THE
SCIENCE OF HISTORICAL THEOLOGY

THE SCIENCE OF HISTORICAL THEOLOGY

ELEMENTS OF A DEFINITION

by
Msgr. John F. McCarthy

TAN BOOKS AND PUBLISHERS, INC.
Rockford, Illinois 61105

Imprimi Potest: Francus Biffi
                  Rector Universitatis Lateranensis
                  Romae, die 25 Iunii 1976

Imprimatur: ✠ Giovanni Canestri, *Vicegerente*
                  Arcivescovo tit. di Monterano
                  E Vicariatu Urbis, die 26 Iunii 1976

Copyright © 1976 by John F. McCarthy.

First published in Rome, 1976, by Propaganda Mariana, as part of "The Roman Theological Forum, Library of Historical Science." Republished in 1991 by TAN Books and Publishers, Inc.

All rights reserved. No part of this book may be reproduced or transmitted in any form or by any means, electronic or mechanical, including photocopying, recording, or by any information storage or retrieval system, without permission in writing from the publisher. Brief quotations may be excerpted without permission.

Library of Congress Catalog Card No.: 91-65352

ISBN: 0-89555-441-0

Printed and bound in the United States of America.

TAN BOOKS AND PUBLISHERS, INC.
P.O. Box 424
Rockford, Illinois 61105

1991

TO JESUS, SACRED OCCUPANT
OF MARY'S HEART ABOVE

TO HIS SUBLIME DIVINITY
POSSESSED IN BLESSED TRINITY
AND SHARED WITH OUR HUMANITY
GREAT MYSTERY OF LOVE

FOREWORD

Within this century, following on the opening up of new theological speculations in the latter part of the last century, much religious research and exegetical writing has been published. In no small part this work has had as its object the 'updating' of the presentation of faith in Jesus Christ for a world that has seen radical changes in culture and human understanding since the days on this earth of Christ. The background of this writing has been the movement of man into the age of technology and the positive sciences. The changes include the way man sees himself and the way in which he considers his surroundings and, above all, his own meaning and destiny.

In brief, the new studies about how modern man thinks and believes have sought to bring the faith into line with man's outlook today and thereby make the faith acceptable. This effort is of course consistent with the general effort of the Church to make the Gospel understood by all men at all times. However, certain aberrations have been noted as the fruits of this recent operation reached the arena of daily life where religion is lived and the faith put into practice. The frequent presentation of highly scientific theological speculation and special exegetical conclusions as if they were the teaching of the Church and, indeed, were identified with the saving faith itself has in fact confused many people, particularly those who have never had reason or occasion to master, as Msgr. McCarthy has done, the fine-honed academic distinctions on which the orthodox scientific theological structure necessarily rests. Sometimes the personal presuppositions of the theologian or exegete himself may get in the way of his presentation and thereby diminish the value of his work for others. At other times, the principles applied to the deposit of the faith by the theologian intent on re-presenting the mysteries of the faith were such that when applied across the board, there was no trace of the 'baby' after the 'bath water' was thrown down the drain.

The world of faith and all the events that are part of it remain a mystery. Revelation has opened to us a glimpse of God's loving kindness and plan for our salvation, but even with that fact of revelation there resides a mystery. God is and remains beyond our intellectual grasp. Christ reveals, yet is never absorbed entirely by our human comprehension. The Holy Spirit enlightens and makes holy, yet not according to man's timetable. In the attempt to make this understood and accepted by men of all ages, the Church studies, meditates on and repeats the Word she received in the beginning.

In the various attempts to penetrate the meaning of the Gospel and make it acceptable today, serious problems inevitably arise. Can man accept today the message that was originally delivered in terms and propositions presumably adapted to another age? Is the message so linked to the original media that the content is irretrievably lost to this generation?

In the attempt to face these questions and arrive at answers to them there has been a lot of theological spin-off in the form of theologizing and popularizing of heavier content-orientated works that has aggravated rather than helped solve the problem of how to make clear, without changing it, the faith to each generation.

Another problem arises when we face the question of adapting the word of God to our age or any specific time. Is there some outside limit to how far one can go with the reformulation of the good news before it ceases to be Christ's message and becomes rather the position and teaching of the individual theologian or the presentation of a specific school of particular theological thought? Is there not also some norm against which one, either individual theologian or school of theology, can check special conclusions and adaptations to see if there remains that fidelity to the revealed word of God without which theological studies may be pleasant, perhaps even stimulating, but ultimately useless, for salvation's sake, personal memoirs?

These questions apply to the world of scientific theological speculation as a whole, but equally pertinent questions can be brought up when we look at the field of exegesis which must supply scientific theology with much of its data. In the attempt to make relevant the Gospel message today, much scientific theology finds itself relying on exegesis that is based as much or sometimes more on the premises and presuppositions of a special, even isolated, school as it is on the Church's understanding of the Word of which she is the custodian and by which she lives and gives life. The multiplication of the problem through the translation of the original revelation into another cultural climate carries with it an increased risk of failure. The failure in many instances can take the form of the emasculation of the original message to make it not only less challenging to the modern culture but less demanding on those within the culture; thus there is born indeed a fatal 'historic compromise'!

To this problem and its world of relative questions Msgr. McCarthy addresses himself in his immensely scholarly work, The Science of Historical Theology. *The work itself is not intended as a Sunday supplement piece and presupposes an intellectual acumen just less than that of the author himself, if it is to be fully appreciated - as may the Lord grant it will be!*

For what makes this work so valuable is not the size of the readership which it may draw but rather the precision with which Msgr. McCarthy presents a specific scientific theological problem and in strict conformity with the rules of that science presents a critique of one of the major exponents of a theological school that for a long time has dominated a growing theological world view. More than that, he offers an approach to this sometimes 'closed shop' world of theology that meets the questions proposed for study by the same school in their own terms and vocabulary while doing so with a freshness of outlook that makes possible a viable rendering of a solution to the very core problems plaguing theological analysis in past decades. What he has done is to challenge some of the presuppositions that have been presented by some exponents of a school of thought that rejects much of the traditional interpretation of the faith, however sincere the intentions of those who have questioned the established tenets. His work looks at some of the framework of thought that has produced the world, the theology and the school of Bultmann, and then confronts its presuppositions with viable answers that trenchantly oppose and offer valid alternatives to the Bultmannian conclusions which have tended to 'take over' among scholars less conscious and convinced of the Church's theological heritage. Msgr. McCarthy has sown a seed here that may yet produce a mighty tree dwarfing the brush and underbrush in the forest of much contemporary theology.

John Cardinal Wright

Vatican City, July, 1976

AUTHOR'S PREFACE

This volume presents a definition of historical theology in terms of its three central concepts: science, history, and theology.

The contrast between the inductive (empirical) and the deductive (Aristotelian) conceptions of science has caused them to be considered mutually exclusive. Actually, they are species of a single genus, but, because their common feature cannot be grasped without an awareness of the scientific medium, the structure of intellectual consciousness must first be brought into view (*Article 2*). The theory of intellectual consciousness broadens and deepens both conceptions of science.

The word 'history' has been fraught with troublesome ambiguities. The search for a univocal concept of history leads to a discovery of the medium of historical science. Once this medium has been identified (*Article 3*), it can be analyzed (*Article 4*), compared with the medium of inductive and deductive science (*Article 5*), and divided into categories (*Article 6*). It becomes clear how historical science uses the knowledge of individual things to arrive at its conclusions, and how final causality functions in historical reasoning. Classical (inductive and deductive) science and historical science turn out to be species of the one comprehensive scientific genus. To reach this concept it is necessary to trace human knowledge down to its psychological roots. The consequent modification of the Aristotelian theory of knowledge is more a development than an alteration.

This development carries over to the notion of faith and theological science. In the light of intellectual consciousness, the factor of infused contemplation is seen to be included in the act of faith and to have a controlling function in valid theological reasoning. An examination of the object of faith from the viewpoint of historical science delineates the field of historical theology (*Article 7*).

The lengthy presentation of W. H. Walsh's theory of history is a tribute to his work as a whole. My critique is intended merely to indicate how certain elements of the theory can be improved by using other principles.

The ample space given to the writings of Rudolf Bultmann is a recognition of his success in formulating problems for historical theology. Since I totally disagree with Bultmann's method of surrendering to difficulties that can be resolved, I also totally disagree with his conclusions. This accounts for the space given to my reply.

The abbreviation *cf.*, as used in this study, is intended in the full sense of 'compare and contrast,' and does not necessarily imply agreement, even with regard to the point being made.

In first footnote references to articles, where a specific page is indicated, I have often included in brackets the pages of the entire article.

I am grateful to Prof. Brunero Gherardini for his wise and erudite advice, to Prof. Claudio Zedda for his critical reading of the manuscript, to John Cardinal Wright, and to all who by their prayer and activity have sustained me during the years of research into the subject. I am particularly grateful to Mary, Mother of Jesus, Mother of the Word Incarnate, Mother of the Church, and our Mother, for having so many times and in so many ways led me to her Son. And, in humble adoration, I am most grateful of all to the Person of Jesus for being, in the unity of the Most Holy Trinity, the divine Reality that this science is about.

SPECIAL ABBREVIATIONS
(with full footnote reference)

BEF	Bultmann, *Existence and Faith* (1.2:2a)
BEPT	Bultmann, *Essays Philosophical and Theological* (1.2:2a)
FaU	Bultmann, *Faith and Understanding* (1.2:2a)
GST	Bultmann, *Die Geschichte der synoptischen Tradition* (8.3:11c)
GuV	Bultmann, *Glauben und Verstehen* (1.2:2a)
HE	Bultmann, *History and Eschatology* (1.2:3a)
HST	Bultmann, *The History of the Synoptic Tradition* (8.3:11c)
IH	Collingwood, *The Idea of History* (3.1:1a)
IPH	Walsh, *An Introduction to Philosophy of History* (3.1:1a)
JCM	Bultmann, *Jesus Christ and Mythology* (1.2:1a)
Jesus	Bultmann, *Jesus* (8.3:5a)
JW	Bultmann, *Jesus and the Word* (8.3:5a)
KaM	Bartsch, *Kerygma and Myth* (1.1:1b)
KuM	Bartsch, *Kerygma und Mythos* (1.1:1b)
S.Th.	Aquinas, *Summa Theologiae* (2.1:1a)
TDNT	Kittel, *Theological Dictionary of the New Testament* (I.32a)
Theologie	Bultmann, *Theologie des Neuen Testaments* (I.1b)
Theology	Bultmann, *Theology of the New Testament* (I.1b)
TWNT	Kittel, *Theologische Wörterbuch zum Neuen Testament* (I.32a)
ZTK	*Zeitschrift für Theologie und Kirche*

TABLE OF CONTENTS

	Page
FOREWORD	vii
AUTHOR'S PREFACE	ix
SPECIAL ABBREVIATIONS	x

Article 1: *Some Distinctions Stressed by Bultmann in the Demythologizing Debate* 1

   1.1. The Demythologizing Debate 1
   1.2. The Goal of Demythologizing 5
   1.3. Two Replies of Bultmann Concerning Demythologizing . . . 9

Article 2: *The Locus of Science in Human Consciousness* 15

   2.1. Intellectual Consciousness 15
   2.2. The Discovery of Truth 21
   2.3. Self-Evident Truth 25
   2.4. The General Locus of Science in Human Consciousness . . 34
   2.5. The Locus of Common Sense in Human Consciousness . . 42
   2.6. The Locus of Specialized Science in Human Consciousness . 46

Article 3: *The Locus of History in Human Consciousness* 57

   3.1. The Locus of History in Prescientific Consciousness . . . 57
   3.2. The Locus of Historical Science in Scientific Consciousness . 60

Article 4: *The Relationship of Past and Present in Historical Science* . 64

   4.1. The Object of Historical Science 64
   4.2. The Present of Historical Science 65
   4.3. The Past of Historical Science 70

Article 5: *The Form of Historical Understanding* 73

   5.1. Historical Causality 73
   5.2. The Form of Historical Reasoning 74
   5.3. The Value of Historical Science 79

Article 6: *The First Dichotomies of Historical Science* 83

   6.1. Historical Method 83
   6.2. The Art of Historical Science 84
   6.3. The Broadest Division of History 85

Page

Article 7: *The Locus of Historical Theology in Human Consciousness* . 87

7.1. Sacred Theology According to St. Thomas 87
7.2. The Act of Faith According to St. Thomas 87
7.3. The Gift of Understanding According to St. Thomas . . . 88
7.4. The Gift of Knowledge According to St. Thomas 89
7.5. Revealed Truth 90
7.6. Faith and Reason 91
7.7. The Object of Theological Understanding 92
7.8. The Theology of Concrete Meanings 97

Article 8: *A Reply to Bultmann's Arguments for the Need of Demythologizing* 101

8.0. The Need of Demythologizing 101
8.1. The Question of Sacred Authority 102
8.2. The Question of Science 103
8.3. The Question of Historical Science 113
8.4. The Question of the Genre of Mythology 119
8.5. The Question of the Scientific Mentality 126
8.6. The Question of the Place Where God Dwells 128
8.7. The Question of Angelic Spirits 135
8.8. The Question of the End of the World 138

Appendix I: *Concerning Rudolf Bultmann's Conception of New Testament Theology* 141

A. The Insight Underlying Bultmann's Theology 141
B. A Tentative Characterization of the Genre of Bultmann's Theological Writing 143

Appendix II: *Concerning W. H. Walsh's Conception of History* . . . 165

Appendix III: *Concerning R. G. Collingwood's Conception of History* . 179

Appendix IV: *Concerning Some Representative Current Conceptions of History* 185

A. The Historical Theory of W. B. Gallie 185
B. Morris Cohen's Theory of History 188
C. Gustaaf Renier's Notion of History 191
D. Bernard Norling's Notion of History 192

BIBLIOGRAPHICAL INDEX 194

Article 1

SOME DISTINCTIONS STRESSED BY BULTMANN IN THE DEMYTHOLOGIZING DEBATE

1.1. The Demythologizing Debate

1. Rudolf Bultmann's articulate need for the demythologizing of the New Testament provides an imposing set of problems for the study of historical theology. The reader who wishes to have a particularized idea of Bultmann's method can do no better than consult the corpus of his own writings, which is abundantly clear for this purpose.[a] The literature of criticism, pro and contra, of the system has grown to immense proportions and now defies the capacity of anyone to digest it thoroughly.[b] Among Bultmann's Protestant confreres the publication of the program of demythologizing has provoked a long and voluminous discussion, partly in the attempt to ascertain exactly what Bultmann means to say, partly in criticism of what he seems to have said. Many theologians have extended more sympathy to his ideas than would have seemed possible. Beneath the surface they think they have caught sight of an evasive something more solid than cynicism, more valuable than contempt. And thus the notion of demythologizing has reached the consideration of a wide circle of thinkers. Protestant criticism has in

---

1.1:1[a] An almost complete bibliography of Bultmann's works published in German and in English up to the year 1965 may be found in C. W. Kegley, ed., *The Theology of Rudolf Bultmann* (London: SCM Press, 1966), 289-310. Some additional works are listed in F. Theunis, *Offenbarung und Glaube bei Rudolf Bultmann: Ergänzung zu Kerygma und Mythos V* (Theologische Forschung 19: Hamburg-Bergstedt, 1960), 133-134; and by José Ewaldo Scheid, *Die Heilstatt Gottes in Christus: Kerygma und Mythos V - Ergänzungsband II* (Theologische Forschung 23: Hamburg-Bergstedt, 1962), 229-234. Especially noteworthy more recent additions are the ET of *GuV* I (*infra*, 1.1:2[a]), *The Gospel of John* (Oxford, 1971), *A Commentary on the Johannine Epistles* (Hermeneia, 1973), and "General Truths and Christian Proclamation" (*infra*, 1.1:2[a]). A guide to the ET of a number of Bultmann's works is given in W. Schmithals, ET, *An Introduction to the Theology of Rudolf Bultmann* (Minneapolis: Augsburg Pub. House, 1968), 325-328. See also the enlarged bibliography in the combined edition of *KaM* I and *KaM* II (London: SPCK, 1972), 357-358.

1.1:1[b] Hans Werner Bartsch, in a supplement to *KuM* I and II, published separately under the title, *Der gegenwärtige Stand der Entmythologisierungsdebatte: Beiheft zu Kerygma und Mythos, I-II* (Theologische Forschung 7: Hamburg-Volksdorf, 1954) (ET, "The Present State of the Debate," in *KaM* II, 1-82), presents a review of the discussion to that date from a viewpoint favorable to Bultmann's basic position and includes a copious bibliography. A handy bibliography of the earlier contributions to the discussion by Protestant and Catholic writers is given by Anton Vögtle in his "Rivelazione e mito," in *Problemi e orientamenti di teologia dogmatica* I (Milan, 1957) [827-960], 954-960. Additional works are listed by René Marlé in his *Bultmann et l'interprétation du Nouveau Testament* (Paris, 1956), 189-195, supplemented by Franz Theunis, in *Offenbarung und Glaube bei Rudolf Bultmann* (*supra*, 1.1:1[a]), 134-141, and by José Scheid in *Die Heilstatt Gottes in Christus* (*supra*, 1.1:1[a]), 229-259. An excellent bibliography on the more recent discussion, compiled by Egon Brandenburger, is presented by Günther Bornkamm, "Die Theologie Rudolf Bultmanns in der neueren Diskussion," in *Theologische Rundschau*, N.F., 29 (1963), 33-46, together with a lengthy survey of the literature to that date (*ibid.*, 46-141). Bartsch continues his work of summarizing the discussion in "Die nicht ausgetragene Entmythologisierungsdebatte," in the volume *Entmythologisierende Auslegung* (Theolo-

large part tended to conclude that the thesis of Bultmann sets up an unthinkable reduction, impoverishment, and emptying out of the kerygma of the New Testament, differing only in degree from the devastation of the New Testament proclamation carried out by the older liberal theology and by the school of the history of religions. But the prolonged consideration and even sympathy which Protestants have found for this conception has assured Bultmann a profound psychological impact in the Protestant world, especially among the younger generation. It is in general a sympathy for the sobre, scholarly, and consistent work of a Lutheran theologian in an age of perplexity.

2. In characterizing the debate which has ensued in Protestant theological circles over Bultmann's program of demythologizing, it is important to keep in mind that brief or lengthy collections of the observations made by the various participants are indeed indica-

---

gische Forschung 26: Hamburg-Bergstedt, 1962), 50-58 [reprinted from *Religion in Life* (1960)]. A useful bibliography on the question of myth and Christian faith is given by Pierre Barthel in his *Interprétation du language mythique et Théologie biblique* (Leiden: E. J. Brill, 1963), 383-389. See also *KaM* I and II (1972 combined edition), 358-361.

The original nucleus of the demythologizing debate was solidified in H. W. Bartsch et al., eds., *Kerygma und Mythos* (1948-), referred to in the present work as *KuM*. This is a special series contained within a larger series published by Herbert Reich — Evangelischer Verlag of Hamburg, under the title, Theologische Forschung: Wissenschaftliche Beiträge zur kirchlich-evangelischen Lehre:

KuM I: *Ein theologisches Gespräch* (1948)

KuM II: *Diskussion und Stimmungen zum Problem der Entmythologisierung* (1952)

KuM III: *Das Gespräch mit der Philosophie* (1954)

KuM IV: *Die Oekumenische Diskussion* (1955)

KuM V: *Die Diskussion innerhalb der katholischen Theologie* (1955)

Selected articles from *KuM*, trans. by R. H. Fuller et al., were published in H. W. Bartsch, ed., *Kerygma and Myth: A Theological Debate* [referred to in the present work as *KaM*], Vol. I (London: SPCK, 1953); Vol. II (London: SPCK, 1962). The 2d ed. of *KaM* I (1961, with slightly rev. trans.) and *KaM* II (1962) have been republished in a single book (London: SPCK, 1972) with an updated bibliography.

A series entitled Studi sulla Demitizzazione, ed. Enrico Castelli (Rome: Istituto di Studi Filosofici), contains the proceedings (in French and Italian) of the international discussions held annually in Rome since 1961 under the sponsorship of the International Center for Humanistic Studies and the Institute of Philosophical Studies connected with the University of Rome. *KuM* VI, which constitutes several titles of Theologische Forschung, represents the German version of some of the volumes of the series published in Rome, whose titles to 1971 are as follows: 1) *Il problema della demitizzazione* (1961) = *KuM* VI-1 (1963); 2) *Demitizzazione e immagine* (1962) = *KuM* VI-2 (1964); 3) *Ermeneutica e tradizione* (1963); 4) *Tecnica e casistica* (1964); 5) *Demitizzazione e morale* (1965); the last two titles combined (1964-1965) = *KuM* VI-3 (1968); 6) *Mito e fede* (1966) = *KuM* VI-4 (ed. F. Theunis: 1968); 7) *Il mito della pena* (1967); 8) *L'ermeneutica della libertà religiosa* (1968); 9) *L'analisi del linguaggio teologico* (1969); 10) *Dibattiti sul linguaggio teologico* (1969); 11) *L'infallibilità* (1970); 12) *Ermeneutica e escatologia* (1971); 13) *Rivelazione e storia* (1971).

Summaries of the discussion published by Catholic scholars include the following: A. Kolping, "Sola Fide: Aus der Diskussion um Bultmanns Forderung nach Entmythologisierung des Evangeliums," in *Theologische Revue*, 49 (1953), 121-134 (reprinted in *KuM* V); R. Marlé, "A propos du projet de R. Bultmann relatif à une 'demythologisation' du Message neotestamentaire," in *Recherches de Science Religieuse*, 41 (1953), 612-632; B. Brinkmann, "Für und gegen die Entmythologisierung der neutestamentlichen Botschaft," in *Scholastik*, 30 (1955), 513-534; A. Vögtle, "Rivelazione e mito," (*loc. cit. supra*), 827-960; R. Marlé, "Bultmann devant les théologiens catholiques," in *Recherches de Science Religieuse*, 45 (1957), 262-272; B. Rigaux, "L'historicité de Jésus devant l'exégèse récente," in *Revue Biblique*, 65 (1958), 481-522. For bibliographical references to works by Catholic writers on the question of demythologizing, see A. Vögtle, *op. cit.*, 955-956; F. Theunis, *Offenbarung und Glaube bei Rudolf Bultmann* (*supra*, 1.1:1a), 134-141.

tive of the discussion but do not epitomize it. What is perhaps most obvious about each respective quotation is its being out of context. Hence, the context must be recognized to be an important factor of the debate. The long array of Protestant commentators on Bultmann's program can be somewhat nebulously divided into those who agree with the proposal and those who disagree, but the division is of little use in the absence of precise distinctions as to the sense in which the respective commentators agree or disagree and as to the depth of disagreement in the latter case. It is easy to show that many writers who claim to disagree are actually in agreement on more fundamental issues; it is no more difficult to show that many who claim to agree reflect differences which are significant.

3. The program of demythologizing may be said to have had its first great impact largely within the circle of Lutheran theological activity, then to have widened out to the circle of Protestant theological activity in general, and finally to have challenged theologians in Catholic and other Christian circles. As the sphere of impact grew, the context of discussion tended to change. This context can be considered from a logical as well as from a chronological point of view. From the logical viewpoint, the first and fullest impact of the program of demythologizing was felt by those whose thinking was most allied with that of Bultmann, in such wise as to give rise to the 'Bultmannian school of thought.' The comments of Bultmannian theologians may be considered to agree with the context implied in Bultmann's teaching, but even here the agreement is almost never complete, and it is surprising how independent Bultmannians have been in their thinking, issues of fundamental importance not excepted. Bultmann did not request rote or slavish acceptance of his *dicta* and he did not receive it. Bultmann's most dedicated followers presumed to disagree on one point or another right from the start, and this divergence also gave rise to what is now known as the 'post-Bultmannian school.'[a]

4. Farther out on the circles are the Lutheran theologians who share the theological tradition from which Bultmann's thought has sprung and who agree with Bultmann's intention of expressing the Lutheran notion of faith in terms which are meaningful to contemporary man. These theologians may be divided into a majority who think that Bultmann has gone too far in his attempt to demythologize the New Testament basic message and a minority who feel that he has not gone far enough.[a] Opinion varies as to the 'fidelity' with which Bultmann has interpreted the Lutheran notion of faith and expressed it in the program of demythologizing. Related to this issue is the question of Bultmann's success or failure in offering the program as a reaction against the more extreme liberalism of the preceding generation in Germany. Many critics are of the

---

1.1:3a Oscar Cullmann (*Salvation in History*, 20) thinks that the expression 'Bultmann School' is accurate, contrary to the objections of Erich Dinkler and other disciples of Bultmann who insist upon the independence of their thought, but that the expression 'post-Bultmannians' is not happily chosen, "since a *true pupil* always begins from the standpoint of his teacher and proceeds from there." A good example of the independence of Bultmann's pupils is G. V. Jones, *Christology and Myth in the New Testament* (London: George Allen and Unwin, 1956), *ad rem*, p. 10.

1.1:4a A representative statement of the view that Bultmann has not gone far enough in his proposal to modernize the New Testament message is that of Fritz Buri, "Entmythologisierung oder Entkerygmatisierung der Theologie," in *KuM* II (1952), 85-101, where he maintains that Bultmann has failed to carry his method to its logical conclusion and that Bultmann in retaining the kerygma is hanging on to what is but a final holdover of the outdated mythology, since there can no longer reasonably be said to be a divine event of redemption or any revelation of God at all. Christian theology, says Buri, cannot preserve its autonomy from philosophy by some appeal to a saving act wrought within human history; its task is rather to help the man of today to understand himself, making use, of course, of the riches of its own tradition (*KuM* II, 96). Cf. F. Buri, "Theologie der Existenz," in *KuM* III (1953), 81-91, and his book, *Theologie der Existenz* (Berne and Stuttgart, 1954); ET, *Theology of Existence* (Greenwood, South Carolina: Attic Press, 1965).

view that it is a poor alternative.[b] In his earlier years, Rudolf Bultmann had collaborated with Karl Barth in working out the general lines of a 'dialectical theology' which would restore the faith-element to a field from which it had been driven by the relentless rationalism of the liberal theological movement.[c] After Bultmann had declared his stand on demythologizing, Barth and his disciples published their opposition to it on grounds that it does not fulfill adequately the essential aim of the dialectical movement. Yet the common 'dialectical' element in both theologies provides a context of considerable importance in this aspect of the debate. Other Lutheran theologians, in opposition to both Bultmann and Barth, have denied the need of modern man to demythologize the New Testament. They do not share the outlook of the higher-criticism from which form-criticism sprang.[d] Some Lutheran commentators agree with the aim of Bultmann to 'derationalize' the Christian interpretation of the New Testament according to the ideal of the Lutheran notion of faith,[e] but others find that the emptying out of the content of the object of faith which results from the application of Bultmann's approach depends upon a notion of faith that is not acceptable according to the principles of Lutheranism.

5. A middle circle of participants in the demythologizing debate includes those Protestant commentators who do not subscribe to the notion of faith from which Bultmannian theology has sprung. It is obvious that at this level of the discussion an important principle of Bultmann's program no longer remains basically unquestioned. The debate over faith and reason, faith and science, faith and history, faith and philosophy assumes a new dimension, and the searching for answers, a new direction. The issue of faith and reason moves from the secondary to the primary order; the fall of man is no longer understood in quite the same way. Modified in corresponding fashion is the conception of modern man, his aspirations, and his history. Critics in this second circle have deeply questioned Bultmann's notion of history and his historical method, his idea of the world, his conception of understanding, his projection of eschatology. Less taken for granted are the results of form-criticism and its pre-history of biblical research. The circle begins well within Lutheranism itself and extends to the Anglican theologians and beyond, with every shade and type of position in between.

1.1:4[b] Thus, e.g., H. Sasse, "Flucht vor dem Dogma," in *Luthertum* (1942); A. Oepke, "Entmythologisierung und existentiale Interpretation," in *Evangel.-lutherische Kirchenzeitung* VI (1952); E. Schmidt, "R. Bultmanns Programm der Entmythologisierung der christlichen Botschaft," in *Zeitschrift für systematische Theologie* (1943). Walter Grundmann (*Die Geschichte Jesu Christi*: Berlin, 1952², with supplement), while admitting as valid some of the results of form-critical research (such as the conclusion that the Infancy Narratives are only Christological affirmations and not historical reports), maintains that in general the Jesus described in the Gospel accounts is a faithful representation of the Jesus whose historical life serves as their model.

1.1:4[c] Cf. J. D. Smart, *The Divided Mind of Modern Theology: Karl Barth and Rudolf Bultmann, 1908-1933* (Philadelphia: Westminster Press, 1967).

1.1:4[d] Thus, e.g., F. Rienecker, *Stellungnahme zu Bultmanns 'Entmythologisierung'* (Wuppertal, 1951); E. Schlink, in *Studium Generale* (1948). Ethelbert Stauffer

("Entmythologisierung oder Realtheologie," in *Deutsches Pfarrerblatt*: 1949) [reprinted in *KuM* II, 13-28] opposes Bultmann's *a priori* exclusion of the miraculous in historical research. Nils Alstrup Dahl ("[Notes concerning Bultmann's] *Die Theologie des Neuen Testaments*," in *Theologische Rundschau*, Neue Folge, XXII [1954], 21-49) opposes Bultmann's existentialist approach. The Scandinavian, Harald Riesenfeld, has published a rather searching criticism of what he considers to be excessive optimism as to what form-critics have succeeded in establishing, in his book *The Gospel Tradition and Its Beginnings: A Study in the Limits of Formgeschichte* (London, 1957).

1.1:4[e] Thus, e.g., G. Gloege, *Mythologie und Luthertum: Das Problem der Entmythologisierung im Lichte lutherischer Theologie* (Berlin, 1953). This is one of the views weighed in the *Memorandum* of the School of Evangelical Theology of the University of Tübingen, published under the title, *Für und wider die Theologie Bultmanns* (Tübingen, 1952).

6. A salient feature of the whole discussion has been the frequency of the distinction made between the method of Bultmann and his principles. This distinction has served to clarify Bultmann's position from the viewpoint of various critics, but it has not tended to remain only on Bultmann's side of the discussion, for it has had the effect of calling into question even in their own minds the premises from which the critics themselves have taken their stand. In this sense the debate over demythologizing has invited all of the participants to make a kind of 'examination of conscience,' in which they might test the validity of the presuppositions of their own thought as a fitting prelude to their evaluation of the presuppositions of Bultmann's thought.

1.2. The Goal of Demythologizing

1. The task of 'demythologizing' the kerygma of the New Testament as proposed by Rudolf Bultmann aims at the attainment of a comprehension of the Christian message (das Kerygma) which is free of every picture of the world (Weltbilt) projected by objectifying thought, whether mythological or scientific.[a] Myth objectifies the other side into this side. Negatively speaking, demythologizing is a criticizing of the mythological conception of the world inasmuch as it veils the true intention of myth. Positively speaking, demythologizing is giving the true existential interpretation of myth so as to render clear its intention "to speak of the existence of man."[b]

2. We should serve our era, wrote the long-time Professor of Marburg, as the primitive disciples served theirs. By demythologizing the Gospels, interpreting them in a manner consonant with the means and data at hand, and thus making them acceptable to a more sophisticated generation of believers, we are performing the same service as the early disciples did in surrounding the crucifixion of Jesus with a mythology attractive to their own generation. The theologian of today must continue the work of composing the Gospels, substituting modern concepts.[a] Modern man cannot be expec-

---

1.2:1[a] R. Bultmann, "Zum Problem der Entmythologisierung," in *KuM* II (179-208), 207; ET (of pp. 191-208 only), "Bultmann Replies to His Critics," in *KaM* I (191-211), 210. Cf. Bultmann, in *KuM* II, 187; *KuM* I, 15-27 (*KaM* I, 1-16); *Jesus Christ and Mythology* (New York: Scribner's, 1958) (hereinafter referred to as *JCM*), 14-21. Cf. *infra*, 8.0: 1-8.

1.2:1[b] Bultmann, in *KuM* II, 184: "Kurz gesagt: der Mythos objektiviert das Jenseits zum Diesseits und damit auch zum Verfügbaren, was sich daran zeigt, dass der Kultus mehr und mehr zu einer das Verhalten der Gottheit beeinflussenden, ihren Zorn abwendenden, ihre Gunst gewinnenden Handlung wird. ... Negativ ist die Entmythologisierung daher *Kritik am Weltbild des Mythos*, sofern dieses die eigentliche Intention des Mythos verbirgt. Positiv ist die Entmythologisierung *existentiale Interpretation*, indem sie die Intention des Mythos deutlich machen will, eben seine Absicht, von der Existenz des Menschen zu reden." Bultmann (*JCM*, 18) admits that the term 'demythologizing' is misleading: "This method of interpretation of the New Testament which tries to recover the deeper meaning behind the mythological conceptions I call *demythologizing* — an unsatisfactory word, to be sure. Its aim is not to eliminate the mythological statements but to interpret them. It is a method of hermeneutics."

1.2:2[a] R. Bultmann, *Glauben und Verstehen: Gesammelte Aufsätze* [Tübingen: J. C. B. Mohr (Paul Siebeck): I (1933); II (1952); III (1960); IV (1965): hereinafter referred to as *GuV*], I, 267; ET of *GuV* I 1966[6], *Faith and Understanding* (New York: Harper and Row, 1969) (hereinafter referred to as *FaU*), 284. The following two essays, published earlier in ET, are omitted from *FaU*: "Das christliche Gebot der Nächstenliebe," in *GuV* I, 229-244; ET, "To Love Your Neighbor," in *The Scottish Periodical* I (1947), 42-56; "Das Bedeutung des Alten Testaments für den christlichen Glauben," in *GuV* I, 313-336; ET, "The Significance of the Old Testament for the Christian Faith," in B. W. Anderson, ed., *The Old Testament and Christian Faith* (New York: Harper and Row, 1963), 8-35. The ET of *GuV* II is entitled *Essays Philosophical and Theological*, trans. by J. C. Greig (London, 1955) (hereinafter referred to as *BEPT*). Some of the essays in *GuV* III have appeared in English: "Der Begriff der Offenbarung im Neuen Testament," in *GuV* III, 1-34; ET, "The Concept of Revelation in the New Testament," in R. Bultmann, *Existence and Faith: Shorter Writings of Rudolf Bultmann*, selected, translated and introduced by S. M. Ogden (Lon-

ted to believe the mythological conception of the New Testament preaching, because thought has been irrevocably reformed by modern science.[b]
3. In the ultimate analysis, God is totally diverse from the world and our conception of it.[a] Our conception of the world is derived from the data of natural science and from mythology. The natural sciences presuppose that the world and its laws constitute a closed and compact system. And as long as the science of nature does not give up the experimental method, it is naïve to desire to make use of the relativity of the principle of causality in the phenomena of the microphysical world in order to prolong the mythological faith in miracles, as though with such a relativity the door could be opened to the intervention of supernatural forces.[b] Bultmann defines mythology as "the conception which makes the non-wordly and divine seem worldly and human, that which pertains to the other side seem to be on this side."[c] What myth truly affirms does not pertain to the realm of objectivity or to the concreteness of verifiable affirmation; it is, on the contrary, ahistorical, not having transpired as a fact, and not demonstrable as such. Jesus does not speak objectively of God, of his nature and attributes; he does not speak of theoretical truths, but only of what God is for men, what God works upon men.[d] There is no Incarnation or Christology or presence or influence of God in the sense of conventional theology.[e] Equally obsolete in the conventional sense is sacred history with its supernatural historical events and external signs of revelation.[f] Man can meet God and experience his hidden action only in faith, whose paradoxical aspect consists precisely in this that it considers as the action of God even an event which can be verified in its natural and historical situation.[g]
4. The final and complete form of all understanding is understanding of one's own self. In fact, all understand-

---

don: Hodder and Stoughton, 1961) (hereinafter referred to as *BEF*), 58-91; "Die christliche Hoffnung und das Problem der Entmythologisierung," in *GuV* III, 81-90; ET, "The Christian Hope and the Problem of Demythologizing," in *The Expository Times*, 65 (1954), 228-230, 276-278; "Geschichte und Eschatologie im Neuen Testament," in *GuV* III, 91-106, Germ. trans. of "History and Eschatology in the New Testament," in *New Testament Studies* I (1954), 5-16; "Allgemeine Wahrheiten und christliche Verkündigung," in *GuV* III, 166-177; ET, "General Truths and Christian Proclamation," in *Journal for Theology and the Church*, 4 (1967), 153-162.
1.2:2[b] *Infra*, 8.0:6. Cf. Bultmann, in *KuM* I, 136 (*KaM* I, 120): "The real *skandalon* of faith in God vis-à-vis modern technology can become clear only when we have abandoned the false view of God which that technology has exploded."
1.2:3[a] R. Bultmann, *History and Eschatology* (New York: Harper and Row, 1957) (hereinafter referred to as *HE*), 96.
1.2:3[b] Bultmann, in *KuM* II, 181: "Für das mythische Denken sind die Welt und das Weltgeschehen 'offen' — offen nämlich für den Eingriff jenseitiger Mächte, also durchlöchert vom Geschichtspunkt des wissenschaftlichen Denkens aus. . . Es tut auch gar nichts zur Sache, wenn man darauf hinweist, dass das Weltbild der Naturwissenschaft von heute nicht mehr das des 19. Jahrhunderts ist; und es ist naiv, die Tatsache der Relativierung des Kausalgesetzes hinsichtlich der atomaren Vorgänge zur Repristinierung des mythischen Wunderglaubens benutzen zu wollen, als ob mit dieser Relativierung das Tor für das Eingreifen jenseitiger Mächte geöffnet worden wäre!"
1.2:3[c] *KuM* I, 22 (*KaM* I, 10).
1.2:3[d] *JCM*, 43.
1.2:3[e] Bultmann, in *KuM* II, 206, n. 1 (*KaM* I, 209, n. 1): "Mich dünkt, die Christologie sollte endlich radikal aus der Herrschaft einer Ontologie des objektivierenden Denkens befreit und in einer neuen ontologischen Begrifflichkeit vorgetragen werden."
1.2:3[f] *JCM*, 78-83. Cf. *HE*, 36: "*The New Covenant* is not grounded on an event of the history of the people as was the Old Covenant. For the death of Christ on which it was founded is not a 'historical event' to which one may look back as one may to the story of Moses. *The new people of God* has no real history, for it is the community of the end-time, an eschatological phenomenon. How could it have a history now when the world-time is finished and the end is imminent! The consciousness of being the eschatological community is at the same time the consciousness of being taken out of the still existing world." Cf. also *HE*, 143-144; 151-152.
1.2:3[g] Bultmann, in *KuM* II, 198 (*KaM* I, 199).

ing is self-understanding. Any understanding which is not understanding of self is not authentic understanding.[a] Real understanding is the developed form of the self-understanding which everyone possesses in its initial stage. Authentic understanding evolves from this initial pre-understanding, which itself can exist in either of two forms: a) a native ingenuous grasp of being, such as friendship before one has come to realize what friendship really means;[b] b) a scientific formulation of the object to be understood. The first form of pre-understanding is called *existentiell*; the other is called *existential*.[c] It would be an *existential* inquest to analyze the significance of existence, but the *existentiell* problem has an answer only in existence.[d] All of the classic historical phenomena should be re-examined, says Bultmann, in the light of the existential comprehension contained in them, as he has done, by way of suggestion, in his study of Christianity in the framework of the Old Testament.[e] Theological hermeneutics (especially) must use concepts through which one can speak of existence in a fitting manner.[f]

5. All men outside of faith are immersed in the world of disposable things. Theirs is an 'unauthentic' existence. The man of faith is possessed of 'authentic' existence, since he lives in the pure realization of existence, in the pure act of obedience and love.[a] Hence, God's revelation of himself in Christ could better be called 'suppressed knowledge.'[b] Since Christian faith is eschatological in nature, it precludes the notion of a "unique and final revelation of God in history," a *revelatum* [*Offenbarungsgeschehen*] which could be observed in the past as from a distance and which would reduce the kerygma to a mere record of something now over and done with.[c]

6. Bultmann sets up a radical difference between history as *Historie* and history as *Geschichte*. The former is composed of causally connected events and relationships between facts which are objectively verifiable and chronologically determinable.[a] The latter possess true historicity (*Geschichtlichkeit*), i.e., the existential constitution of the being which necessarily exists in history (*das Dasein*).[b] Das

---

1.2:4[a] R. Bultmann, "Das Problem der Hermeneutik," in *GuV* II (211-235), esp. 227-235; ET, "The Problem of Hermeneutics," in *BEPT* (234-261), esp. 252-261. Cf. Bultmann, in *KuM* II, 192 (*KaM* I, 191-192); *HE*, 117; *JCM*, 48-50.
1.2:4[b] R. Bultmann, "Die Geschichtlichkeit des Daseins und der Glaube: Antwort an Gerhardt Kuhlmann," in *ZTK*, N.F. 11 (1930) [339-364], 351; ET, "The Historicity of Man and Faith," in *BEF* [9-110], 99-100.
1.2:4[c] "... *existential* is used normally as the general term applied to the principles of existentialistic philosophy, whereas *existentiell* has reference to particular situations that are handled in a way compatible with or in agreement with these principles." The equivalents 'ontological' and 'ontical' have been suggested by English-speaking writers as covering this distinction." (R. Bultmann, *Essays Philosophical and Theological* [*supra* 1.2:2[a]], 236, translator's note.) *Existentiell* refers to "that which has to do with the individual's own unique situation and responsibility." (S. Ogden, "Preface" to R. Bultmann, *Existence and Faith*, 7.) Cf. Bultmann's clarifications in *JCM*, 66 and 74.
1.2:4[d] Bultmann, in *KuM* II, 192 (*KaM* I, 193).

1.2:4[e] R. Bultmann, *Das Urchristentum im Rahmen der antiken Religionen* (Zurich, 1949); ET, *Primitive Christianity in Its Contemporary Setting* (Cleveland: World Pub. Co., 1956).
1.2:4[f] Bultmann, in *KuM* I, 124 (*KaM* I, 104); *JCM*, 55.
1.2:5[a] Bultmann, in *KuM* I, 29 (*KaM* I, 19-20).
1.2:5[b] Bultmann, in *KuM* I, 126 (*KaM* I, 107).
1.2:5[c] Bultmann, in *KuM* I, 129 (*KaM* I, 111).
1.2:6[a] *HE*, 143-144.
1.2:6[b] Bultmann bases his position on that of M. Heidegger, *Sein und Zeit* (Tübingen, 1927); *Being and Time*, trans. by J. Macquarrie and E. Robinson (London: SCM Press Ltd., 1962), as he himself admits (*KuM* I, 33; *KaM* I, 24-25; cf. *KuM* II, 192). In *KuM* II, 191 (*KaM* I, 191-192) he states: "I think I may take for granted that the right question to frame with regard to the Bible — at any rate within the Church — is the question of human existence. I am driven to that by the urge to inquire existentially about my own existence. But this is a question which at bottom determines our approach to and interpretation of all historical documents. For the ultimate purpose in

Dasein, i.e., the human being, ultimately signifies to exist, to be confronted with non-being, to be able to be and ever to decide anew.[c] Genuine human existence is an historical existence constituted at bottom by encounters.[d] Hence the terms 'subjective' and 'objective' should be eliminated from philosophical and theological discussion, or, at any rate, should not be used in their proper sense when one is treating of historical matters.[e]

7. Time as 'flux' (*Verlaufzeit*) is distinguished from time as 'now' (*Jetztzeit*). The eschatological 'now' is the present separation which is always made in faith between the absolute future of the new era and the present era, which, in face of this 'absolute future,' is rated as 'absolute past.' Life is attained where this separation is made.[a] Meanwhile, the determined past of sin is retained as pardoned. Pardon is the manner in which the separation is made in the eschatological instant. The expression, πλήρωμα τοῦ χρόνου (fullness of time),[b] signifies that "history (*Geschichte*) has arrived at its mortal end and its nonsense has now become obvious."[c] It is realized time after time in the 'now' of the decision of faith.[d] The instant of the eschatological decision is not to be considered as an infinitely small part of the line of mundane time, but as a point without dimensions. The past, in this eminently 'historical' (*geschichtlich*) sense, is not simply that which was, but that which is subjected to the 'law of the past.' The future of mundane time is a part of this 'past.' But the 'future' is not something which, in the eminently 'historical' sense, will ever come about in the flow of time. It will never be in any way disposable.[e] The mythological concept of a Christ who has come 'once and for all' is to be changed to the concept of 'each time in the individual now.'[f] Bultmann attains this new concept of ἐφάπαξ through the decisive rôle of the kerygma. The Jesus of St. John is basically the appeal of God which becomes 'event' in the eschatological 'now.' The 'history of salvation' thus becomes limited to the kerygmatic sit-

---

the study of history is to realize consciously the possibility it affords for the understanding of human existence."

1.2:6c Bultmann, in *KuM* II, 193 (*KaM* I, 193-194): "Sie [*die Philosophie*] zeigt ihm, dass menschliches Sein im Unterschied von allem anderen Sein eben Existieren bedeutet, ein Sein, das sich selbst überantwortet ist und sich selbst zu übernehmen hat. Sie zeigt ihm, dass die Existenz des Menschen nur im Existieren zu ihrer Eigentlichkeit kommt, sich also nur immer jeweils im konkreten Hier und Jetzt verwirklicht. Sie meint aber nicht, durch existentiale Analyse das existentielle Verständnis des Hier und Jetzt zu beschaffen; sie nimmt dieses dem Menschen nicht ab, sondern schiebt es ihm gerade zu."

1.2:6d Bultmann, in *KuM* II, 199 (*KaM* I, 200): "Ist menschliches Sein in echtem Sinne als ein geschichtliches Sein verstanden, das sein Erleben in seinen Begegnungen hat...."

1.2:6e R. Bultmann, "Das christologische Bekenntnis des Ökumenischen Rates," in *GuV* II (246-261), 258; ET, "The Christological Confession of the World Council of Churches," in *BEPT* (273-290), 287.

1.2:7a "This is exactly what gives the event of Christ its eschatological significance. It anticipates that future event in which the time process is destined to culminate, while to speak of it as happening 'first' is to speak in terms of the empirical time-process, not in terms of the eschatological event." (*KuM*, I, 130; *KaM* I, 112.) Again: "[the 'now' of the New Testament] is not the timeless 'now' of the mystics, stoics, or idealists. It is not as if all events in time were but parables of eternity. Rather, the 'now' of the New Testament implies that timelessness (*Zeitlosigkeit*) becomes an event for each particular individual only by virtue of an encounter in time; it has itself the character of an encounter... I cannot regard the reproduction of the events of the years 1-30 in memory as the equivalent of the eschatological encounter." (*KuM* I, 132; *KaM* I, 115.)

1.2:7b Gal 4:4.
1.2:7c *GuV* I, 7 (*FaU*, 34).
1.2:7d *HE*, 143, 151-152.
1.2:7e *HE*, 36.
1.2:7f "... Damit ist das ἐφάπαξ erst in seinem wahren Sinne verstanden als das Einmal des eschatologischen Geschehens. Denn es meint nicht die datierbare Einmaligkeit eines historischen Ereignisses, sondern lehrt — höchst paradox — ein solches als das Ein-für-alle-Mal des eschatologischen Geschehens glauben. Als das eschatologische ἐφάπαξ hat dieses Geschehen ständig Präsenz im verkündigenden Wort." (*KuM* II, 206; *KaM* I, 209.)

uation, i.e., to the situation of the encounter of man with the eschatological appeal of the message of Christ, and the event of salvation does not exceed the limits of the decision which takes place in the existential 'now.'[g]

(T)o the Christians the advent of Christ was not an event in that temporal process which we mean by history today. It was an event in the history of salvation, in the realm of eternity, an eschatological moment in which rather this profane history of the world came to its end.[h]

1.3. *Two Replies of Bultmann Concerning Demythologizing*

1. In a general reply to accumulated criticism of his notion of existence and faith,[a] Bultmann seeks to clarify his proposal of demythologizing. He states that the presence of the conflict between the biblical mythological world-picture and the modern scientific world-picture merely testifies to the fact that faith has not yet discovered the proper terms in which to express itself, it has not realized that it cannot demonstrate itself by logic, it has not clearly understood that its ground and object are identical, it has not clearly apprehended the transcendental and hidden character of the divine activity. The conflict shows that faith has failed to perceive its own 'nevertheless,' and thus has tried to project God and his acts into the realm of the world. When demythologizing starts from the modern world-picture and calls into question the biblical mythology and the traditional teaching of the Church, it is presenting a new kind of criticism which is performing for faith the supreme service of recalling it to radical reflection upon its own nature.[b]

2. Bultmann reaffirms that no interpretation can take place in the absence of the framing of specific questions, even though the interpreter may not have formulated these questions explicitly or consciously. The Bible takes the form of a word which addresses me personally, but this fact I cannot anticipate or take into account as a systematic principle for my exposition of the Bible, since it is, in traditional language, the 'work of the Holy Ghost.' But, if faith in the Word of God can be the work of the Holy Ghost only as operating through intelligent decision, it follows that an understanding of the text can be reached only in systematic interpretation. Now, the terminology which governs this understanding can be acquired only from profane reflection; it is the business of the philosophical analysis of existence to provide this terminology.[a] It has become vital not to accept traditional terminology without critical reflection, but rather to take into consideration its nature and its source. What this means is that we must be concerned to use a terminology that is based upon the 'right' philosophy.[b]

3. Bultmann maintains that existential analysis provides the 'right' philosophy for the exegesis of the Bible. Apart from the decision to be human in the sense of being a person who accepts the responsibility for his own Being as shown by existential analysis, not a single word of Scripture is intelligible as having existential relevance.[a] But, if we cannot speak of an act of God without speaking simultaneously of our own existence, if an act of God cannot be posited apart from its existential reference, does divine activity have any objective reality (*Realität*) at all? Is not faith reduced to pure experience? Is God no more than a subjective experience in the soul, despite the fact that faith makes sense only when it is directed towards a God with a real (*wirklich*) existence outside the believer? By no means. This would follow only if faith and experience are interpreted in a psychologizing sense, as bare psychic phenomena. Human Being is properly understood

---

1.2:7[g] *GuV* I, 292-293 (*FaU*, 310-312).

1.2:7[h] Erich Frank, *The Role of History in Christian Thought*, 74-75 (quoted by Bultmann in *HE*, 153).

1.3:1[a] R. Bultmann, "Zum Problem der Entmythologisierung," in *KuM* II, 179-208; ET (of pp. 191-208 only), "Bultmann Replies to His Critics," in *KaM* I, 191-211.

1.3:1[b] Bultmann, in *KuM* II, 207 (*KaM* I, 210).

1.3:2[a] Bultmann, in *KuM* II, 191-192 (*KaM* I, 192).

1.3:2[b] Bultmann, in *KuM* II, 192 (*KaM* I, 193).

1.3:3[a] Bultmann, in *KuM* II, 193 (*KaM* I, 194).

only as historic Being, whose experiences consist of encounters. So the encounter of faith with the act of God cannot defend itself against the charge of illusion: the encounter with God is not objective like a worldly event. If faith tried to refute that charge, it would be misunderstanding its own meaning.[b] The true strength of faith lies in the fact that it is not demonstrable with regard to its object. If faith were susceptible of proof, we could know and ascertain God apart from faith, and that would place him on the level of the world at hand and accessible to objectifying thought. Faith is a listening to Scripture as the Word of God, not as a compendium of doctrines or a document enshrining the beliefs of other people, but as a word addressed personally to ourselves — as kerygma. Scripture becomes the Word of God only in the here and now of the encounter: the Word of God is not a fact that can be determined objectively.[c]

4. Bultmann avers that the invisibility of God excludes every myth which tries to make him and his acts visible, every conception of invisibility and mystery that is formulated in terms of objective thought. "God withdraws himself from objective view; he can be believed in only in defiance of all outward appearance, just as the justification of the sinner can be believed in only in defiance of the accusations of conscience."[a] Miracles, in the sense of acts of God, are not visible or ascertainable like worldly events, they are not interferences in worldly happenings. Miracles take place *in* worldly events in such a way that the closed weft of history as it presents itself to objective observation is left undisturbed. This is why the act of God is hidden to every other eye than the eye of faith.[b] The paradox of faith is precisely this: faith understands as the act of God an event which is ascertainable in its context in nature and history, which is recognizable as a unit in the series of worldly happenings that continues to be unbroken both on the level of science and on the level of everyday life. Mythological thought imagines the series to be torn asunder, but faith transcends the entire picture when it speaks of the activity of God. In the ultimate analysis, this picture is already transcended when I speak of myself, for I myself, my authentic self, am no more visible or ascertainable than is an act of God.[c]

5. Bultmann insists that in existential self-understanding there is also an understanding of the person or environment encountered. If, for example, my encounter with the love of another should afford me a new understanding of myself, this happening is not restricted to consciousness. But consciousness is here being considered as an existential phenomenon, not as a psychic phenomenon, as Thielicke and others wrongly suppose. "By understanding myself in this encounter I understand the other in such a way that the whole world appears in a new light, which means that it has in fact become another world; I acquire new insight into and new judgment regarding my past and my future, which means that they have become my past and future in a new sense."[a] Clearly such an understanding cannot be possessed as an abiding truth, since its validity depends upon its being constantly renewed and upon an understanding of the imperative which it involves.[b]

6. Hence, for Bultmann the nature of the other is apprehended *in* the love and *in* the trust of the existential encounter and this nature necessarily includes an element of risk. The ground and the object of faith are identical. What God is in himself becomes identical with what he does to us. Faith becomes a new understanding of ex-

---

1.3:3[b] Bultmann, in *KuM* II, 198-199 (*KaM* I, 199-200).
1.3:3[c] Bultmann, in *KuM* II, 199-200 (*KaM* I, 201).
1.3:4[a] Bultmann, in *KuM* II, 207 (*KaM* I, 210).
1.3:4[b] Bultmann, in *KuM* II, 196 (*KaM* I, 197).
1.3:4[c] Bultmann, in *KuM* II, 198 (*KaM* I, 198-199).
1.3:5[a] Bultmann, in *KuM* II, 201-202 (*KaM* I, 203): "Indem ich mich in dieser Begegnung verstehe, verstehe ich den Andern, und damit erscheint die ganze Welt in einem 'neuen Licht', d.h. aber: sie ist faktisch eine andere geworden; ich gewinne über mein Vergangenheit und Zukunft neue Einsicht und neues Urteil, d.h. aber: sie werden in neuen Sinne meine Vergangenheit und meine Zukunft."
1.3:5[b] Bultmann, in *KuM* II, 202 (*KaM* I, 203-204).

istence, and the act of God becomes that which affords to us a new understanding of ourselves.[a] Bultmann rejects a theoretical distinction between an act of faith which sees in the event of the cross the revelation of the love of God and a second act of faith — a radical change of self-understanding — for which the prior act sets us free. We cannot see and believe the revelation of God's love without at the same time being set free for a new understanding of self.[b] God encounters us in His Word — i.e., in a particular word, in the proclamation inaugurated with Jesus Christ. It is the Word of God only to the extent that it is a word which happens on definite occasions, and not in virtue of the ideas it contains. "It is only in this way that it is really the *verbum externum*: it is not a possession secured in knowledge, but an address which encounters us over and over again. This is why it is a word addressed *realiter* to me on a specific occasion, whether it be in the Church's proclamation, or in the Bible mediated through the Church as the Word of God addressed to me, or through the word of my fellow Christian. That is why the living Word of God is never a word of human wisdom but an event encountered in history. The fact that it originates in an historical event provides the credentials for its utterance on each specific occasion. This event is Jesus Christ."[c] The Word of God is such only in event, and the paradox is that this word is identical with the word which originated in the apostolic preaching, fixed in Scripture and handed on by men in the Church's proclamation. When a man just like myself speaks to me the Word of God, the Word of God becomes incarnate in him. The Incarnation is not a datable event of the past; it is an eschatological event which is being continually re-enacted in the event of proclamation. And the Church is really the Church only when it too becomes an event. "For the Church is the eschatological congregation of the saints whose identity with a sociological institution and a phenomenon of the world's history can be asserted only in terms of paradox."[d] The acts of God do indeed have "a social and corporate reference,"[e] but the Church, being an eschatological community, does not have a "social and corporate" nature. Hence, says Bultmann, the integration of faith which he has proposed is not guilty of excessive individualism.[f]

7. Bultmann says that he does not consider his concept of myth to be a question of any great importance.[a] A myth is an account of an event or happening in which supernatural, superhuman forces or persons are operative. Such accounts are often simply referred to as stories of the gods.[b] Scientific thought has to do with a closed nexus (*Zusammenhang*) of cause and effect. It is the fundamental complement or counterpart of mythical thought, because it assumes the unity of the world and the presence of an order, a law which everything existing and coming into the world must obey. Scientific thinking takes its departure from the question of the ἀρχή, that is, of the unifying source of the multiplicity of the world; it does not seek the origin of the world in an extramun-

---

1.3:6[a] Bultmann, in *KuM* II, 200 (*KaM* I, 202): "Es gibt kein Vertrauen und keine Liebe ohne Wagnis. So fallen — wie auch schon Wilh. Herrmann lehrte — Grund und Gegenstand des Glaubens nicht auseinander, sondern sind ein und dasselbe, und zwar eben deshalb, weil wir von Gott nicht sagen können, wie er an sich ist, sondern nur, was er an uns tut."

1.3:6[b] Bultmann, in *KuM* II, 202 (*KaM* I, 204).

1.3:6[c] Bultmann, in *KuM* II, 204-205 (*KaM* I, 206-207).

1.3:6[d] Bultmann, in *KuM* II, 206 (*KaM* I, 209-210).

1.3:6[e] Bultmann, in *KuM* II, 206 (*KaM* I, 210), quoting Amos Wilder, *Eschatology and Ethics in the Teaching of Jesus* (New York: Harper, 1950), 65, who uses this expression against what he sees to be the "excessive individualism" of Bultmann's interpretation.

1.3:6[f] Bultmann, in *KuM* II, 206-207 (*KaM* I, 210).

1.3:7[a] Bultmann, in *KuM* II, 180. Bultmann says that he accepts the concept of myth as commonly used in the sciences of history and religion (*Geschichts- und Religionswissenschaft*), and he refers the reader to W. Nestle and B. Small.

1.3:7[b] Bultmann, in *KuM* II, 180: "Mythos ist der Bericht von einem Geschehen oder Ereignis, in dem übernatürliche, übermenschliche Kräfte oder Personen wirksam sind (daher oft einfach als Göttergeschichte definiert)."

dane power or divinity, for it sees that source as immanent and ever-present.[c] To this unity of the world there corresponds the unity of scientific thought, due to the rational argument it must provide for every conclusion, and in this, again, it is distinguished from the incoherence of mythical narration. Scientific thought is indeed open, not to the intervention of supernatural powers, but to the new discoveries that increase its knowledge of the world. It is true that the present knowledge of science can change and even be reversed, but that does not open the door to the intervention of powers from the beyond. It is the method that is definitive in science, not the conclusions.[d]

8. The man who is liberated from mythical thinking, Bultmann explains, sees himself to be a unity in his feeling, his thinking, and his willing: he does not countenance the intervention of any outside powers, diabolical or divine. It follows from this that scientific man does not contemplate interventions of God, of the devil, or of demons.[a] Myth expresses in language that is no longer suitable the awareness that the world in which man must live is filled with puzzles and mysteries. Myth is therefore the expression of a definite manner of understanding human existence. It knows another reality (*Wirklichkeit*), different from the reality (*Wirklichkeit*) that science has in view. It knows that the world and human life have their ground and confines in a power which lies on the other side of all the things that can be brought within the reach of human calculation and disposition, but it speaks of this reality of the beyond in the inadequate terms of spatial distance.[b] Demythologizing is based on the awareness that we cannot say anything about what lies beyond the world or about God; we realize that we cannot say what the other side and the world are 'in themselves,' because in trying to do so we objectify God and the other side into worldly phenomena of this side.[c] Demythologizing is therefore a hermeneutical method: it does not eliminate the mythological expressions; it merely provides them with an existentialist (*existential*) interpretation.[d]

9. Bultmann points out that Helmut Thielicke and others who claim that mythological language is indispensable to the human mind attempt nevertheless to escape from mythical thought by understanding the mythological elements as 'images' or 'symbols.' In doing so they are refuting their own position, because they are embarking on a process of demythologizing without even realizing the fact.[a] But it is not a case of an either/or choice between science and mythology. Just as one seeks the existential meaning of myth, so could one seek the existential

---

1.3:7c Bultmann, in *KuM* II, 180. He adds that it is absurd to call this scientific understanding of the world a 'myth' or to appeal to the 'myth of the twentieth century.'

1.3:7d Bultmann, in *KuM* II, 181. He takes occasion here to deny that the relativizing of the laws of causality in atomic processes makes an opening for the intervention of forces from the beyond (*jenseitiger Mächte*), since scientific thinking consistently retains the need of the λόγον διδόναι.

1.3:8a Bultmann, in *KuM* II, 181-182. He observes also that the non-materialist historiographer does recognize the existence of spiritual forces and of the persons who are their bearers. For the distinguishing of a mythological from a real event, he cites the rules formulated by C. Hartlich and W. Sachs ["Einfürung in das Problem der Entmythologisierung," in *Für Arbeit und Besinnung*, 4 (1950), 436]: "That is mythological which cannot really have happened (*was nicht wirklich geschehen sein kann*) because: 1) it cannot be ascertained according to the general rules of science: miracles are impossible; 2) it conflicts with the assumed unity of personal life: the only means that God can use is spirit as understandable; what is not spiritual cannot transmit the divine; 3) it contradicts the rules of morality; 4) it has no redemptive meaning in a sense that is meaningful for the personal life of the individual person (*KuM* II, 182)

1.3:8b Bultmann, in *KuM* II, 183.

1.3:8c Bultmann, in *KuM* II, 184. In *KuM* II, 185, Bultmann quotes W. Hermann [*Die Wirklichkeit Gottes* (1914), 9]: "Gottes Wirklichkeit liegt jenseits alles dessen, was die Wissenschaft beweisen kann. Sehen wir das ein, so wird dadurch unser Glaube nicht geschwächt, sondern an das Verborgene erinnert, das seine Stärke ist."

1.3:8d Bultmann, in *KuM* II, 185.

1.3:9a Bultmann, in *KuM* II, 185-186.

meaning of science, once the methodology of science has been clearly recognized.[b] In fact, expressions like 'I love you' and 'Please excuse me' are neither scientific nor mythological. There is a language in which existence is expressed ingenuously; there corresponds to this language a science that speaks of existence without objectifying it.[c] It is absurd to reproach demythologizing for seeking a scientific content in the Gospel, where by science is meant the science of objectifying thought, because what demythologizing seeks is an understanding of Scripture that is free of every worldly image, be it that of myth or that of science.[d] Demythologizing means posing the problem of the conceptuality according to which the underlying intention of the biblical statements can have found expression. It is the problem of a language that preaching and faith can speak without occasioning misunderstandings. It is, of course, evident that the preaching of faith does not presuppose reflection on its own conceptuality, but when the situation becomes such that the interpretation of Scripture is no longer certain — is, in fact, a matter of dispute — and preaching is misunderstood to the point of non-comprehension, then there arises an absolute need to reflect, to ponder over the conceptuality that pertains to exegesis, to preaching, and to the profession of faith. Since a reflection of this kind is a task of theology, more precisely, of hermeneutics, it is therefore a task of science, at least to the extent that science serves to express in words the thing that is being treated.[e] What kind of science is it? Bultmann is thinking of a science which is nothing other than the clear and methodical development of the understanding of existence given with existence itself, in contrast with objectifying science, which is nothing other than the methodical development that proceeds from thought put to work for the purpose of gaining domination of the world. As long as faith recognizes itself as self-understanding, there is no danger of its falling under the tyranny of a science that is essentially foreign to it. Nor is existentialist interpretation a foreign science ruling faith, for the original *existentiell* rapport of the believer with the Word of God in Scripture is a *prescientific* understanding that antecedes all existentialist interpretation. This means that existentialist interpretation does not provide the basis of faith. What such interpretation does is enable faith to preserve itself from a mistaken understanding of itself.[f]

10. In a later general reply concerning the problem of demythologizing,[a] Bultmann explains that by 'demythologizing' he means "a hermeneutical method which examines mythological statements and texts for their reality-content."[b] It is understood that myth speaks in an inadequate manner of some reality (*Wirklichkeit*), and in this a particular understanding of reality (*Wirklichkeit*) is also presupposed. 'Reality,' however, can be legitimately understood in two different ways: in the common manner as the objective representation of the world in which man finds himself; [c] in a special way

---

1.3:9b Bultmann, in *KuM* II, 183.
1.3:9c Bultmann, in *KuM* II, 187: "Es gibt doch wohl eine Sprache, in der sich Existenz naiv ausspricht, und es gibt entsprechend eine Wissenschaft, die ohne die Existenz zum welthaften Sein zu objektivieren, von der Existenz redet."
1.3:9d Bultmann, in *KuM* II, 187. Four pages earlier Bultmann says that science is objectifying thought: "...one could call mythical thought primitive scientific thought... above all to the extent that it is objectifying thought, as is the thought of science." (*KuM* II, 183.)
1.3:9e Bultmann, in *KuM* II, 188. He maintains that it is deceitful for the preacher to pass off improvised mysteries like the Trinity as though they were the genuine mystery of the act of God. The mystery of the Trinity, he says, is a mere *x* (a sheer unknown quantity) in the light of which neither preacher nor hearer can distinguish anything thinkable (*KuM* II, 190).
1.3:9f Bultmann, in *KuM* II, 189.
1.3:10a R. Bultmann, "Zum Problem der Entmythologisierung," in *KuM* VI-1, 20-27.
1.3:10b Bultmann, in *KuM* VI-1, 20: "Unter *Entmythologisierung* verstehe ich ein *hermeneutisches Verfahren*, das mythologische Aussagen bzw. Texte nach ihrem Wirklichkeitsgehalt befragt."
1.3:10c Bultmann, in *KuM* VI-1, 20: "...die ein objektivierenden Sehen vorgestellte Wirklichkeit der Welt, innerhalb deren sich der Mensch vorfindet..."

as the reality of the historically existing man.ᵈ If a person, according to the more common usage of the word, makes himself an object, as is done in sociology and in 'explanatory' psychology (as distinguished from the 'understanding' psychology of Dilthey), he thereby reduces his ownmost reality to the reality of the world. Contemporary thinkers are realizing more and more that the historian cannot place himself opposite history in a subject-object relationship, since the act of viewing the historical process is itself an historical process and the apparently objective image of an historical event bears always the imprint of the individuality of the viewer.ᵉ

11. Bultmann says that what we call 'existence' is the specifically human way of existing proper to man. Since man must take possession of himself, his life is history. According to existentialist interpretation, history is the field of human decisions. The possibility of being either authentic or unauthentic belongs to historicity, that is, to human reality. The reality of man is his history, which stands always in front of him to such an extent that being for the future (*Zukunftigsein*) is the reality in which man stands. The historical meaning of a happening in the history of mankind can be understood only in terms of its future. The future is thus an essential element of the event, but because the end of human history is not in view, the full meaning of human events cannot be grasped except as the meaning of the moment in the sense of the place of decision.ᵃ

12. Bultmann admits that reality is one and that there is only one true proposition regarding a single phenomenon. But the one reality can be viewed under either of two aspects, corresponding to man's possibility of existing genuinely or not: of understanding unauthentically in terms of what is at one's disposal, or of understanding authentically in terms of the future that is not at one's disposal (*unverfügbar*). Correspondingly, man can consider the history of the past (*die Geschichte der Vergangenheit*) in a merely objectifying way or he can regard it as a call to responsible decision.ᵃ The existentialist interpretation of history must function in dialectical relationship with the objectifying view of the past that accompanies it. Existentialist interpretation cannot grasp the historical meaning of an event apart from a careful ascertaining of the facts bearing on the case. Since existentialist interpretation operates on facts, it is not a product of fancy.ᵇ

13. Bultmann contends that it is the intention of myth to speak of a reality beyond the reality that is observable and objectifiable. Demythologizing seeks to bring to light this genuine reality of man. God is not an objectively perceivable phenomenon of the world. Therefore, we can speak of an act of God only inasmuch as we are speaking of our existence as encountered through the act of God.ᵃ This way of speaking of the act of God may be called 'analogical.' It is in this manner that faith recognizes God to be the Creator and Ruler of everything that happens in nature and in history. This confession of faith can never have the character of a universal truth, valid for everybody, like a thesis of science or of the philosophy of history, for that would inevitably objectify the act of God. The affirmation of God's creating and ruling can therefore have its legitimate basis only in the understanding that a man has of himself.ᵇ

1.3:10ᵈ Bultmann, in *KuM* VI-1, 21: "... als die Wirklichkeit des geschichtlich existierenden Menschen."
1.3:10ᵉ Bultmann, in *KuM* VI-1, 20-21.
1.3:11ᵃ Bultmann, in *KuM* VI-1, 21-22.
1.3:12ᵃ Bultmann, in *KuM* VI-1, 22-23.
1.3:12ᵇ Bultmann, in *KuM* VI-1, 23.

1.3:13ᵃ Bultmann, in *KuM* VI-1, 25: "Da Gott kein objektiv feststellbares Weltphänomen ist, lässt sich von seinem Handeln nur so reden, dass zugleich von unserer Existenz geredet wird, die durch Gottes Handeln betroffen ist."
1.3:13ᵇ Bultmann, in *KuM* VI-1, 25-26.

ARTICLE 2

THE LOCUS OF SCIENCE IN HUMAN CONSCIOUSNESS

2.1. Intellectual Consciousness

1. What is characteristic of man in that vital act of awareness called consciousness is the power of intelligence.[a]

The normal person in his waking consciousness is immediately and continuously aware of himself as conscious and of the reality by which his conscious self is surrounded.[b] It is a

---

2.1:1[a] The word *conscientia* (*cum+scientia*) was used by classical Latin writers sometimes to denote knowledge shared by more than one person but more often simply to signify definite and unwavering personal knowledge, especially of something good or evil (cf. A. Forcellini, *Lexicon totius Latinitatis* [Patavii, 1940], "Conscientia"). One was described as 'conscious' (*conscius*) who shared knowledge with someone or especially who had personal knowledge of a moral act affecting himself. Even in the special sense the adjective *conscius* came to replace the original *scius* because it was more euphonic (cf. Forcellini, *op. cit.*, "Conscius"). St. Thomas teaches that the word *conscientia* in its original sense means knowledge referred to something else (*cum alio scientia*) and thus implies the application of knowledge to something else (Aquinas, *Summa Theologiae* [Turin: Marietti, 1962] [hereinafter referred to as *S.Th.*], I, q. 79, art. 13). The question is whether those who have traced the origin of the word *conscientia* considered the possibility that it signified an awareness of the double character of intellectual consciousness with particular reference to the subjective element. John Locke was the first to use (in any language) the abstract noun 'consciousness' (cf. "Consciousness" in the *Oxford English Dictionary*). In his *Essay Concerning Human Understanding* (1st pub. 1690; repub. in 2 vols., Everyman's Library: London, Dent, rev. ed., 1965) he speaks of "that consciousness which is inseparable from thinking and, as it seems to me, essential to it" (*ed. cit.*, I, 280). Again he says that "since consciousness always accompanies thinking, and it is that that makes everyone to be what he calls *self*, and thereby distinguishes himself from all other thinking things: in this alone consists *personal identity*, i.e. the sameness of a rational being" (*ibid.*, 281; cf. *ibid.*, 283-291). He means "that consciousness whereby I am my *self* to *myself.*" (*Ibid.*, 290.) Not all modern languages have distinct words for 'consciousness.' For instance in Italian *la coscienza* is "*la consapevolezza* [knowledge or awareness] *e la giusta valutazione che uno ha di se stesso e dei propri atti o pensieri.*" (F. Palazzi, *Novissimo Dizionario della lingua italiana* [Milano, 1939].) But these languages often allow the distinction between 'moral conscience' and 'psychological conscience.' Thus in French (and in Italian) 'psychological conscience' is "*l'aperception par laquelle l'homme se connaît lui-même dans une vue intérieure.*" (A. Chollet, "Conscience," in E. Amann, ed., *Dictionnaire de Théologie catholique*, Vol. III-1 [Paris, 1923; cols. 1156-1174], 1158.)

In classical Latin the verb *intelligere* (most commonly thought to have been derived from *intus legere*) denotes "mental perception" (Forcellini, *Lexicon*, "Intelligo"). In English the word 'intelligence' denotes a single broad mental power, a cognitive capacity which is inborn in human beings (R. B. Cattell, "Intelligence," in *The Encyclopedia Britannica*, XII [1970], [345-347], 345).

2.1:1[b] "The prime fact of all experience is consciousness. Roughly, we mean by consciousness that state in which we are when we are said to be wide awake, in contrast to a profound, dreamless slumber. We all know the difference, in spite of the fact that behaviorists profess they cannot understand what certain psychologists talk about when they refer to consciousness ... We may describe normal consciousness as a state of the human mind, in which it can perceive and interpret its surroundings and in which the past experience of the individual is fully available [not patho-

curious but obvious fact that human consciousness is always thus divided into a knowing subject and the object of his knowledge. Because of this division, the knowing self as such is never in human consciousness an *object* of knowledge except indirectly, but it pertains to human consciousness for the knowing self to know that he is knowing.[c]

2. The reason for this characteristic division of human consciousness into a subjective and an objective area lies in the nature of human intelligence.[a] While the entire range of consciousness is composed of multiple elements and degrees, the division of consciousness into a subjective and an objective part seems to be the act which makes consciousness properly human,[b] for it enables the human person to distinguish himself from the objects of his knowledge and to gather the latter around the pole of objective intelligibility.[c]

3. These two essential components

---

logically faded or blocked] for the adjustment of the individual to the needs of the present moment." (T. V. Moore, *Cognitive Psychology* [Chicago: Lippincott, 1939], 3.)

2.1:1c The knowing *subject* cannot, in its primary and original function, be an *object* of knowledge, but in a secondary and derived manner it can. While the term 'consciousness' has been extended by William James and others to cover all of the contents of experience of which an individual may be aware at a given moment, the present discussion is centered upon the area of consciousness which we have called 'intellectual consciousness.' We are assuming that rational solutions to problems of consciousness must be situated within the unique kind of awareness that is intellectual consciousness. If, in fact, intelligent solutions become visible only within a proper focus, it is then only within the focus called intellectual consciousness that the relevant questions will arise, and that is the sense in which Edward Holloway's definition of the mind of man as "that centre of unity and meaning within himself, through which he focuses everything else" seems to be valid (*Catholicism: A New Synthesis* [Bedford, Eng., 1969], p. 4). Since a visual focus is a medium between the observer and his object, we may now begin to ask whether the *intus legere* behind the word intelligence (*supra*, 2.1:1b) was limited to a supposed 'reading into' the objects of sensory perception, or whether it implied also and even more fundamentally a 'reading inside,' in the sense of the vision of intellectual objects presented in the focus that is the human intellect. Cf. T. Aquinas, *In libros I et II Posteriorum Analyticorum expositio* (Turin: Marietti, 1955[2]), Lib. II, Lect. XX, No. 14: "Sic enim, scilicet per viam inductionis, sensus facit universale intus in anima, in quantum considerantur omnia singularia." For St. Thomas intelligence is the second stage (*actus secundus*) in the actuation of the possible intellect, and it consists in the act of 'intellecting,' or considering. Now, we may ask whether 'insight' pertains more directly to the active potency or the passive potency of the intellect. It seems to the present writer that insight is at least as much the proper act of the *intellectus agens* as it is the *actus secundus* of the *intellectus possibilis*. Indeed, the grasping of what insight is must derive from the recognition of the divided character of intellectual consciousness and from the intent to include that subjective element which seems to characterize human intelligence (as distinct from the intuition of brute animals), so that the very problem of insight becomes one of finding the place and function of objectivity within the realm of human subjectivity and vice versa.

2.1:2a It should be clear that the present essay is an expression of the dichotomic theory of intelligence, according to which dichotomy is posited as the most fundamental and characteristic act of human intelligence as such. The notion of dichotomy is used both to elucidate the basic composition of human intelligence and to locate it within the field of general consciousness. Cf. T. Aquinas, *In Aristotelis librum de Anima commentarium* (Turin: Marietti, 1959[4]), Lib. III, Lect. XI.

2.1:2b "... consciousness of anything implies and depends on an ability to differentiate an 'I' from a 'non-I' world..." (L.S. Kubie, "Psychiatric and Psychoanalytic Considerations of the Problem of Consciousness," in J. F. Delafresnaye, ed., *Brain Mechanisms and Consciousness* [Oxford, 1954], 454). It seems that it is precisely the act of distinguishing the 'I' from the world of objects that constitutes the crossing of the threshold of intellectual consciousness.

2.1:2c "An individual who is equipped with multiple receiving and responding apparatus can gradually differentiate be-

of the primordial dichotomy of human intelligence — the knowing subject and the object of his knowledge — are at least implicitly known to all human beings who are actually conscious. The awareness of objects pertains in some way to other animals as well; the awareness of self derives from an *act* which is distinct from the 'self' of which the conscious person is aware; it is at first identified with the power to think and later with the power of reason. The act of intelligence is not adverted to at the very beginning of human conscious life, but it is there from the beginning as the ground of intellectual consciousness. The growth of human understanding from the initial awakening of intelligence takes place always in the presence of the three primordial components of intellectual consciousness.[a] The viewpoint of the adult is not the same as it was when he was an infant. He has grown in experience, knowledge, and understanding. His view of himself has changed as has also his view of the world. But in the normal instance the primordial factors will in an important sense have remained constant. As regards the awareness of self, the knowledge of oneself should have increased, crises may have been undergone, conversion and even mystical enlightenment may have been experienced, but the initial awareness of one's own identity will not have changed.[b] As regards the awareness of the other than the self, an enormous succession of objects will have passed in review before the conscious subject, but the essential relationship of subject to object will not have changed. And the native intelligence of the person will be the same. Yet development will have taken place.

4. It is an easily recognized fact that intelligence does not function without the presence of sensory data.[a] If one is unconscious, he will not resume his thinking until he has 'come back to his senses.' And he needs a succession of sensory images in order to be able to pursue his thinking. The factween an 'I' world, i.e., between a somatically and psychologically internal world and a somatically and psychologically external world. Out of this learning process comes the self-aware component which is always either latent or explicit in the process of consciousness." (Kubie, *op. cit.*, 455.) We may say that out of the learning process comes the *explicitation* of the self-awareness already vaguely present from the beginning of intellectual consciousness. We place the dividing-line between the characteristically human and the subhuman in the concept of interiority, at least in the minimum conceptual sense of the vague awareness that the meaning of sensory objects present is in the focus of one's own mind. Thus the vision of sensory objects which is intelligence stands on the subjective side of a dichotomy which does not even exist for animal intuition.

2.1:3a The three primordial components of intellectual consciousness are all indefinite at the beginning of intellectual development. Last of the three to be recognized explicitly is the act of knowing, but it is present from the beginning both as the passive power to know and the active dynamism of knowing. If we assume that the act of knowing has both an active and a passive *aspect* (rooted in the initial dichotomy of intellectual consciousness), we can locate the act of knowing somewhere *between* the poles of pure objectivity and pure subjectivity in such wise that the passive aspect of knowing (the *intellectus possibilis* of Aristotle and Aquinas) is tied to the pole of objectivity and the active aspect (the *intellectus agens*) is tied to the pole of subjectivity. A valid theory of knowledge will both bring into view the objective side of the act of knowing and provide the insights that constitute knowing what it means to know.

2.1:3b "...consciousness implies and requires a differentiation between an 'I' and a 'non-I' world, with a parallel evolution of consciousness of self and of things outside of self." (Kubie, *op. cit. supra*, 2.1:2b, p. 456.) Underlying the dichotomized perception of meaning which is intelligence is the initial division of the known from the knower. It is upon this fundamental dichotomy that all subsequent acts of intelligence are based. While the knowing subject as such remains in consciousness no more than a mere 'existent' devoid of all intelligibility, it nevertheless constitutes the center around which intelligence develops and thus is the direction towards which the thoroughly intellectual area of interiority comes into view as objects lying between sense experience and the knowing subject are elucidated.

2.1:4a *S.Th.*, I, q. 84, art. 7.

ulty of the intellect can be illustrated by the example of the faculties of sense. While any sense might be used, the one which lends itself most to the illustration is the faculty of sight.[b]

5. According to the analogy of sight, intelligence is the power of 'seeing' the intelligibility of things. (By the parallel analogy of feeling, one could be said to 'touch' or to 'grasp' this intelligibility.) To know is to 'see' the meaning of things. This is to say that things 'appear' in the 'light' of their meaning and are collected by the 'light-gathering power' of the human intellect. When one understands, he says that he 'sees' what is meant, and this act of 'seeing' is a valid and meaningful experience, even though it is expressed through analogy with an act of sensory perception.

6. By an illustration based upon the analogy of sight, we can describe the psychological self as situated at the proximate edge of intellectual consciousness, looking through the 'window' of intelligence at the world of objects illuminated by their intelligibility. This illustration is analogously true of the objects of intellectual vision. There is the subject who sees and there is the illuminated field of what he sees.

7. The question arises of the validity of the viewpoint of the "self standing at the window" and of the world he

---

2.1:4[b] Cf. *S.Th.*, I, q. 12, art. 2; I, q. 67, art. 1. Jean Piaget maintains that intelligence is not a faculty, not one form of structuring among others; it is rather "the form of equilibrium towards which all the structures arising out of perception, habit and elementary sensorimotor mechanisms tend." He holds that there is "a radical functional continuity between the higher forms of thought and the whole mass of lower types of cognitive and motor adaptation," in the sense that "intelligence itself does not consist of an isolated and sharply differentiated class of cognitive processes," but is "only a generic term to indicate the superior forms of organization or equilibrium of cognitive structurings." (*La psychologie de l'intelligence* [Paris, 1947]; ET, *The Psychology of Intelligence* [London: Routledge, 1950], pp. 6-7.) It is Piaget's studied conclusion that intelligence is perhaps identified with or characterized by the whole of behavior or with one of its general aspects (*La naissance de l'intelligence chez l'infant* [Paris, 1935], pp. 375-376; ET, *The Origin of Intelligence in the Child* [London: Routledge, 1953], p. 374). The basis of this conclusion is the 'theory of assimilation' (*La naissance*, 412-425; *The Origin*, 407-419), which Piaget accepts in contrast to 'vitalistic intellectualism' and to other theories of knowledge. He refuses to look upon intelligence as a faculty in the sense of "a mechanism complete in its structure and in its operation," for this would be to posit as irreducible, not intellection as such, but "a certain reification of that act in the form of a given mechanism in the completely formed state." (*La naissance*, 373; ET, 372.) But he admits that the same reasons which support vitalism in biology favor intellectualism in psychology: "the difficulty of accounting for intelligence, once it has been achieved, by anything else than its own organization considered as a self-sufficient totality... From that to making intelligence a power *sui generis* (just as vitalism makes the organism the expression of a special force) is only a step." (*La naissance*, 371; ET, 369-370.)

There is a current of imprecision in Piaget's exposition. What the faculty psychologist regards intelligence to be is not self-sufficient but self-explanatory. Piaget does not attempt to face the problem which he himself mentions that intelligence cannot be accounted for in any other way. In his exposition he must repeatedly have recourse to the specific intellection which his theory rejects. For instance, when he says that intelligence is "only a generic term," he overlooks the fact that in making the statement he must depend upon the other generic terms in the same sentence: 'superior forms,' 'organization,' 'equilibrium,' and 'cognitive structuring' (cited above). To disparage one generic term on the ground that it is only a generic term calls into question the set of generic terms on which his own theory is woven. The oversight lies in not having determined the value of the generic terms in his own train of thought. Curiously, it is the tendency to dissolve otherwise specific concepts into formless generic terms that seems to be the chief error in Piaget's method. In approaching the whole subject of his study he immediately dissolves the rather specific concept of intellection into "the form of equilibrium towards which all the (lower) structures.... tend," thus substituting an empty goal-idea which tells us nothing about what the act of intel-

claims to see. Is he aware of a real self, substantially existing in a real world, or could he not be but a tenuous character in a dream of a god or of an angel or of a demon? Is the world outside his mind even remotely similar to the image of it in his mind? Is the world of intelligible being really divided into the dichotomies presented by his intellect in its effort to understand?

8. It is true that the self at the window could not demonstrate his own reality to anyone except a character in the same imaginary dream, but he has no real need to, since his own reality is self-evident in consciousness. And he need not fear that he himself is dreaming, for when he awakes from a dream he knows that he is awake. The world of his waking consciousness is derived from an internal reconstruction of stimuli from without, but he has no reason to fear that the world without is that much different. Indeed, the telescope is constructed by an imitation of the optics of the human eye, and whoever surveys the world through a telescope is looking at a reconstructed image. Yet he knows and he can verify that the image is a faithful reproduction. There is no reason to think less of the eye. And the division of the self from the otherness of intelligible being which is the first act of intelligence does indeed suggest that simpler beings might know it in another and better way. But this is the condition of human consciousness, and man only contradicts his nature by obstructing its development in the way that it must develop or by entertaining doubts about what is given in consciousness. It is, in fact, a contradiction of human reason to use what is self-evident to

---

ligence is in itself. Furthermore, he limited his observations to human offspring, thus assuming in advance the specific difference which he subsequently denies. He cannot, therefore, tell us why the "mass of lower types of cognitive and motor adaptation" in other animals does not tend towards intelligence as its form of equilibrium. He overlooks the probability that the tendency towards intellection in humans is governed by a capacity of intellection present and functioning at least as a natural dynamism over the whole gamut of his observations. Again, whoever says that intelligence does not consist of a distinct and sharply differentiated class of cognitive processes is thereby simply admitting that he has never succeeded in reaching a clear and distinct concept of intelligence, because a clear and distinct class of cognitive processes can be clear and distinct only in the mind of one who knows it clearly and distinctly. The same is true of the "lower types of cognitive and motor adaptation," since a 'type' is the same as a 'class.' Piaget adheres to a whole set of semi-distinct generic ideas of his own choosing, while rejecting in theory the unique intellectual base from which all concepts receive their scientific validity. To say that intelligence is nothing more specific than a form of equilibrium is merely "giving a name which stops investigation." (T. V. Moore, *Cognitive Psychology*, 5-6.)

Piaget sees no need to admit that "an invariable structural mechanism of intelligence exists," seeing that it does not differ distinctly from behavior in general. Thus, "adaptation — intellectual and biological, hence adaptation of intelligence to 'things' as well as of the organism to its 'environment' — always consists in a balance between adjustment and assimilation... In short, the biological interpretation of intellectual processes based upon the analysis of assimilation does not result at all in the epistemological realism of vitalistic intellectualism." (*La naissance*, 375-378; ET, 374-376.) The error lies in confusing a general feature of cognition with the specific character of cognition. Physical sight is also a balance between adjustment and assimilation, as is any other living act. Yet between physical sight and the external world there lies the 'invariable structural mechanism' of the eye. Piaget's argument for the exclusion of the intellect would exclude just as much the organs of sense. From the genus of cognition to the species of intellectual cognition is only one step, but Piaget fails to take that step. He does not succeed in realizing that "the biological interpretation of intellectual processes" is itself an intellectual act standing wholly within the order of intellection and not at all in the biological order — an act based upon the prior intellectual distinction between the biological and the psychological.

Piaget defines perception as "the knowledge we have of objects or of their movements by direct and immediate contact, while intelligence is a form of knowl-

the conscious subject in order to deny that the same self-evident things are valid. The only alternative to playing the game of life by the rules of human consciousness is the frustration of one's life. The drama of life is lived in the little theater of one's own conscious experiences. Outside the intelligible area of this theater are the unperceived *sources* of evident experience: the non-conscious substance of man, the aspects of the external world which are not accessible in any way to human sensation, the human intellect itself, as the unperceived unitary source and ground of that divided mode of being which is intellectual consciousness. Man is implicitly conscious that his world is but a tiny illuminated area in the midst of the surrounding darkness. But his light is his hope. Much of the darkness he will not penetrate in the short span of his mortal existence. But he can move his candle about in the darkness, and by doing so he can bring an enormous amount to light, so much that even the terms of his original problems may change into challenging new questions and unexpected answers.

9. Doubts about the human perspective pertain to the same question of the viewpoint of man. It is true that the outlook would be somewhat different, if man were a thousand times bigger than he is, so that he could eat a whole cow in a sandwich, or a hundred times smaller, so that he would

edge obtaining when detours are involved and when spatio-temporal distances between subject and objects increase." (*The Psychology of Intelligence*, 53.) But the evidence is to the contrary. Just as there is in fact no direct contact with the objects of sense perception, since all sense knowledge comes *through* the organs of sense, so the objects of intellectual perception may have no greater spatio-temporal distance than have the objects of sense perception. Does the intellectual perception of the square root of 25 have greater spatio-temporal distance from the knowing subject than has the physical perception of the star Arcturus? It seems that intellectual objects are *closer* to the subject in spatio-temporal terms, since they stand between him and sensory objects. Intellectual objects are indeed more abstract and therefore more distant from purely material conditions. Piaget's description of the two levels of intelligence does not therefore seem to be factual. "There exists a sensorimotor or practical intelligence whose functioning extends that of the mechanisms of a lower level: circular-reactions, reflexes and still more profoundly, the morphogenetic activity of the organism itself." (*La naissance*, 357; ET, 357.) "From the viewpoint of the structural mechanism, elementary sensorimotor adaptations are both rigid and unidirectional, while intelligence tends towards reversible mobility," but from the viewpoint of the functional situation, "we can say that behavior becomes more 'intelligent' as the pathways between the subject and the objects on which it acts cease to be simple and become progressively more complex." (*The Psychology*, 10-11.) Is there a factual basis for saying that the perception of a rainbow or the unlocking of a door involves "simpler pathways" between subject and object than does the division of 30 by 6? Piaget confuses psychological distance (abstraction) with physical space, and he reifies Thought in order to provide a vehicle for a journey through the physical universe. "Thought alone breaks away from these short distances and physical pathways, so that it may seek to embrace the whole universe including what is invisible and sometimes what cannot be pictured; this infinite expansion of spatio-temporal distances between subject and objects comprises the principal innovation of conceptual intelligence and the specific power that enables it to bring about operations." (*The Psychology*, 121.)

Piaget insists that "from the viewpoint of function" he sees no need to admit a fundamental difference between sensorimotor and conceptual co-ordinations, since "functional continuity in no way excludes diversity or even heterogeneity among structures." (*The Psychology*, 6-7.) But this radical functional continuity of the intellectual with the biological is scientifically useless to the extent that Piaget's distinction between the 'structural viewpoint' and the 'functional viewpoint' is not accounted for either in its origin or in its validity. This distinction implies a viewer and the stable frame of reference which sets Piaget's discourse in a distinctively intellectual milieu. When he uses the 'biological viewpoint' to discount the functional uniqueness of intelligence, he is simply failing to see the uniquely intellectual character of his 'biological viewpoint,' and is thus constructing his whole theory on a misunderstanding.

have to flee for his life at the sight of a mouse. The implications of such alternatives have been treated with suitable art and sobriety by writers like Jonathan Swift.[a] But the search for the viewpoint that a man would have, if he were as big as the Milky Way or so small that he could live on the face of a proton, as he now lives on the face of the earth, loses all meaning, for the terms of the question are absurd, since the components of man could not be put together in those dimensions. Nor shall we ever know what the world seems like from the point of view of a grasshopper or a fly, though the study of their instincts can be rewarding. Man is endowed with his own point of view; his task is to develop it.

2.2. The Discovery of Truth

1. How does the human mind develop? If human knowledge is a form of vital awareness whose unity and totality always include a knowing self, an object known, and an act of knowing, then the development of knowledge must either be the result of a development of these components and of their relationship to one another or the addition of some new component.

2. It is an interesting and verifiable fact that, although the knowing self and the object of his knowledge are essential components of consciousness, the mind can move by concentrating attention upon one component or the other, in such wise as to put its counterpart out of focus while not forgetting it entirely. Such concentration brings into focus new areas and new dichotomies. It is the gradual bringing into intelligible focus of new objects and new areas by means of these dichotomies that constitutes one principal factor in the dynamism of the mind. The other principal factor is the activity of reuniting in mental unity the extramental being that it has brought into focus by its act of division. This work of organization and unification is called synthesis; it is joined to its counterpart, analysis, in dynamic and dialectical union as the reciprocal components and mutual contributors to that process which we call learning.[a] The synthesis of the two parts of the fundamental dichotomy of intellectual consciousness (the subjective and the objective areas) yields the notion of the mind. Hence, the putting together of that fundamental synthesis might well be styled the discovery of the mind.[b]

2.1:9[a] J. Swift, *Gulliver's Travels*.
2.2:2[a] Cf. Aquinas, *De Anima*, Lib. III, Lect. XI, Nos. 746-764.
2.2:2[b] Many empirical psychologists, while plainly rejecting the image of a 'mind-stuff,' nevertheless uncritically entertain the notion of a vague thing which they call 'the mind' or 'Thought,' atttributing to it the enclosed identity that belongs properly to the intellect. The mind is not a self-standing psychological 'bag' floating in an ocean of experience; nor is thought a 'bubble' of some kind on an amorphous sea of energy. The mind is better regarded as the abstract concept resulting from the unification of the objective and the subjective aspects of consciousness in a single identity, while the intellect is the concrete source from which the capacity to be a mind originates. Cf. F. H. Bradley, *Appearance and Reality* (Oxford, 1893[1], 1897[2]), 76, 144-145.
William James maintains that 'mind' in the terminology of the psychologist "is only a class name for *minds*... To the psychologist, then, the minds he studies are *objects*, in a world of other objects." (*The Principles of Psychology* [London:

Macmillan, 1910] [1st pub. 1890], Vol. I, 183.) "The mental state is aware of itself only from within; it grasps what we call its own content, and nothing more. The psychologist, on the contrary, is aware of it from without, and knows its relations with all sorts of other things. What the thought sees is only its own object; what the psychologist sees is the thought's object, plus the thought itself, plus possibly all the rest of the world." (*Ibid.*, 196-197.) In rendering this tribute to the psychologist, James overstates the rôle of his own field of study. In dismissing the concept of 'mind' as a mere class name, he fails to distinguish the *kind* of object that other things in the world may be, and thus he falls into the trap of reifying 'minds' into tenuous organisms vaguely existing on their own. He does indeed warn the psychologist against "the confusion of his own standpoint with that of the mental fact about which he is making his report," as well as against "the assumption that the mental state studied must be conscious of itself as the psychologist is conscious of it." (*Ibid.*) But two relevant factors are omit-

3. The development of the mind proceeds by concentration of the attention of the intellect upon the subjective or the objective area. It is the usual case that an infant is centrally aware of the subjective area, and only occasionally and peripherally aware of the objective area. The feelings of the infant are of paramount importance, while objects are related to his comfort or pleasure. At the beginning of the intellectual life the appetites are strong and demanding of attention; only gradually does the light of intelligibility exert its influence amidst the crowd of emotions and become strong enough to attract attention. In the meanwhile, objects become known, but their value is proportioned mostly to the satisfying of appetites, of which curiosity is one. The awareness of self is the dominating element. But in the normal course of events, under the influence of parents and of society, the appeal of objectivity is accentuated, and thus begins the struggle within the mind of the child between two rivals for his love. The first is the intelligibility of reality, the fruition of that power of understanding that was first manifested in the isolation of objectivity from the self; the other is the satisfaction of appetite, the blind gratification of the will in the conscious ability to enslave the independence of objectivity to the aggrandizement of the self. The outcome of that emulation spells the verdict of success of failure upon one's life.

4. The concentration of mind upon the subjective area is an ambiguous notion requiring clarification. That form of concentration of the mind which consists in subjection (after the initial stages of infancy) to the gratification of appetite represents a frustration of the natural function of the intellect and is called a 'concentration

---

ted from James' notion of 'mental fact': the first is the formative influence of the psychologist's own standpoint upon what he considers to be a mental fact; the other is the formative influence of the standpoint of the mind being observed (a factor defying complete observation from without) upon what the outside observer is describing as mental fact. The psychologist's error, therefore, lies not merely in confusing his own standpoint with that of his subject, but also and especially in forgetting that *two* elusive standpoints are operative within the field of observation. In other words, the difference between a mind and any other kind of object is precisely the operative factor of the knowing subject. To ignore systematically the subjective factor of the mind, on the ground that it is empirically unverifiable (that is, not identifiable with any *object*) means to falsify the observations and produce pseudo-science. The science of psychology must begin from an intellectual concept of the human mind as being one of the class of things composed of a subject aware that he knows objects — as being therefore itself a kind of object that by very definition is partly unobjectifiable but whose unintelligible subjective element has its own identity in functional relationship to the world of objects.

To put the fact another way, the psychologist can never grasp or adequately observe the content of another's mind, but by proceeding scientifically he can identify objects within the mind of another which become intelligible objects within the frame of reference in the mind of the psychologist and may or may not have the same meaning in the mind of the subject. The psychologist does not see the object of another's thought "plus the thought itself," since there is no 'thought itself' apart from the object of thought and the invisible knowing subject. What James calls the 'thought itself' is confused with the focus or standpoint of the knowing subject, and it is the aim of the present article to show, not only that this standpoint belongs ultimately to either the objective or the subjective factor of the mind, but also to identify the goal of all science (including psychology) with the objectification of that standpoint. A psychologist observing the mind of Einstein or of Aristotle flatters himself too much about seeing the 'thought itself' to the extent that the medium of his own thought is not sharpened to grasp the meaning in the focus of his subject's mind. Similarly, a mind that is not morally pure will miss a whole realm of intelligibility (and mental fact) in the mind of a chaste person. What James calls "the thought itself" is therefore better called the *focus* of the thought, and this focus is not scientifically seen until it has been *identified* primarily in the consciousness of the psychologist and secondarily in the consciousness of the psychologist's subject.

of the mind' only in the sense that the area of objectivity is blurred out of focus. But there is another form of concentration on the subjective, consisting in attention to objects proximate to the knowing subject within the field of consciousness and fulfilling the natural inclination of the intellect to know. This area of objectivity proximate to the knowing self is called morality.

5. In the intellectual dichotomy of the mind, the knowing self stands at the proximate edge of the field of knowledge, marking the boundary of intelligibility. Behind the self in the darkness of the unintelligible lie the driving forces of human nature, the appetitive faculties.[a] These forces are known directly as gratification alone, and this gratification or satisfaction lies in an order inaccessible to understanding. In the intellectual dichotomy into cognitive and appetitive functions, the appetitive drives are *represented* in intellectual consciousness by the intellectual self, which is directly experienced as the knower, but is not subject to direct examination as the known.[b]

6. The self is conscious of a power to control his appetites to a certain extent, but not to dominate them completely. The appetites of sense are unruly, but they can be domesticated. There is also intellectual appetite; it is called the will. The self is conscious of a certain power of choice. He can concentrate attention upon himself or upon the other. He can let his appetites run freely to every excess, or he can seek reasons to bring them under control. He can let his intellect open to the truth, or he can turn its attention away. He can admit the meaning of reality into his consciousness, or he can block its access. We call this power the freedom of the will, and the self-gratification derived from the shutting out of truth and meaning from their deserved place in one's consciousness is called pride. Truth offers pure and rational satisfaction to the will in the fulfillment of its nature; pride offers impure and irrational gratification to the will in the frustration of its nature. If truth prevails, the region of objectivity develops, and the self withdraws into a tiny area within consciousness. The state of withdrawal is called humility. If pride prevails, the self exults in its lustful exercise of freedom and its primacy over everything it knows.[a]

2.2:5a Cf. T. V. Moore, *The Driving Forces of Human Nature* (London: Heinemann, 1948).

2.2:5b Cf. F. H. Bradley, *Appearance and Reality*, 81, 93-96.

2.2:6a William James contends that "ideas preserve their own several substantive identities as so many several successive states of mind," and he bases his psychology upon this premise. "The only identity to be found among our successive ideas is their similarity of cognitive or representative function as dealing with the same objects. Identity of being there is none." (*Principles of Psychology*, I, 174-175.) James drew this erroneous conclusion because he overlooked the abiding presence of the knowing subject in the successive acts of thinking. James avers that 'mind' is "only a class name for *minds*" (*supra*, 2.2:2b) and that minds have no other identity than the identity of the objects which they represent. But the continuity of identity of the knowing self is obvious to everyone who is intellectually conscious. James attributes to each successive thought its own instantaneous "substantive identity" in opposition to the obvious fact that thoughts adhere to the substance man and have no identity apart from a knowing subject. James seeks to escape from unverifiable metaphysical entities, but in doing so he makes a metaphysical substance out of 'thought,' personifies it, and eventually endows it with a capital letter so as to become Thought (*ibid.*, 340ff.).

James defines the 'spiritual Self' as "a man's inner or subjective being, his psychic faculties or dispositions, taken concretely," while the 'pure Ego' is limited to "the basic principle of personal Unity." (*Ibid.*, 296.) He holds that from what he calls the concrete point of view "the spiritual self in us will be either the entire stream of our personal consciousness, or the present 'segment' or 'section' of that stream." (*Ibid.*) He insists on this concrete approach, rather than the old abstract viewpoint by which the spiritual self was divided into faculties isolated from one another and identified in turn with the intellect or the will. (*Ibid.*) He agrees with Wundt that "the images of feelings we get from our own body, and the representations of our own movements distinguish themselves from all others by forming a *permanent* group,"

7. It is said that truth resides fundamentally in things, but formally in the mind, for truth is the conformity of a judgment concerning a thing with the way the thing really is.[a] Thus, a statement is true if the judgment it expresses does not contradict reality; otherwise it is an error or a lie. But judgment is not simply an act of the intellect recognizing the identity of terms. It is also an act of the will, permitting the intellect to recognize the identity of the terms, and thus allowing the truth to assume its rightful place within consciousness. The will is free to accept the truth or to exclude it. The story of the exclusion of truth is a major theme in the history of thought and of science. It is essentially a tale of the artifices of pride in the workshop of knowledge and of the almost incredible lengths to which it will go to preserve its hegemony.[b] The proponents of the geocentric theory of

so that our coming to conceive this permanent mass of feeling as being subject to our will is what we call the consciousness of self. Wundt sees this self-consciousness to be completely sensational at the outset, with the element of subjection to the will gradually emerging and becoming dominant, the abstract Ego (self-consciousness reduced to the process of apperception) emerging at the end of the process of development. But bodily feeling remains as the basis of self-awareness. "The most speculative of philosophers is incapable of disjoining his ego from those bodily feelings and images which form the incessant background of his awareness of himself." (Ibid., 303, quoting Wundt.) "The sense of our personal identity . . . is a conclusion grounded either on the resemblance in a fundamental respect, or on the continuity before the mind, of the phenomena compared." It must not be treated as "a sort of metaphysical or absolute Unity in which all differences are overwhelmed." (Ibid., 334-335.)

James does not explain how there can be "continuity before the mind" in the absence of an abiding knowing subject as a factor of the mind. In the ambiguity of his 'Thought' the mind both is and is not identical with successive objects of thought; it both is and is not a state of itself. Furthermore, if thoughts are not within the substance man, how can James reasonably speak of "a man's inner or subjective being"? In preferring his 'concrete view' of the 'spiritual self' to the 'abstract view' of philosophical tradition he uses an abiding frame of reference located in the intelligence of the knowing subject as such (himself) and antecedent to the object he is describing. Thus his 'stream of consciousness' flows in an intellectual medium which he theoretically excludes but uncritically uses. The scientific thinker does not attempt to *disjoin* his ego from the "permanent mass of bodily feelings" that lie behind it; he simply frees his ego from emotional attachments by using his intellect to find and adhere to objective reality. In this way a thinker can *distinguish* objective truth from subjective feelings. Neither Wundt nor James discovered the place and function of viewpoints in his own consciousness. The failure to scrutinize the terms which constitute the overview of their psychology in this respect limits their conclusions to the level of the pre-scientific.

James could not refer to "man's inner or subjective being" except in terms of the prior intellectual distinction between subject and object, specified here as 'inner' and 'outer.' His expression "taken concretely" is therefore a *petitio principii*: he professes to look at man's inner being purely as a concrete object — from across a vacuum as it were — but the very notion of 'inner being' can come into view only in the intervening focus of the distinction between subject and object, that is, only in an intellectual milieu whose abstract nature is wrongly called a 'concrete viewpoint.' The fact that the knower sees objects in an abstract focus is what leads us to posit the intellect as the principle of unity underlying knowledge and linking it to the vital unity of the whole man. This ultimate vital unity underlying consciousness is what Aristotle calls the *entelecheia*, or substantial form (soul), and which, according to Thomas Verner Moore, "not only organizes growth but lives a life of sensory and intellectual experience." (*Cognitive Psychology*, 89.) The bare principle of this vital unity of man is what we have called the 'metaphysical person,' because it transcends the data of consciousness which here constitute the 'physical' level of observation.

2.2:7a Cf. T. Aquinas, *Quaestiones disputatae* I: *De Veritate* (Turin: Marietti, 1953 [ET, *Truth* (Chicago, 1953,)], q. 1, art. 1 (see *infra*, 2.4:2b).

2.2:7b Bernard Lonergan epitomizes the process of ignorance in *Insight*, xiv.

the universe before the seventeenth century devised ingenious systems of cycles and epicycles of the planets to keep the earth at the center of the world. In similar fashion the followers of the egocentric theory of knowledge have in every century been able to devise most clever and intricate systems to maintain the stature of their psychological self at the center of their psychological world. So common is this theme that not a few thinkers even in our day have been led to conclude that self-assertion is the goal of life, whose fulfillment requires victory of the Self over the Truth.

8. A bias is a propensity influencing the rational activity of the intellect. The exclusion of the truth is a biased act of the will, based either upon the sheer pride of the will or upon a felt conflict with another supposed truth that the will has irrationally admitted or devised for its own gratification. This is bias in the negative sense, and its contrary is that happy propensity of the well-disposed will to seek and accept the truth. Ignorance, error, and truth are mixed to some degree in every human mind, whether the conscious subject be of the best will or of the worst, but the quality of the mixture tends to differ. Persons whose minds are relatively closed to truth by a proud disposition of will [a] have the tendency to admit those portions of truth that appear to them neutral, not challenging the hegemony of their pride, and they will even be apostles of enlightenment in those remote regions of consciousness, while dismissing the more proximate truths as pietistic and mythological. They will pity and patronize the humble selves of persons who have opened the proximate regions of their consciousness to the entry of truth; they will exclude the mortified wills of the devout from qualified membership in the fashionable world of self-assertion. This conflict of dispositions regarding the rôle of truth will not be completely resolved short of the tribunal of God, so deep-seated is the cause.[b] The bias of pride will have its parties and its factions, its causes and its crusades. It will always be shocked or amused at the beliefs of the humble. The answer to the problem is to seek humility, not to predicate it of oneself, and to strive in every way possible to subdue the inordinate will of an exaggerated self in order to find the fulfillment of self in the realized possession of the truth. The acceptance of the truth is the truest act of freedom, for it is the truth which truly makes one free.[c]

2.3. *Self-Evident Truth*

1. It has now been noted that the concentration of the mind upon the area of the knowing self can be healthful or harmful. To the extent that such concentration consists in giving free rein to the lower appetites or even to the unenlightened will it is irrational and foolhardy. The will by its very activity does not necessarily promote the good of man or his finality. Because it is free, it can also promote his destruction, and because in itself it is blind it can act reasonably only in collaboration with the intellect. Such collaboration must take place as a service to truth and of the meaning it contains; the usurpation by false systems of thought of the place in consciousness deserved by truth amounts to an enslavement of the intellect to the self-interest of the will and a frustration of human nature. The intellectual self of human consciousness promotes its own welfare and fulfills its purpose by subordinating the non-rational drives at the end of intelligibility to the motives presented by the intelligible objects of the mind.

2. To find such motives the mind must concentrate upon the objects of intellectual consciousness. The first objects known are the presentations of the senses. An infant will begin to distinguish colors, sounds, tastes, odors, feelings, such as hunger, thirst, pain. Then he begins to recognize more complex objects, like visible bodies, kinds of food, more precise feelings. All of these things are objects, if we define an object in its broadest sense as "a recognized unit of experience." [a] A color or a noise is the vaguest kind of object, but it can be recognized as a unit of experience. A toothache is a different ex-

---

2.2:8[a] Cf. Aquinas, I-II, q. 77, art. 2; I-II, q. 84, art. 2.
2.2:8[b] Cf. Jn 12:35-36; 18:37-38.

2.2:8[c] Jn 8:32.
2.3:2[a] Cf. T. V. Moore, *Cognitive Psychology*, 237.

perience from a sore foot; when it returns it can be recognized as a unit of experience. Thus it is that while the cognitive faculties of sense are the chief source of objects of the mind, even the experiences of the appetitive faculties can be a source as well. In this case the directly unintelligible is rendered indirectly intelligible as an object of intellectual regard.

3. As the mind concentrates upon the field of objectivity, which at first is a meaningless conglomeration of sensory presentations, its attention is focused on particular material things, which is to say that it isolates things in their individuality and sees them as meaningful wholes. The 'synthetic sense' constructs the meaningless sensory impressions into pre-intellectual things that have meaning on a level common with the higher animals, and the intellect derives rational meaning from these sensible images in keeping with its characteristic mode of operation.[a]

4. It seems that when an object is recognized in intellectual consciousness it is not known in a merely sensory way, but the mind begins to derive intellectual meaning from the images presented. It does this by the process of intellectual analysis and synthesis.[a] The focusing of the mind upon an object of sense is already an advance from the fundamental dichotomy of intelligence to a series of subsequent dichotomies by which the object becomes known. The focus upon the object puts other objects and sensory impressions into the background at the periphery of the field of vision, and an identity that is more than sensible is sought amidst the dichotomies summoned forth by the sensible qualities. Thus, the object may be hot or cold, heavy or light, bright or dark, hard or soft, moving or still, and the location of the object on one side or the other of these divisions suggested by the sensible qualities leads the knowing subject gradually to establish the identity, which he may later have to revise more accurately. It is thus that the qualities of material things enable the knower to locate them within categories and give them general names.

5. We call this location of things within general categories 'conceptualization.' The recognition with which it terminates is called a 'concept.' We may define a concept in its most general sense as "a synthesis of two recognized intellectual objects within the opposite parts of the most proximate dichotomy." For instance, what is recognized to be characteristic of man is his intellect, or reason. The concept of man resides within his definition as a 'rational animal,' because it is a synthesis of the two parts of the proximate dichotomy into rationality and animality. It is the function of intellectual awareness to make the relevant distinction.

6. But are such dichotomies valid? To answer this question we must return to the primordial dichotomy of the human intelligence. In the original opposition of the knower and the known, the validity of the opposition is supposed in the very act of knowing. We may say that the validity of the knower and of his world of sensory impressions as well as the validity of the act of knowing are presuppositions to all knowledge and to any discussion about knowledge, since there cannot be any act of knowing without them. But the interpretation that we give to these components is subject to correction and revision.

7. With regard to the objects known, we must distinguish the assumed validity of the object seen or otherwise sensed from the unquestionable validity of the sensation itself. Thus, for example, the visible star Theta the First (θ_1) in the constellation of Orion is easily resolved through a small telescope into four smaller stars rather widely separated.[a] The impression of light received by the naked eye is valid; the recognition of a unified object is shown by the telescope to be invalid. Doubt can be cast upon the validity of interpretations, the construction of images and of syntheses, when a closer look yields stronger evidence to the contrary and divides what seemed to be a single thing.

8. The same may be said of the knowing subject. The reality of the knowing subject as such is given in consciousness and cannot be excluded,

---

2.3:3[a] Cf. *S.Th.*, I, q. 78, art. 4; I, q. 79, art. 3; Moore, *Cognitive Psychology*, 241-242.
2.3:4[a] Cf. Aquinas, *De Verit.*, q. 1, art. 3.

2.3:7[a] Cf. D. Menzel, *A Field Guide to the Stars and Planets* (London: Collins, 1966), 188-189.

but the seeming unity of the *idea* of the self admits of doubt to the contrary. Here, however, we must distinguish between unity of nature and unity of function. A baseball is a thing recognizable in its unity as materially continuous and discrete from the matter around it. A baseball diamond has unity of function, or purpose, and is also materially continuous, but is not discrete from its surroundings except by mental designation. A baseball team is not materially continuous and has unity of function alone. Yet the mind ascribes unity and identity to each of these things.[a] We may thus distinguish in the individual man between his metaphysical and his intellectual self. The metaphysical self is the identity of the man in his entire being, material and spiritual. The intellectual self is the identity of the man as he emerges within his own intellectual consciousness.[b] Both are things, if we take a 'thing' to be "any element of being," as in its broadest sense a 'thing' is known to be,[c] for we speak of 'everything' and 'anything' as the broadest differentiation of being itself. But not all things are real, and the reality of the self can be questioned.

9. There are some who deny the 'metaphysical self' as a product of primitive philosophy long since outmoded by science. There are some who regard the 'intellectual self' as a figment of the prescientific imagination, claiming to have discovered that from the scientific viewpoint the unified image of the self divides into the elements of the stream of consciousness. Thus would they consign to the archives of antiquity belief in the human spirit.[a]

10. But even those who claim to be scientific sometimes use words and ideas without stopping to examine their import. In treating the question at hand, they may speak of the scientific 'viewpoint' and of various 'aspects' of the problem without taking time to ask themselves what they mean by a 'viewpoint.' All of us tend at first to make this mistake. We can determine something about viewpoints from the primordial dichotomy. Within the primordial division of the mind, the intellect may concentrate its attention on one part or the other. This concentration is a viewpoint, and hence we may define a 'viewpoint' as a concentration of the mind upon one side or the other of a dichotomy that it has already realized.[a] From every dichotomy two viewpoints arise, and the acceptance of one requires the rejection of the other within the context of thought.[b]

2.3:8a Cf. Aquinas, *De Anima*, Lib. III, Lect. XI, No. 755.
2.3:8b Cf. Bradley, *Appearance*, 76-77.
2.3:8c Bradley falls into the common error of restricting 'thing' to self-standing substances and their physical parts: "A rainbow probably is not a thing, while a waterfall might get the name, and a flash of lightning be left in a doubtful position." (*Appearance and Reality*, 61.)
2.3:9a William James calls "a perfectly wanton assumption" the belief that the reflective consciousness of the self is essential to the cognitive function of thought. "As well might I contend that I cannot dream without dreaming that I dream, swear without swearing that I swear, deny without denying that I deny, as maintain that I cannot know without knowing that I know." (*Principles of Psychology*, I, 274.) James here confuses the primary function of the knowing subject as *subject* with its secondary function as an *object* of thought. The knowing subject can know without making 'I know' or 'that I know' the object of his awareness, but he cannot know without knowing, and he cannot humanly know without the concomitant awareness (on the subjective side of the primordial dichotomy) that he knows.
2.3:10a The application of dichotomy to form and privation is shown by St. Thomas in his exposition of Aristotle's *De Anima*, Lib. III, Lect. XI, No. 759.
2.3:10b Cf. T. Aquinas, *In Aristotelis libros Peri Hermeneias expositio* (Turin: Marietti, 1964^2,) Lib. I, Lect. XI, No. 149. William James rejects any factual distinction between sensory and intellectual activity. "The contrast is really between two *aspects*, in which all mental facts without exception may be taken; their structural aspect, as being subjective, and their functional aspect, as being cognitions... From the cognitive point of view, all mental facts are intellections. From the subjective point of view all are feelings." (*Principles of Psychology*, I, 478-479.) We admit, with Aristotle and Aquinas, that all cognition has an appetitive aspect, since the will is the intellectual appetite and sensory perception is accompanied by feelings. But James

11. It should be clear that a 'point of view' in the sense just considered is an intellectual term based on the analogy of sight. A 'viewpoint' in the physical sense is a geographical term, expressing the way in which a given object will be seen by the eye when it is located at such a position. But there is no geography of the mind in exactly the same sense. How the mind moves from one viewpoint to another needs to be understood. The mind moves forward by concentrating its attention in more restricted areas brought into focus by the recognition of dichotomies. Within the field of attention objects arise, and they provide new dichotomies. The parts of dichotomies are somehow combined into syntheses that may be called 'concepts.' And concepts somehow provide 'viewpoints.' These are the basic elements of the solution, but we have not yet unified them into an organized whole.

12. The question most immediately at hand is that of the validity of the elementary 'viewpoint' according to which the mind concentrates upon the psychological self, as distinguished from the objects which it knows. Is the 'self' that immediately emerges as the object of attention a unified thing with its own undeniable identity, or is it an 'illusion' like Theta the First of Orion, which loses the autonomy of its identity and dissolves into a plurality of more solid things under the scrutiny of a closer examination? Indeed, just as the 'solid' band of the Milky Way exchanges its continuity in the telescopic field for the isolated beauty of myriad points of light, so also has the recurrent claim been made by students of our mental states that the seeming continuity of the 'stream of consciousness' will under more critical examination dissolve into a countless number of successive moments and simultaneous phenomena of biological life. It is not to our present purpose to review or appraise the data of clinical psychology, but only to examine the place of the 'self' in the nature of knowledge and of its acquisition.

13. A distinction must first be made regarding the 'psychological self.' While consciousness is divided in its every moment between the knower and what he knows, it is not the 'psychological self' that constitutes the knower, for the 'psychological self' is a subsequent construct of the mind, a kind of object formed by the abiding presence of past subjective experience. It is the 'private' and 'personal' area unclaimed by objectivity, the region immediately surrounding the consciousness of free will. It is my suspicion that a systematic and unbiased study by the knowing subject of his psychological self will indeed dissolve the construct into a number of separate parts: feelings and habits of the emotions, the awareness of power and of weakness, the experience of the past, the awareness of one's freedom and how he tends to use it. The consciousness of self in this sense is seen to be a construct of the mind as the rhythm of life leads it again and again to concentrate upon

---

does not determine the scientific validity and function of *aspects*. An aspect presupposes a viewer and therefore stands wholly within the cognitive order. What James is trying unsuccessfully to say is that, in the intellectual distinction between the cognitive and the appetitive, mental objects may be viewed either as objects of cognition or as objects of appetition. But objects viewed in the framework of this intellectual distinction are directly and most intelligibly cognitive, and only paradoxically and less intelligibly appetitive. James's failure to recognize the fact of this structural framework in his own mind leads him to identify falsely the structural aspect of cognition with feeling, whereas mental structure is in fact directly identified with cognitive objectivity (as the intelligible medium of understanding) and only paradoxically identified with feeling. For only one of the two aspects arising from an intellectual dichotomy will fit logically into the context of thought as the direct basis of an insight, the other aspect yielding only paradoxical insight. James's conclusion that the sensory is intellectual and vice versa results from an illogical twisting of the paradoxical insight, that the sensory is *in some sense* intellectual, into the direct line of reasoning about intellectual activity. His observation that the structural aspect is always subjective is a forcing of the structural side of his dichotomy 'structural-functional' into the subjective side of the primordial dichotomy operating unrecognized in his mind, and the unfortunate result is that he thus suppresses the structural side of human cognition, in which all science, including the science of psychology, is located.

the area of the knowing subject. This concentration upon one's 'self' by the feeling of well-being or of distress, by hunger or by pleasure, by the satisfaction of desires or by the need to resist them, by countless 'reasons' of the corporal sensibility, but above all by that feeling of awareness that we call 'love,' is the object of our present attention. We need to know the place of appetition in that image of concentration that is the 'psychological self.' I would suggest that the place of love is dominant, that the 'psychological self' is principally composed of the awareness of freedom attached to the knowing subject by vital and inseparable union. For it seems that the area of feeling and appetite which the mind automatically associates with the knowing subject is dominated by the will and its desires, except to the degree that its function is diminished by the demands of bodily appetite. And thus, for instance, an addiction to alcohol or narcotics, or the growth of any vice, has its influence upon the functioning of the will. But the will has the primacy of rule in a normally functioning man, and it is the loves of the will that are most closely associated with the mature image of the self.

14. I do not look for the unity of awareness in that object of concentration that is called the 'psychological self,' for this is but the image of a feeling.[a] It is what is left of intellectual consciousness, when the intellect concentrates upon the feeling of its lower nature or upon the blind desire of the will, and blurs the rest of objectivity out of focus. I look for the unity of awareness, rather, in the unquestionable consciousness of the knower that he is knowing, and this is no object or construct, but is self-evident subjectivity. And I look for the continuity of function, not only in the operation of consciousness as a psychological whole, but also and especially in the way in which the knowing subject, the 'intellectual self,' allows his mind to develop through the conscious use of his will. The choice is radical and immediate. Either he will use his freedom to concentrate the mind upon 'himself' by a studious attention to undisciplined appetites and the 'blind passion' of the will, bending objectivity itself to serve the purposes of his pleasure, or he will use his freedom to concentrate the mind upon the intelligibility of objects in such wise that his understanding may develop in the pursuit of truth, and free

---

2.3:14a William James reduces what we have called the 'psychological self' to the permanent mass of feelings we derive from our body. (*Principles of Psychology*, I, 303.) "The past and present selves compared are the same just so far as they *are* the same, and no farther. A uniform feeling of 'warmth,' of bodily existence (or an equally uniform feeling of pure psychic energy?) pervades them all; and this is what gives them a *generic* unity, and makes them the same in *kind*. But this generic unity coexists with generic differences just as real as the unity. And if from the one point of view they are one self, from others they are as truly not one but many selves." (*Ibid.*, 335.) He finds no other identity within the 'stream of consciousness' than "resemblance among the paths of a continuum of feelings (especially bodily feelings)," professing adherence to the doctrine of the empirical school of the Associationists and the Herbartians that the Self is "an aggregate of which each part, as to its *being*, is a separate fact." (*Ibid.*) He overrides the objection raised by common sense that the unity of the successive selves involves "a real belonging to a real Owner, to a pure spiritual entity of some kind" (*ibid.*, 337), on the supposition that the 'title' of a collective self may rather be passed from one Thought to another. "Each pulse of cognitive consciousness, each Thought, dies away and is replaced by another... Each later Thought, knowing and including thus the Thoughts which went before, is the final receptacle — and appropriating them is the final owner — of all that they contain and own." (*Ibid.*, 339.)

In evading the evident unity of the human person, James is constrained to personify Thought, thus creating an unstable quasi-substance which will not stand up under logical analysis. He merges thought with an unscientific image. In reducing thought to momentary pulses of consciousness he fails to account for two important elements of thought which are not included in these pulses: the constant awareness of the knowing subject as such and the abiding state of his understanding. James's notion of Thought is thus a step backwards from the scientific analysis of thought initiated by Aristotle.

will may become an act of acquiescence to the truth. If, happily, he chooses the latter course, in keeping with the higher interests of his being, the 'psychological self' will tend to disappear until not much else is left but the 'intellectual self,' fully engrossed in the self-fulfilling sight of pure self-appropriated objectivity.[b]

15. But, as the undeveloped mind begins to discover the world of objects, the self-evident world of the self yields the center of focus to a universe filled with other selves. There are the table itself, the chair itself, the room itself, the house itself; there are warmth itself and coldness itself, the inside itself and the outside itself, the bottom itself and the top itself; there are beauty itself, truth itself, goodness itself; there are objectivity itself, nonselfhood itself, and even nothingness itself. Every object of thought or observation has its 'self' in opposition to the self of the knowing subject. What is the credibility of the 'self' residing in other things?

16. The answer is to be sought in the primordial dichotomy of consciousness. When the mind begins to concentrate upon the area of objectivity, it immediately makes a distinction between the things of its attention and its way of knowing these things. There arise in a vague and implicit manner those principles of thought which we call 'self-evident' and which will themselves become objects of explicit regard only after the mind has advanced to a considerable degree of maturity. But they are present and operative from the beginning of its activity. When the mind looks at an object like a table or a chair, it looks 'beyond' the impressions of sense for an identity which will characterize the thing and set it apart from everything else. This identity is called the self. The identity of the self is located within the parts of the most proximate dichotomy ascribed to the concrete intuition of the particular.

17. We know that the mind ascribes identity to things. We know too that

---

2.3:14[b] William James seems to be aiming at this conclusion where he says that 'consciousness' does not stand for an entity but it does stand for a function of experience which thoughts perform, namely, the function of knowing. ("Does 'Consciousness' Exist?" in *Essays in Radical Empiricism* [London, 1912], 1-38.) His thesis is that "if we start with the supposition that there is only one primal stuff or material in the world, a stuff of which everything is composed, and if we call that stuff 'pure experience,' then knowing can easily be explained as a particular sort of relation towards one another into which portions of pure experience may enter. The relation itself is a part of pure experience; one of its 'terms' becomes the subject or bearer of the knowledge, the knower, the other becomes the object known." (*Ibid.*, 4.) In this way for James 'consciousness' is "but the logical correlative of 'content' in an Experience of which the peculiarity is that *fact comes to light* in it, that *awareness of content* takes place." Thus 'self' and its activities are said to belong to the content. "To say that I am self-conscious, or conscious of putting forth volition, means only that certain contents, for which 'self' and 'effort of will' are the names, are not without witness as they occur." (*Ibid.*, 5-6.) I would agree with James's reduction of the 'self' to a content of consciousness to the extent that he means the proximate surroundings of the psychological self, but his reduction of the knowing self as such (the 'pure Ego') to the primal stuff of 'pure experience' is scientifically defective.

We cannot start with the supposition that the one primal stuff in the world is 'pure experience,' because experience is by definition a subjective event and therefore stands in its original meaning on the subjective side of the subject-object dichotomy, which is the first act of human intelligence. Now, every experience is a subjective event related to something objective, and every experience of knowing cannot be reduced to anything simpler than the two factors of subject and object. James tries to reduce this act to the primal stuff of 'pure experience,' but in doing so he does not find an *intelligible common element* which they share. Knower and known are not parts of a recognizable stuff that could be called 'experience.' No object belongs to the experience as far as its own nature is concerned. It is irrational violence to force the whole universe of objects known under the heading of 'experience.' To do so destroys the clear notion of what an object is. And the fact tht the proximate surroundings of the knowing subject can be discovered to be objects of knowledge does not eliminate the knowing subject as such

the mind can err in this operation. The fact that the morning star and the evening star are always seen at opposite ends of the sky could lead one to believe that they are different stars. Yet, a closer look will show that they are identical and that what seems to be a star is actually the planet Venus. The two selves of the two stars are the object of an illusion. Hence, we cannot start out by assuming that the identity ascribed to objects is not an illusion of the mind. But we should recognize the fact that it is the natural function of the mind to seek identities and to proceed by a process of precision from what seems to be so to what becomes known to be so. This process has a unity of function, and the growth of certainty which it produces, rather than destroy confidence in the power of the mind, becomes its greatest confirmation.

18. When the mind begins to concentrate within the area of objectivity, it brings with its concentration the awareness that *a thing is itself.* The thing may not turn out to be what at first it seemed to be, but at any stage or degree of knowledge a thing is known to be itself. The morning star is the morning star, the evening star is the evening star, whether or not both turn out to be the planet Venus. This 'principle of identity' is part of the essential equipment of the mind; it cannot be denied without its being used in the very process of denial. The principle is already at work at the level of the primordial dichotomy. The knowing subject knows that in his essential identity as the knowing subject he cannot be the object of his knowledge. He knows that he cannot step outside of himself to look at himself objectively (that is, without constructing an image of himself, and this is not himself in the same sense). This principle of identity as applying to the knowing subject is self-evident to the knower, and it is perhaps from this recognition that he can apply it to

or make him a part of the content of knowledge, except to the extent that the word 'content' is extended to include both the subjective and the objective sides of the primordial dichotomy. Even the consciousness of breathing might be identified confusedly with one's 'self,' but scientifically only until it has been recognized as an object. "I am as confident as I am of anything that, in myself, the stream of thinking (which I recognize emphatically as a phenomenon) is only a careless name for what, when scrutinized, reveals itself to consist chiefly of the stream of my breathing." (James, *ibid.*, 36-37.) Once he had identified the stream of his breathing as an object, James was required by scientific method to place this object into the framework of objectivity and to notice that in the stream of his thinking about the stream of his breathing his knowing self was on the subjective side of the dichotomy looking at his breathing on the objective side. Thinking not distinguished from breathing is prescientific. Our thesis is that there is a *per se* dualism in the act of knowing. Human intelligence is experience based on the awareness of this dualism. To 'witness' a content of consciousness called 'self' is to confirm this dualism. The science of psychology cannot eliminate the knowing self; it can only provide the means by which the knowing self knows the truth about himself in the field of intellectual objectivity. Knowing cannot be *explained* as a particular sort of relation to one another of portions of pure experience, namely, of the portion of pure experience called 'subject' to the portion of pure experience called 'object,' in the absence of two clarifications omitted by James: 1) an intelligible concept of 'pure experience' which presents something to the inquiring mind after the specific differences of 'subject' and 'object,' which are its species, have been removed; 2) an intelligible concept of 'relation' as applied to the idea of knowing. Since James avers that this relation is itself a portion of pure experience, he sees it as a sort of third species of the elements of pure experience contained in knowing, and he conceives it as being prior to the 'subject' and 'object' which are its terms. Now we readily agree that knowing is some sort of relation between a subject and an object, but to define knowing in such vague terms has no scientific value. A valid scientific explanation would have to show *where* in human consciousness the experience of relation is located and *how* this kind of experience is related to experience in general. But such an inquiry will quickly reveal that 'subject,' and 'object,' are not species of a more general concept ('pure experience') having any intelligibility either as regards objectivity or as regards the relation of terms to one another.

everything else.[a] When the knowing subject knows that *he cannot both be and not be the knowing subject at the same time and in the same way* within the state of consciousness, he has recognized another principle native to the operation of the mind, the 'principle of non-contradiction' Other inescapable principles are implicit in the mind as well. The knowing subject knows that the extension of his consciousness is fully divided between himself as the knowing subject and the object as what he knows. As an intellectual dichotomy, there is nothing in between. While the nature of the act of dichotomy excludes a third alternative to its two most evident parts, still there is an entire potential field, or 'middle ground,' between the knowing subject and the initial object of his knowledge, a field which develops into the objective content of the mind as the intellect itself develops outside of the field of vision. The function of the intellect is only implicit within the state of consciousness until the mind has developed to the point at which it can distinguish between the act of knowing and its content, and at that point the act of knowing becomes in a secondary way an object of knowledge, entering 'visibly' into the area of objectivity. The act of knowing thus becomes the subject of such disciplines as logic and epistemology.

2.3:18a According to the theory of Aristotle, as expounded by St. Thomas, the mental dispositions (*habitus*) of *per se* evident first principles arise with the knowledge of universals, which in turn are still objects in the soul standing outside of the flow of inductive experience (*experimentum*). Sense images, he says, can remain in the memory of the higher animals, including man, and, from repeated memories of the same kind of image as it is presented by different individuals of the same class, man is able by a kind of reasoning in which one particular thing is related to another to grasp something (common) in the many individuals remembered. This inductive reasoning of experience provides the condition in which intelligence can know universals. We must, however, presuppose, he says, a soul that is capable of having universal knowledge and of actuating intelligibles by the abstraction of universals from singulars, which could not be possible if there were not in some way sense knowledge also of universals. The senses know Callias, not only as Callias, but also as *this man*, and the senses know Socrates, not only as Socrates, but also as *this man*, and thus it becomes possible for intelligence to go on to consider *man* in both individuals. "Si autem ita esset quod sensus apprehenderet solum id quod est particularitatis, et nullo modo cum hoc apprehenderet universalem naturam in particulari, non esset possible quod ex apprehensione sensus causaretur in nobis cognitio universalis." (Aquinas, *Post. Analyt.*, Lib. II, Lect. XX, Nos. 11-14; quotation from No. 14.)

The explanation of Aristotle-Aquinas does not fully resolve the issue. While based on the principle that intelligence is of universals, it is constrained to posit a kind of sense knowledge of universals. In the midst of this vagueness in the distinction between intelligence and sense, there is posited also a kind of reasoning, called 'inductive,' which is neither intellectual nor sensitive, but belongs somehow to man. Furthermore, the process by which intellectual knowledge arises in consciousness is not explained, except as the emerging of quiescent universals in the soul, and the relation of *per se* evident principles to the emerging of universals is not clarified.

The lacunae in Aristotle's explanation of the origin of *per se* evident principles can be filled from the notion of intellectual consciousness. A puppy can recognize its mother, a cat, some milk. A human infant can recognize its mother, a cat, some milk. But only the infant can recognize *that this* is its mother, or a cat, or some milk. In order to recognize *that* this is something, the infant must be aware of objectivity. As intellectually conscious, the infant is spontaneously aware of himself as distinct from the sensory object confronting him at the moment; he knows himself subjectively as a concrete *identity*. From this awareness proceeds the knowledge *that* the object with which he is confronted is distinct from his own identity and that each successive object is in like manner distinct from his own identity. The knowledge *that* is the first universal, and it contains under it every succeeding object of intellectual knowledge. The next step is for the infant to recognize that *this* is its mother, or a cat, or some milk. By a simple transfer of the knowledge that an object is distinct from *his* own identity, the infant distinguishes the object from

19 Knowledge is "the consciousness of objects present or remembered," and it is distinguished from pure experience, which is its source.[a] The consciousness of the knowing subject as such is not knowledge, because it cannot be an object of consciousness. It is a self-evident source of knowledge, and it has its important function in the operation of the mind. The act of knowing is another self-evident experience which enters into knowledge only in a secondary sense, when it is made an object of consciousness. Sensory experiences become elements of knowledge, when they are isolated as units and constructed into images. While such constructions can be erroneous, there is no reason to assume that the construction of images from the experience of external stimuli is not a faithful reconstruction, granted that the image is valid only for the perspective concerned.[b] But knowledge does not consist only in the consciousness of objects immediately derived from the experience of present external stimuli. There is the factor of memory (which can be faulty). And, moreover, there is the whole world of fanciful objects, manufactured from the imaginative powers within the consciousness of man, having no authentic counterpart in the external world, and yet in an undeniable manner pertaining to the knowledge of their possessor. It is possible to know what Alice saw on the other side of the looking glass, even though the adventures of Alice are pure figments of the imagination.

20. Objects arise in consciousness through the process of 'objectification,'[a] but it is important to note that the factors in the process are not always the same. The construction of external stimuli into an object by the 'synthetic sense' involves an automatic factor of recognition and a creative factor of interpretation.[b] The recognition of the identity of a sensible thing involves a given image of sense and the recognition of the proximate dichotomy. The derivation of concepts involves a process of analysis and synthesis which is partly automatic and partly creative.[c] The fabrication

---

its own identity. In the judgment, 'this is my mother,' the identity of the mother, represented by the demonstrative *this*, is distinguished from the *sensible form* of the mother, which can therefore be applied to other identities at a later time in order to form a universal in the Aristotelian sense. The separation of form from matter thus involves a recognition of the 'self' in the object. In the case of the infant's mother, the self is like the infant's own self. In the case of the cat it is only somewhat similar, and in the case of the milk it is not very similar at all, but the normal infant will soon learn the difference. Implicit in this process is the principle of identity, namely, *that* (step one) a thing is its *self* (step two).

The explanation from intellectual consciousness is basically in agreement with the Aristotelian theory that only intelligence can know *that* (cf. Aquinas, *Peri Hermeneias*, Lib. I, Lect. III, No. 9) and that the use of *per se* known first principles arises with the knowledge of universals. Because the dichotomic theory of intellectuality does not posit the awareness of Aristotelian universals as the essential feature of intelligence, it leaves room for the possibility both of intellectual knowledge of particular objects and of a kind of knowledge of universals even on the pre-intellectual level of higher animals.

The universal whose awareness characterizes intelligence is *that*. The knowledge *that* is broader and more fundamental than the open classes of things which Aristotle presents as examples of intellectual knowledge, and the fact that this primordial form is operative from the very beginning of intellectual consciousness accounts for the ability of man to reason from particular to particular without the need to postulate a power midway between intellectual knowledge and sense knowledge.

2.3:19a Cf. Moore, *Cognitive Psychology*, 237-238.

2.3:19b Cf. H. Margenau, *The Nature of Physical Reality* (New York: McGraw-Hill, 1950), 46.

2.3:20a Cf. Moore, *Cognitive Psychology*, 237.

2.3:20b Cf. *ibid.*, 350.

2.3:20c "The act of reification of data involves more than integrations; it involves *construction*, construction in accordance with rules. Objectivity emerges as a result of this procedure; to assert objectivity is our way of acknowledging the success of the transition from data to the rational wholeness of constructed objects ... (O)bjectivity, if it attaches to experience, is brought into the scene during our passage from the immediately given to what may

of imaginary objects involves the creative use of the imagination upon the raw materials of remembered experience. From this complex process of objectification emerge within consciousness two distinct worlds: the 'real' world and the 'imaginary' world. By the presence of the self-evident principles the knowing subject knows that the imaginary world cannot be the real world and the real world cannot be the imaginary world. An object of the one world cannot cross over into the other, and there is no third alternative world. The knowledge of the knowing subject is cleanly divided, at least in concept, between the world of reality and the world of imagination, although the borderline may be indistinct and one can be deceived into placing in the realm of the real what is actually imaginary and vice versa.

2.4. The General Locus of Science in Human Consciousness

1. Science is a division of knowledge within the consciousness of man. Its discovery dates from the moment at which the knowing subject recognizes the difference between the real and the unreal among the objects of his knowledge. The notion of the real is almost self-evident in the consciousness of the normal person, because it arises immediately from the awareness of the personal freedom to let one's thought follow the attraction of the creative imagination or to resist that attraction. Even little children soon become aware of the conceptual difference between the real and the imaginary. They dream and they build castles in the sky. They are deceived sometimes by their senses and sometimes by the deliberate intent of other persons. Thus the *notion* of the real is grasped and provides the dimension of a field of knowledge, the medium of a focus of the mind according to which one's consciousness is differentiated and the objects of real knowledge are separated from objects pertaining to the illusory, the deceptive, the imaginary, the aesthetic, and the satisfaction of non-intellectual appetite.[a]

2. In its initial stage as a reference to the real, science seems to add no new positive specification to what one knows. Its influence seems merely negative in that it serves to exclude whatever falls short of the 'real' as verified in the data of sense, but in fact this exclusion implies an at least vaguely conscious recourse to verification in intellectual experience as well.[a]

---

here loosely be called concepts or ideas." (Margenau, *The Nature*, 60.)

2.4:1a Francis Bradley admonishes, "It is a mere superstition to suppose that an appeal to experience can prove reality. ... Any deliverance of consciousness — whether original or acquired — is but a deliverance of consciousness." (*Appearance and Reality*, 182.) We admit that sensitive experience is given, and merely given. But the door to science is opened by the recognition of the *notion* of reality, so that the pathway of science leads over the development of an ever deeper understanding of the meaning of reality.

2.4:2a Henry Margenau maintains that "It is wholly unwarranted to *start* a theory of knowledge with the ontological premise characterizing the spectator-spectacle distinction. If experience, on proper analysis, invests this distinction with meaning, we are ready to accept it, but even then only as a property of the contents of experience, actual or possible." (*The Nature*, 46-47.) It is Margenau's position that the given is to be sought within experience, and not within mind, so that

"it is a fatal error to lift (the mind) out of context at the beginning of all inquiry and thus to convert it into an intractable singularity in the field of experience, without use or meaning." (*Ibid.*, 47.) "Location is not one of the properties of bare experience, though elements within experience may or may not have location. It is therefore never necessary to say *where* experience is. Nor is there anything *external* to experience, for such a spatial attribute can at best be only a metaphor." (*Ibid.*, 48.) We admit that the subject-object dichotomy is one element of experience among many and that location is not one of the properties of bare (i.e., sensitive) experience. But it is not unwarranted to begin the exposition of a theory of knowledge from the spectator-spectacle distinction, because this distinction is the first and fundamental element of *intellectual* experience, having from its inception at least a vague location with regard to every element of sensitive experience. As used here, the same distinction is also a unit of *theoretical* experience which provides a frame of reference for

The objects in its field and the conclusions of its thought must surmount simple tests or comparisons by which they qualify to belong in reality. That is so which can be *shown* (to the senses and to reason) to be so, or at least which cannot be disproved in a confrontation with the evident experience of sensation and intellection. Science is, therefore, mediate knowledge: no element can be admitted or permitted to remain within its field, if it is not contained in the focus of reality.b The notion of reality brings into focus an area of intelligibility which otherwise would remain vague, confused, undeveloped, and misleading. It provides a field of thought which is differentiated from the simple pursuit of pleasure, from the wanderings of an undisciplined imagination, from the creative production of fiction and fantasy, of music and fine art. It thus determines for itself a separate 'universe of discourse.' Once the knowing subject has recognized the difference between reality-oriented thinking and 'creative' thinking, the two modes of discourse become divisions of the realm of objectivity, and the value of each is enhanced precisely to the degree that the thinker is able to distinguish clearly between them and make them complementary parts of his unified knowledge. Yet the parts are not equal. He who has discovered science has found a value which exceeds that of the prescientific areas of consciousness and which invites the knowing subject to organize his consciousness in terms of the realistic insight it bestows.

---

the fitting of all other elements of experience into an organized totality, and that is what a theory of knowledge purposes to do. Location, therefore, *is* a vague property of all intellectual experience and a precise property of that kind of intellectual experience which we call cognitive theory. The object of the subject-object dichotomy is not external to experience, but it is external to the subjectivity which is left behind as the intellect concentrates on the object as such. Nor is the mind the only singularity in the field of experience, but it is the object of special attention for theory of knowledge.

2.4:2b According to St. Thomas the true is the object of intelligence and the fulfillment of understanding (*In Peri Hermeneias*, Lib. I, Lect. III, No. 7). Truth is the conformity of intelligence to the thing (*adaequatio rei et intellectus*). (*De Veritate*, q. 1, art. 1.) This means, says St. Thomas, that truth is primarily (*per prius*) in intelligence composing and dividing (*in intellectu componente et dividente*), because "intelligence first begins to have something of its own when it begins to judge concerning an apprehended thing, that is, when it says that something is or is not, and truth is only secondarily in the true or false composition of the definitions of things. (*De Veritate*, q. 1, art. 3.) Being and the true are convertible (*ens et verum convertuntur*), because every natural thing is conformed to the divine intellect. Human intelligence knows truth when it knows that it is conformed to the thing, and to know this disposition of conformity is nothing other than to judge that it is or is not thus in the thing (*iudicare ita esse in re vel non esse*). (*In Peri Herm.*, Lib. I, Lect. III, Nos. 8-9.)

St. Thomas seems to intend by *adaequatio rei et intellectus* the conformity of the understanding to *reality*. (Cf. Moore, *Cognitive Psychology*, 352.) When St. Thomas says that truth resides in the judgment of intelligence that a thing *is* or *is not*, he is restricting the *esse* of the copulative to the *esse* of extramental existence. Now it should be clear that, if *ens* is the broadest of all categories, it must include, not only real beings, but imaginary beings as well, for otherwise it would be necessary to have a broader term to cover both. When, therefore, St. Thomas restricts the true to the judgment that a thing *is*, he is applying the copulative *is* within the perspective set up by the interposition of the concept of reality and thus identifying the true with the real.

The explanation from intellectual consciousness is in agreement with the position of St. Thomas in the sense that it identifies 'real knowledge' with the knowledge of reality. It is the concept of reality which provides the viewpoint for the recognition of truth. The realistic point of view is implicit in intellectual consciousness itself, but it has to be *chosen* in order to provide the focus of real knowledge. It is in the focus of reality that the self-evident principles begin to produce the knowledge that we call science. Outside of the reality-medium human intelligence can distinguish the form of a thing from its identity, whether the thing be real or fantastic. Hence the principle of

3. Science may be defined as "the knowledge of the real as such."[a] Its method is characterized by constant recourse to reality as the criterion of verification. Its boundaries are coextensive with the limits of reality itself.

4. The question of the nature and limits of reality has been studied and controverted since ancient times. Materialists have tended to limit reality to bodily things perceived by the senses, that is, to the origin of the sensations themselves. Idealists have tended to deny the validity of sensory perceptions as the criterion of reality. These two positions have constituted the extremes of the spectrum of graduated opinions. The earliest tendency of science-oriented consciousness is toward verification merely by the senses; images and objects in contradiction to such verification are removed from the sphere of reality. But, as scientific consciousness develops, it becomes clear that reality includes more than the qualities of material things as such. For the mind has a natural tendency to admit concepts not included in matter and to derive conclusions by way of reasoning which are removed from the material as such.

5. In the modern era empiricists and positivists have tended to stress the inductive method of verification by the senses to the point of almost excluding the validity of essences and the use of the deductive method. Yet they have tended to admit the generalizations of mathematical principles and scientific concepts, as well as the use of mathematical deduction. Hence

---

identity does not of itself cause real knowledge. But, as intelligence finds or does not find objects within the reality-medium that it has chosen, so it judges that a thing *is* or *is not* and is no longer free to imagine that within the given frame of reference the same identity can both be and not be. It is therefore within the concept of reality that the *per se* known first principles achieve unity and meaning.

2.4: 3a Science may just as well be defined as "the knowledge of reality as such," according as the remote or the proximate formal object of science is considered. "La scienza, infatti, è sempre, anche oggi, conoscenza del reale, dell'essenza, della verità; e la sua intenzionalità è sempre la scoperta della struttura intima della realtà e delle cause dei fenomeni." (F. Selvaggi, "Evoluzione del concetto di scienza e dell'epistemologia," in *Seminarium*, N.S., XIV [1974], [491-515], 512.)

St. Thomas, expounding Aristotle, sustains the theory that science is the knowledge of things through causes (*In Post. Analyt., passim*). "Tunc scimus cum causas cognoscimus." (*Ibid.*, Lib. I, Lect. IV, No. 15.) By the term 'science' he means knowledge in the proper and complete sense, and he limits the verb 'to know' (*scire*) to scientific knowing. "Oportet igitur *scientem*, si est perfecte cognoscens, quod cognoscat causam rei scitae." This means the perfect, and not partial, apprehension of the *truth* of things, for the principles of the being of a thing and the principles of the truth of a thing are the same. Science is certain (not dubious) cognition of a thing, and therefore what is scientifically known cannot be otherwise than it is. Science is not the knowledge of mere effects, but of effects as caused; it is the knowledge of causes as effecting the thing (*ibid.*, Lib. I, Lect. IV, No. 5). But science is the result of syllogistic demonstration; it is the comprehension of the truth of a conclusion which has been demonstrated, and a demonstration is a syllogism making someone know (*ibid.*, Lib. I, Lect. IV, No. 9). Demonstration must be based upon true premises (*ibid.*, Lib. I, Lect. IV, No. 13) and upon what is necessarily so (*ibid.*, Lib. I, Lect. IX, No. 2). Every demonstrative syllogism proceeds from propositions which are either necessary or statistically probable (*verae ut frequenter*) (*ibid.*, Lib. I, Lect. XLII, No. 2). Science does not consist in sense cognition, for sense does not know what a thing is, but only sensible qualities which affect particular substances existing in a definite place and time. Science consists in the knowledge of the universal, and we call universal that which is always and everywhere (*ibid.*, Lib. I, Lect. XLII, Nos. 5, 7).

St. Thomas says that the knowledge of *per se* known first principles is not science in the complete sense (*ibid.*, Lib. I, Lect. IV, No. 8); it is rather understanding (*intellectus*), to which pertains the knowledge of the universal, and the universal is the principle of science. Understanding is science only in the sense that it is the font (*caput*) and origin (*principium*) of the sciences, whose own object is even more certain and better known *quoad se* than is the object of science. But there cannot be a science of first principles,

they have not been able to make consistent use of their concept of reality, but have resorted to the suspicious expedient of using the concepts, principles, and deductions that satisfy their desires, while maintaining the general position that concepts, principles, and deductions do not pertain to the field of science.

6. Their inconsistency is forced by necessity. As the scientific mind advances, it cannot help recognizing the important rôle played by objects of consciousness that cannot be reduced to the level of sensory perception. Among these objects are concepts, principles, and the reasoning process itself. Realists have tended to include such objects within the realm of reality. The extreme realists of earlier times went so far as to ascribe to essences (ideas) a substantial existence apart from the mind of the knower. Moderate realists have accorded them an

---

because every science arises from demonstrative reasoning, and first principles, standing as they do at the beginning of the reasoning process, are not themselves the result of reasoning (*ibid.*, Lib. I, Lect. XX, No. 15). The first general principles are known of their very nature to be true. They cannot be confirmed by exterior reason (*ratio exterior*), but they are rooted in the interior reason which is in the soul, because by the light of natural reason they immediately become known and they are so utterly true that their opposites cannot even be conceived by the intellect. They can be denied by exterior words but not by interior thought (*ibid.*, Lib. I, Lect. XIX, Nos. 2, 3).

The definition of science as the knowledge of reality in the reduplicative sense does not differ substantially from the theory of Aristotle as expounded by St. Thomas. What St. Thomas calls "knowledge of the causes of the *thing*" is here simply translated into "knowledge of the causes of *reality*," and this is what St. Thomas and Aristotle ultimately mean (*supra*, 2.4:2b). It is our contention that what St. Thomas calls 'truth' (*rerum veritas*) is identical with the concept of reality (*supra*, 2.4:2b), the genus which gives unity to the whole of science. The concept of reality is the general medium in which extramental being is reflected. This extramental being may be divided into the four causes of Aristotle (cf. *In Post Analyt.*, Lib. II, Lect. VII, Nos. 2-3), and subdivided into more restricted species, giving rise to the specific branches of science, but all of these causes have a common name, and that name is 'meaning.' In the medium of the concept of reality is reflected the meaning of extramental things, and this meaning corresponds to being that exists in extramental things. The material object of science is the object constructed from the data of sense. The immediate formal object of science is the meaning which intelligence sees in (and beyond) the sensory image, because the intelligibility of that meaning is picked up in the concept of reality. The mediate and remote formal object of science is the intelligible being in the sensory object, which human intelligence cannot see directly, but which it can by the power of reason affirm to correspond to the conceptual representation in the medium of understanding.

Science, then, is the knowledge of the meaning of reality, and it may be divided into the knowledge of the various kinds of meaning of reality. Intelligence can distinguish between the identity of a sensible object and its form. It can, for instance, distinguish between the cow and its whiteness or blackness. Again, it can distinguish between the sensible form of the cow and what it has narrowed down to be the 'essence' of cows as such (*supra*, 2.3:18a). Identity is an intelligible concept associated with the 'matter' of a thing. Science concentrates upon the identities of things in order to confirm its reasoning in the data of sense, which guarantee the extramental existence of things. This is the 'material cause' of Aristotle. Or science can concentrate upon the universal forms of things, and this is the 'formal cause' of Aristotle. Science can also concentrate upon meanings that are 'extrinsic' to a given sensory object. If it studies the relation of another identity to the given identity, it is treating the 'efficient cause' of Aristotle. And, if it directs attention to successive forms of the same identity, it is examining the 'final cause' of Aristotle. To work according to any of these causes, the scientist must rely upon both deduction and induction, but the elaboration of formal causality is especially dependent upon the deductive method of classical scientific reasoning.

What result from reasoning according to the various causes of things are not completely different sciences but different branches of science as a whole. St. Thomas apportions the sciences according to the different genera of their subject-mat-

extramental, but not substantial, mode of existence.[a]

7. We say that science begins with a recognition, however vague and implicit, of one's medium of thought and tends to advance towards an ever more precise and explicit apprehension of that medium. Just as consciousness is the medium of all thought, and reality is the medium of all scientific thought, so each specialized science has its special medium within reality by which it can come into being. The function of the special medium of each specialized science is related to the function of reality as the general medium.

8. Real physical things are sources of knowledge acting upon consciousness from outside the mind in order to bring knowledge into being, but the 'reality' by which they are covered is a psychological object, a general meaning existing remotely in extramental things, proximately in the mind that recognizes it, and only fundamentally in the physical things which are the sources of experience.[a] The content of

---

ter, basing each science upon a generic concept which provides the common medium (*In Post. Analyt.*, Lib. I, Lect. XLI, No. 11). He does not explicitly affirm that all of the sciences are unified under the common concept of reality, but he does maintain that they all adhere to truth, and he traces all of the sciences back to the same self-evident principles of understanding (*ibid.*, Lib. I, Lect. XLIII, No. 11). It follows that the sciences are united in the first principles of reason, and I have attempted to show above (2.4:2b) that the concept 'reality' is based immediately upon the principle of the excluded middle, not as the species of a genus, but as the act of acceptance of the principle. This acceptance is implicit in the Aristotelian notion of judgment.

2.4:6a Margenau identifies both existence and being with physical reality. "The only alternative to a denial of meaning in the word being is to identify it with [physical] reality. ... The term existence will be dealt with in the same way." (*The Nature*, 4.) I prefer, on the contrary, to preserve the distinctive meaning of each of the three words. *Existence* is the uncomprehended (and *per se* unintelligible) residue of a thing that is grasped intellectually; it is identified with the singularity and individual identity of the source of an experience. *Reality* is first and foremost a concept in the mind identified with that portion of mental objects which cannot be recognized to be illusory, deceptive, or fantastic, and referring to their sources as known by intellectual inference to exist extramentally. *Being* is the general term which includes both the conscious and the extra-conscious modes of existence, including illusions, deceptions, and fantasies. Thus, Gibraltar has real existence, while Oz has only fantastic existence. Gibraltar is a source of sensitive experiences that can be verified over and over again by the same person and confirmed by other persons; it is an external source in the sense that one must go to it by traversing a tract of physical space, lined with predictable sense experiences of its own. Oz can be brought before the imagination at will, and its features can be changed at will. The number five does not have extramental substantial existence; it can be brought before the mind at will. But it has a real place within the human intellect. Five really is the whole number which stands between four and six. The *real* square root of twenty-five is five. The number five fits within the concept of reality as a real conscious feature of the concretely existing intellect standing outside of consciousness and constituting a part of the substance man. Reality, then, is not identified with physical reality; it is rather a genus whose meaning becomes clear as it is divided into the two species of physical reality and intellectual reality.

2.4:8a The concept 'reality' may be defined as "the totality of extramental being." If a concept is a mental being, how can there be a concept of what is totally outside the mind, and what could possibly be the content of such a concept? I think that the concept of reality arises with the distinction between the existence and the meaning of things and that its content consists in the grasp of meaning. After a person has come to know that there are objects and that the identity of objects is distinct from their sensible form (*supra*, 2.3:18a), he is able to go on to distinguish between the identity of objects and their meaning. In this third step, the sensible image of an object is associated with its identity to form the existence of the object, and the intellect grasps the meaning that emerges distinct from the existence. But certain conditions are necessary for the emergence of meaning. The knowing subject must realize that there *is* meaning in the object

'reality' as a concept lies in a reference to things as they are in themselves. It therefore implies a native trust in the ability of sensation and intellection to apprehend things as they are in themselves, since the immediate reference is to the data of sensitive and intellectual experience, while the remote reference is to the extramental existence of the meanings of things.[b]

and that this meaning is rooted in the extramental existence of the object. It is not the meaning that human creativity is free to inject into objects from the fancy of the imagination. This meaning requires the knowing subject to test and verify his grasp in the existences presented by the senses. It requires him also to be aware of the instrumentality of his mind, because he must distinguish constantly between the images of fact and the images of fancy. And this attention to facts draws him to search for further meaning.

The grasping of the notion of reality is therefore the step to understanding. The wholeness of each object of intelligence has become a universe of objects presented to the mind, and now there arises the realization that there is meaning in this universe. There is the awareness that only real objects can help or hurt the subject, and there is the inclination to face reality. The notion of reality sustains a simple reference to the data of sense, but not without the awareness that these data have meaning. The notion of reality is universal in the sense of containing the implicit awareness that the universe of objects presented by the external senses is a meaningful whole and also in the sense that the idea of meaning applies to each and every object presented by the data of sense. But the idea of meaning is a mental event; it is a development of the intellect of the man who grasps it.

I think the concept of reality is expressed in the principle of the excluded middle: a thing either is or is not. This principle does not have meaning except as indentified with the concept of reality. And there is no concept of reality except as identified with this principle. But the identity of the concept of reality with the principle of the excluded middle means that whoever uses the notion of reality is thereby constrained to adhere to this principle and to whatever this principle implies. Reality does not, then, mean merely verification in sense experience; it means also verification in intellectual experience. Just as a scientist who uses a telescope must attend, not only to the objects in the medium, but also to the character and condition of the medium itself, so does the one who would survey reality have to attend also to the medium of his knowledge. Reality is the experience of the intelligibility of things. It imposes itself upon the mind, not only as the existence of sensory objects, but also as the meaning which lies behind them.

2.4:8[b] Margenau says that science "takes experience for granted." He maintains, however, that "reality is not the cause but a specifiable part of experience." (*The Nature*, 289.) We find also that 'reality' is a part of experience, since it is a concept realized in intellectual experience, but it refers to extramental sources which are not in themselves a part of experience. Margenau denies that the physical world is in that sense outside of experience (*ibid.*, 455). He cautions that, "if we once succumb to the temptation of acknowledging the existence of the *one* additional metaphysical dimension which realism demands, there is no way of avoiding a silly encumbrance with a nestling infinite hierarchy of dimensions, much like Zeno's nest of spaces.... The attitude of realism, or idealism, or indeed any attitude that projects a single phase of experience beyond the confines of experience relinquishes control over reality to agencies of doubtful competence. And it leaves science without defense against fairies, ghosts, and goblins. For, as we have tried to show, unless it be found in the counterplay of construction and verification, there is no available criterion to give reality its warrant and to set it apart from the unreal." (*Ibid.*, 456-457.) We agree that the one criterion of reality is experience, precisely, intellectual experience rooted in sense experience. We therefore do not admit into the system of reality (our personal experience of reality) what has not been validly constructed by the intellect and verified (as regards objects that can be imagined) in sensory experience. But realism is correct in locating the physical world outside of experience, for it is the source and cause of experience. If this source were not outside of experience, it could not cause experience in anyone but myself; and there could not be anyone but myself, for experience is a subjective event. Therefore all others would be in every sense within the limits of my experience. But this in obviously false, for the reality of the physical world and of other

9. The act of isolating intelligible objects from the data of sensory experience is known as abstraction.[a] These objects often have the form of universals, or classes of things, but they may also be concrete. The knowing subject sees the abstract object as intellectually interposed between himself and the individual sensory object. The recognition of the intellectual object (objective medium) as an object is called 'total abstraction.' The (concrete) development of understanding in the intellect of the person who experiences the recognition is called 'formal abstraction.'[b]

10. Any object of knowledge may be called an 'idea.' The word 'idea' comes from the Greek verb 'to see,' and it is often applied to sensory as well as to intellectual sight. When a person says that he has an idea of what a people imposes itself upon me. We do not, nevertheless, maintain that the physical world, as it stands outside of experience, 'looks' the same as when it is constructed in the imagination, for how it looks depends upon the receptive capacity of the observer. Nor does our position set up a "nestling infinite hierarchy of dimensions," since the concentration of intellectual experience is upon the dimensions it knows and upon the scientific criterion of reality. The realistic view does not indulge in worry or superstitious fantasy concerning the world beyond experience, but it remains open to the possibility of new experiences from extramental sources as yet unknown to it.

2.4:9a F. H. Bradley denies that ultimate reality could be multiple or that there could be independent reals. He sees the Absolute to be an individual and a system whose empty outline is filled up by experience, where experience is taken to mean something much the same as given and present fact. "Sentient experience, in short, is reality, and what is not this is not real. We may say, in other words, that there is no being or fact outside of what is commonly called psychical existence." (*Appearance and Reality*, 127.) "For if, seeking for reality, we go to experience, what we certainly do *not* find is a subject or an object, ... (but rather) a whole in which distinctions can be made, but in which divisions do not exist. ... I mean that to be real is to be indissolubly one thing with sentience. ... Being and reality are, in brief, one thing with sentience; they can neither be opposed to, nor even in the end distinguished from it." (*Ibid.*, 128-129.) It was a mistake in method for Bradley to go to sense experience in order to find reality and thus to immerse intellectuality in sentiency, for the true path of science is from sentiency to intelligence. Grasping reality is an intellectual experience carrying with it the insight that there are real things independent of any given act of experience. There is no intelligence where the distinction between subject and object is not operative, and there is no science where reality is not distinguished from sentiency.

2.4:9b According to Henry van Laer the basis of the difference between formal and total abstraction lies in the 'twofold aspect' of the universal, namely, its ontological vs. its logical aspect. He sees formal abstraction as seizing the datum of experience in its intelligible content, divesting it of the concrete and individual form in which it appears. "The result is an intelligible content which represents the real datum with respect to its specific character and therefore, in point of fact, is universal in the ontological sense. Total abstraction, on the other hand, strips the sense datum of its concrete individual appearance — which really amounts to dematerializing it — and seizes that which material things have in common, namely, the specific essence. The result is a universal concept in the logical sense ...." (*Philosophy of Science*, Part One, *Science in General* [Pittsburgh: Duquesne Univ., 1956], 26.) "Formal abstraction yields the universal *qua* intelligible, and total abstraction yields the universal *qua* universal (i.e., *qua* communicable)." (E. D. Simmons, "In Defense of Total and Formal Abstraction," in *The New Scholasticism*, 29 [1935], [427-440], 434.) It is the opinion of the present writer that the difference between the two kinds of abstraction, as described by Simmons and van Laer, is not sufficient for a clear concept. The needed clarity will appear only in the subject-object frame of reference. Total abstraction is a kind of intelligibility adhering to an intellectual object as such. Formal abstraction is a movement of development of the understanding of the knowing subject as such. As the understanding develops, the whole (concrete) intellect of the man becomes abstracted from the materiality of its original situation. Total abstraction is the view resulting from a dichotomy; formal abstraction is the stage of development of the intellect that possesses this dichotomy.

violet looks like or smells like, he is using the word 'idea' to represent a phantasm of sensory experience. An idea partly composed of sense imagery and partly of intellectual apprehension is a 'conception.' A conception is a vague intellectual idea. A clear and precise intellectual idea is known as a 'concept.' The achievement of clarity is called a 'definition.'[a]

11. When we speak of the 'notion' of reality, we are referring to a conception not clearly defined in consciousness. A clear *concept* of reality is the insight of a mature mind, while the *notion* of reality is one of the earliest insights of any normal mind. When the definition of an idea reaches sufficient clarity to satisfy the requirements of specialized science, it is given the status of a concept.

12. Notions are preconceptual ideas. When the mind focuses upon reality, it has a natural tendency to proceed from mixed conceptions to clear conceptions, from notions to concepts. The *notion* of reality lies in the idea of constantly referring to what is real and non-imaginary. In that sense, it might seem to effect an unbreakable bond with the data of sensory experience, as opposed to the conceptions of intellectual thought; it would seem to require the monopoly of observable sense phenomena as the object of scientific thought. But this interpretation is an illusion. The fact is that the very act of recognizing reality as such, which is the *sine qua non* of all scientific thought, is itself a development of the intellect not verifiable in the data of sensory experience. It is the declaration of independence of the intellect, when it emerges from bondage to sense experience and the vagaries of the imagination to recognize itself and demand the rights of citizenship in the world of reality. The notion of reality appears in consciousness as the higher or more valued part of a division from the merely imaginary, so that the esteem one gives to reality is based upon the implicit acceptance of the validity of the dichotomy. Through this dichotomy, mind, as distinguished from observable matter, first comes into focus as an object of thought. It is not the whole mind or the whole intellect that it seen, but the notion of reality as an intellectual object separated from the *material things* to which it refers.[a] As a mixed conception, reality in its universality never becomes totally independent of the matter to which it refers, but the mental field brought into focus by the notion of reality will tend to develop intellectually, with the result that the intellectual portion of the mixed conception will gradually increase and the material imagery will gradually become more subtle.

13. From this point on, the position of the behaviorist and of the psychological materialist becomes either irrational or dishonest. The thinking method of both types of observers depends upon an implicit recognition of the difference between the sensory perceptions to which they refer their judgments and the intellectual apprehension which gives them a motive for so referring. Moreover, they accept the validity of at least some of the principles of reason. But because of adherence of their wills to some negative bias, be it mere emotional attachment or a canny awareness that an acknowledgment of mental reality leads towards philosophical and religious conclusions that they have decided to avoid, they stunt the growth of intellect towards a grasp of fuller intelligibility and clap it in the irons of materialism. Within these irons, the intellect of the materialist can still

---

2.4:10[a] Margenau distinguishes between 'epistemic' definitions, connecting constructs with immediate experience, and 'constitutive' definitions, relating constructs to one another (*The Nature*, 220-244). Actually, there are various kinds of definitions, according to the frame of reference or basic dichotomy employed by the intelligence.

2.4:12[a] The fact that reality is an intellectual *object* allows the intelligence to study it *as an object*, and not as identified with one's own subjectivity. Failure to grasp this fact led Jakob Friedrich Fries, Friedrich Eduard Beneke, John Stuart Mill, and other proponents of 'psychologism' to maintain that the only instrument at the disposal of philosophical inquiry is self-observation, or introspection (cf. N. Abbagnano, "Psychologism," in *The Encyclopedia of Philosophy* VI [New York, 1967], 520-521). The path of all science, including philosophy, is away from introspection towards *introversion*, which is a concentration upon mental *objects*.

grow to some degree, but only into the grotesque figure of a mind focused firmly upon his subjective prejudice, with its area of objectivity unfolded by a series of material insights but immediately inverted by subjection to the bias. Such 'objectivity' is not scientific, for its medium of thought is not reality within objectivity, but reality inverted into subjectivity.

2.5. The Locus of Common Sense in Human Consciousness

1. Another name for common science is common sense. Historically the term has been used to denote knowledge on both the sensory and the intellectual levels. In ancient Greek and medieval scholastic cognitional theory the term refers to what is common either in the perception of the special organs of sense or in the thinking of all mankind. Thus Aristotle identified the 'common sense' as a general centralizing faculty by which are apprehended the 'common sensibles' of motion and rest, figure and size, number and unity. He and other Greek thinkers recognized also certain common elements in the thought-pattern of all men which established a common and spontaneous agreement concerning certain convictions.

2. In contemporary discussion the term 'common sense' is sometimes used negatively to denote a naive view of reality as contrasted with the viewpoint of empirical science, and it is sometimes used positively to denote "a set of attitudes and assumptions presumed to be held by plain men who are untutored in a conscious philosophy." [a] While the second of these two denotations stresses the realism of common sense as opposed to the sophistic idealism of surrounding philosophies and scientific theories, neither identifies common sense in its full proportions. Common sense, fairly defined, is "the right reason of sound judgment not involving technical concepts." [b] As such it is related both intrinsically and extrinsically to specialized science. It is related intrinsically as the general and initial phase from which specialized science springs, the source of self-evident and incontrovertible truths, and the rough instrument of realistic thinking.[c] Common sense is thus related to science as arithmetic is related to mathematics, to use an example from specialized science itself. As the initial form of science, common sense is vague and imperfect, subject to correction by the more precise thinking and more exact definition of its higher stages. But the basic principles of common sense are intrinsic to all science; they cannot be flatly contradicted or substantially left behind by any science that will remain truly scientific. Science which violates common sense in its premises or in its thinking thereby turns itself into pseudoscience.[d]

3. Common sense is related to specialized science as the potentially perfect to the essentially incomplete. Specialized science by its specialization loses sight of the total picture which common sense retains. What is lost from the field of view by specialization is not merely a residue of vague ideas and irrational biases maintained in the minds of those who have not achieved the same specialization. It includes also the areas of thought capable of developing into other specialized sciences, as well as that area included within the formality of reality which cannot be apprehended by any specialized science because it is the ground of wisdom. But wisdom is not simply the eventual conclusion of a general theory of specialized science. It is rather the maturity of the disciplined mind that combines with a

---

2.5:2a Arthur Danto, "Common Sense," in *Encyclopedia Americana* (1967), Vol. VII, 415.

2.5:2b Common sense is "good sound ordinary sense: good judgment or prudence in estimating or managing affairs especially as free from emotional bias or intellectual subtlety or as not dependent on special or technical knowledge." (Sub verbo, *Webster's Third New International Dictionary*, unabridged ed. [Springfield, Mass., 1966].)

2.5:2c Cf. R. Garrigou-Lagrange, *Le sens commun* (3rd rev. ed., Paris: Nouvelle Librairie Nationale, 1922), 106-108.

2.5:2d "Experimental science can be thought of as an activity which increases the adequacy of the concepts and conceptual schemes which are related to certain types of perception and which lead to certain types of activities; it is one extension of common sense." (James B. Conant, *Science and Common Sense* [New Haven, 1951], 32.)

certain scientific proficiency the domination of the unruliness of those areas of consciousness occupied by emotion, volition, and the subjectivity of the self. The very decision to think scientifically depends not upon specialized science but upon that harmony of intellect and will called total science, and it is this total science that serves to eliminate such scientific enigmas as the brilliant mathematician who is also a moral degenerate, or the profound philosopher who is also an alcoholic. The total answer to the partial questions which specialized science poses for itself is to be found in that absolute abstraction of the intelligence which is identified with wisdom, while in the distinction which scientists sometimes make between their specialty and common sense, this perfection of the whole man is relegated to the lower level of the latter.

4. It is a mistake to define common sense by its defects alone.[a] While it is true that its defective use will lead the unsophisticated to consider it omnicompetent in practical affairs and blind to the long-term consequences of policies and courses of action, its correct use will instill in the simple an esteem for the capabilities of the educated and will preserve the specialist from falling into that form of unrealistic snobbishness that prevents him from seeing the limitations of his own specialty as viewed from the bigger world of common sense. It will instill even in the specialist that humility and sobriety which keep him from exaggerating the importance of his specialized conclusions and keep his mind open to the reality of truth transcending the specialized field. It sets his personality on a course of development concurrent with the growth of specialized insight, so that the scientist does not emerge from his study as a stunted person. It will fit the deepening awareness of the fact and finality of human

---

2.5:4[a] Bernard Lonergan defines common sense as "that vague name given to the unknown source of a large and floating population of elementary judgments which everyone makes, everyone relies on, and almost everyone regards as obvious and indisputable." (*Insight: A Study of Human Understanding* [London: Longmans, Green, and Co.; N.Y.: Philosophical Library, 1958²], 289.) He holds that the field of empirical science is to be reached only by abstracting from the empirical residue, while the field of common sense includes the empirical residue and views things in their individuality, their accidental determinations, their arbitrariness, their continuity. He describes empirical science as dealing with the comprehensive, universal, invariant, non-imaginable domain whose object is the thing-itself, with differences in kind defined by explanatory conjugates and with differences in state defined by ideal frequencies, while common sense is said to deal with the experiential, particular, relative, imaginable domain whose object is the thing-for-us, with differences in kind defined by experiential conjugates and with differences in state defined by expectations of the normal. Hence, common sense is seen as concerned with things as related to us, while science is concerned with things as related among themselves. (*Insight*, 293-294.) Lonergan states that the "method of common-sense eclecticism" allows uncriticized and unchastened common sense to settle by its practicality the aim of philosophy as well as to measure ingenuously the resources which the philosopher has at his disposal. He observes that it denies to philosophy its vital growth on the assumption that "problems are immutable features of the mental landscape, and syntheses are to be effected by somebody else who, when he has finished his system, will provide a name for merely another viewpoint." (*Insight*, 418.) Thus he is led to conclude that the proper domain of common sense is the realm of particular matters of fact, while the correct meaning of such terms as reality, knowledge, and objectivity cannot be attained by appealing to what common sense finds obvious. (*Insight*, 419, 421.) He admits that every specialist runs the risk of turning his speciality into a bias by failing to recognize and appreciate the significance of other fields, but he adds that common sense does so almost invariably, "for it is incapable of analyzing itself, incapable of making the discovery that it too is a specialized development of human knowledge, incapable of coming to grasp that its particular danger is to extend its legitimate concern for the concrete and the immediately practical into disregard of larger issues and indifference to long-term results." (*Insight*, 226.)

Lonergan very plainly declares the independence of scientific thinking from the 'naive realism' of the "familiar world of poetry and common sense" on the ground

love into a 'cosmic' context which empirical science can never recognize, and present within an intelligible focus the moral priority of the ethical mind over the remote reasoning of the empirical mind. It will stay on guard against that 'spiritual concupiscence' which has been the bane of science and philosophy since the beginning of history. Since the differentiation of common sense lies in its general awareness of reality, when rightly used it will insist upon homogeneity only in the sense that it will not allow theory to propose conclusions based upon dubious premises which contradict what common sense already knows, thus preserving specialized science from idealistic fantasy. Theory and common sense comprise a single universe of discourse; theory which violates common sense is pseudo-science. Common sense will recognize that technological advance is partly the fruit of specialized scientific theory but is also due to experimentation on a purely common-sense level. It will remain open to the possibility of recognizing

that the two levels of thought constitute two entirely separate or distinct universes of discourse together comprising a single knowledge of a single world. (*Insight*, 297-298.) Common sense "commonly is unaware of the admixture of common nonsense in its more cherished convictions and slogans" (*Method in Theology* [London: Darton, Longman & Todd, 1972], 53), so that the austere thought of creative thinkers in philosophy and science is soon fused by popularizers and lower-level teachers "more with common nonsense than with common sense, to make the nonsense pretentious and, because it is common, dangerous and even disastrous." (*Method*, 98.) Lonergan distinguishes three stages of the world mediated by meaning. "In the first stage the world mediated by meaning is just the world of common sense." In the second stage it splits into the realm of common sense and the realm of theory. (*Method*, 93.) In the third stage the sciences have become ongoing processes whose special aim is not to state the truth about this or that kind of reality but ever closer to approximate the truth by an ever fuller and more exact understanding of all relevant data. "In the second stage, theory was a specialty for the attainment of truth; in the third stage scientific theory has become a specialty for the advance of understanding." (*Method*, 94.) He views the sciences as autonomous: they accept as scientific only those questions which can be settled by an appeal to sensible data. (*Method*, 94.) The worlds of theory and of common sense are regarded as two different universes of discourse whose unification could never be effected by logic. (*Method*, 94-95; *Insight*, 294.)

According to Lonergan's exposition, the undifferentiated consciousness of common sense insists upon homogeneity, declaring null the conclusions of theory which conflict with its own validity, while the unity of differentiated consciousness depends, not upon logical homogeneity, but upon "the self-knowledge that understands the different realms and knows how to shift from any one to any other." (*Method*, 84; cf. *ibid.*, 304-305.) But not even the realm of common sense is in the ultimate analysis considered to be a single world mediated by meaning, for "there are as many brands of common sense as there are differing places and times. What is common to common sense is, not its content, but its procedure," and this consists of a characteristic self-correcting process of learning, always with reference to the immediate, the concrete, the particular. (*Method*, 303, 328-329.) Lonergan is thus constrained to distinguish between two languages and two social groups resulting from two worlds mediated by meaning: the world mediated by common-sense meaning and the world mediated by systematic meaning. The former group is limited to common-sense language, but the latter can use both ordinary and technical language. The educated classes may on occasion use one or another technical term or logical technique, but "their whole mode of thought is just the commonsense mode." (*Method*, 304.) He observes that practical people are guided by common sense and are therefore not disposed "to sacrifice immediate advantage for the enormously greater good of society in two or three decades." He sees churchmen as having been particularly afflicted with this shortsightedness. The series of fundamental changes that has served during the past four centuries and a half to modify man's image of himself in his world, his science, his history, his philosophy, and his conceptions of the same, Father Lonergan notes, has in general been resisted by churchmen for two reasons: they have had no real apprehension of the nature of these changes and they have been disturbed by the lack of intellectual

in the ancient and medieval traditions forms of specialized science whose value is not totally contained in modern empirical science, and which therefore belong to the area of common sense, not as common, but as specialized science distinct from what modern empirical science is prepared to recognize as science. It will thus avoid an overly restricted definition of truth and an overly narrow conception of what understanding is.

5. As the ground of the virtue of wisdom, common sense will remain immersed in the particular and the concrete, but without precluding degrees of abstraction that exceed and transcend the physical and mathematical abstraction of the empirical sciences. When common sense flowers into wisdom, it is capable of making greater sacrifices of immediate goods for values that are more absolute and ultimate than those which empirical

---

conversion accompanying them. (*Method*, 317.) "One might as well declare openly that all new ideas are taboo, as require that they be examined, evaluated, and approved by some hierarchy of officials and bureaucrats; for members of this hierarchy possess authority and power in inverse ratio to their familiarity with the concrete situations in which the new ideas emerge; they never know whether or not the new idea will work; much less can they divine how it might be corrected or developed; and since the one thing they dread is making a mistake, they devote their energies to paper work and postpone decisions." (*Insight*, 234-235.) Just as science is a different universe of discourse from common sense, so also theology is independent of the judgments of ecclesiastical officialdom: using the basically irreformable method now at last worked out by Father Lonergan, "each theologian will judge the authenticity of the authors of views, and he will do so by the touchstone of his own authenticity." (*Method*, 331.)

REMARKS. The "touchstone of one's own authenticity" is an unintelligible, antiscientific, and self-centered criterion intrinsically incapable of helping its user to know either what the truth is or in which direction genuine intellectual conversion lies. It merely expresses the folly of attempting to base theological method upon the contradiction of that sound old adage of common sense, "No one is the judge in his own case," which everyone knows, everyone relies on, and almost everyone regards as obvious and indisputable. Scientific judgment requires a fair and exact presentation of the facts. While what Lonergan says about the limitations and the characteristic dangers of common sense contains a grain of truth, his description of common sense almost entirely in terms of its negative and imperfect features is a caricature rather than a factual exposition.

According to Lonergan's division of science and common sense into two different universes of discourse, the universe of common sense is the inferior world of those who have not attained the esoteric world of science. The common sense of this division is defined by its imperfection in the sense that it is but the initial phase of an intellectual development that becomes more perfect on the level of science. And it is defined by its negative features in the sense that it is associated "almost invariably" with an admixture of bias and "common nonsense." This definition is, however, inadequate on both counts. Since common sense contains the 'empirical residue' of Lonergan's basic dichotomy, it includes not only the imperfection that is left behind by the focus of the mind upon empirical science, but also every other science not included in the focus of empirical science. To assume that what stands outside of the focus of empirical science is necessarily dross, as Lonergan does, is unscientific and betrays the lack of even a minimum of investigation. In similar fashion, to make common sense inferior on the ground that it is almost invariably mixed with bias and nonsense, while science is usually not, it to make a gratuitous assertion that will not stand up under any scientific investigation. Bias and nonsense are so common in both fields that to base one's case on their absence from either is an 'exercise in futility.' Lonergan admits that every specialist is tempted to turn his specialty into a bias by failing to recognize the significance of other fields. Common sense, as the 'empirical residue,' contains within it, actually or potentially, the whole of intellectual consciousness not contained within the restricted field of empirical science. By failing to recognize and appreciate the significance of common sense in its positive value, by depreciating it in an unscientific dichotomy, by making this assumed inferiority a theme of a systematic exposition, Lonergan succeeds only in monumentalizing the professional bias that he has set out to overcome.

science can ever know. Having at hand a fund of past experience, it will view with healthy distrust the fatal tendency of the undisciplined reason to dissolve its integrity into irrational pluralism. It will allow areas of its competence to develop into specialized fields to the extent that this development is consonant with the intuition of reality and the correct use of reason. It will not be persuaded by false argumentation that its primary grasp of the world of reality is but a cosmological myth to be burst open into two universes of meaning, when it has been ripened by the growth of empirical science, or that the analogy of sight underlying its view of the world is scientifically an illusion. Thus it will not admit as real what can be shown to be in violation of sensible data, but neither will it admit the appeal to sensible data as the only criterion of the validity of scientific conclusions, as though there could be no intellectual intuition of reality. And it is this fundamental conservatism of common sense which provides the basis for a unified concept of science and the possible unification of the conclusions of the specialized sciences with all of the data of consciousness in terms of intelligible principles.

2.6. The Locus of Specialized Science in Human Consciousness

1. All science must have a static as well as a dynamic aspect. It must be dynamic, for it exists in the consciousness of living human beings and is, moreover, a developmental thing that comes into being in graduated stages as the mind itself develops. Science is, in one sense, a product of the growth of the individual intellect. It is also a product of the activity of many individual intellects over successive generations, so that it has its own interpersonal history and process of development. But science must also be static, for it is knowledge of the real as such, and there could be no knowledge of the real as such, if there were not something stable and abiding to which the intellect could refer as a criterion and according to which the idea of reality would have some consistency and identity in itself. Moreover, science as an interpersonal thing could have no history or development unless there were something in it essential and unchanging so as to make the changing aspect a function of the same identical thing.[a]

2. There has been a tendency in modern times to concentrate so heavily upon the dynamic aspect of science as to ignore the static aspect that gives it unity and full intelligibility.[a] The tendency has been to identify science itself with the 'scientific method.'[b] The result has been an almost exclusive attention to the method of acquiring knowledge and an almost complete neglect of the kind and meaning of the knowledge acquired, except in functional relationship to the further acquisition of the same kind of knowledge within a narrow frame of reference. Now, to declare that science is not knowledge, but a method of acquiring knowledge, leads immediately to the question: "What is the nature of the knowledge that is so acquired?" It is the lack of an answer to this ques-

---

2.6:1[a] Cf. Margenau, *The Nature*, 288-289.

2.6:2[a] "The dynamic view in contrast to the static regards science as an activity; thus, the present state of knowledge is of importance chiefly as a basis for further operations. ... My definition of science is, therefore, somewhat as follows: Science is an interconnected series of concepts and conceptual schemes that have developed as a result of experimentation and observations. ... Thus conceived, science is not a quest for certainty; it is rather a quest which is successful only to the degree that it is continuous." (Conant, *op. cit. supra*, 2.5:2[d], 25-26.) This definition is based upon a confusion of the in-

trinsic and extrinsic purposes of sciences. To say that a calf is of importance chiefly as something getting bigger does not tell us what a calf is. Similarly, arithmetic is something in itself, over and above the fact that in this individual it may be growing into algebra or geometry.

2.6:2[b] "Note that science describes the how, but does not explain the why of nature; it makes no attempt to establish the true and absolute 'nature of things.' This latter activity belongs in the province of religion. Science, then, is a method, not a subject. It is a method for the organized investigation of nature." (G. Abell, *Exploration of the Universe* [New York: 1964], 4-5.)

tion, the placing out of focus of the knowledge acquired so that it never becomes a direct object of consideration, that has constituted a weakness in modern scientific expression.[c]

3. It seems obvious that knowledge acquired by the scientific method must have some identity and significance of its own, just as any other kind of knowledge has. The conclusions of science constitute a doctrine, whether its exponents call it that or not. Any textbook of science is an exposition of doctrine, whether its writer will admit the fact or not. Therefore it is more realistic and scientific to recognize and describe this doctrine than studiously and systematically to ignore it.[a]

4. That body of knowledge of the real that is derived from the organized investigation of nature is implicitly identified with science, even by those who claim that science is not a subject or a body of knowledge. The organized investigation of nature requires a special viewpoint suggesting the special way in which the investigation is to be organized and providing some anticipation of what is being looked for. Physical science, as specialized science, shares the common approach of all science and possesses in addition a proper approach of its own.[a]

5. An approach, in the dynamic sense, is constituted by the method; in the static sense it is constituted by its proper fundamental concept, whether or not this concept has been fully and precisely recognized. The fundamental concept of science in general is the concept of reality. Science begins when the thinker first becomes aware of reality and decides to think in terms of it, although at this point he is seldom reflectively aware *that* he is doing so.[a] There is, in fact, a continual reference to the notion of reality during such periods of thinking, but the thinker does not explicitly advert to the fact that reality is his medium of thought within consciousness. It is only when his mind has become more mature that he is able to reflect upon this fact and realize explicitly what science really is. In the meanwhile, the initial recognition will have developed into a knowledge of reality.

6. The concentration of the mind upon the 'flat field' of reality,[a] that is, of the presentations of sensory experience, will lead the impartial observer to perceive with growing clarity the existence of depth within that field. There is another dimension of reality not contained in the data of sense experience alone. Thus, the nature of reality can be illustrated from the science of mathematics. Just as the point is the principle of the line, even though a line is not composed of any number of points, so the presentations of sensory experience are the principle of reality, even though reality is not composed of any number of sensory experiences. Just as the science of geometry would dissolve into nothing if its lines and figures were never recognized in themselves but only in the point from which they are derived, so is reality dissolved into nothing, when the intellectual objects of which it is composed are not recognized except to the extent that they are reduced to the sense percepts to which they adhere. Science begins from deliberate reference to the things verified by sensory experience as such; it cannot contradict the data of sensory experience at any step along the way; but the component that gives to science its depth and proper character is the developed recognition of the intellectual dimension of reality.[b]

2.6:2c "At the opposite pole from the universalists are those who would sum up the whole of science as nothing but a method. Such professed empiricists .... generalize one feature of the scientific enterprise so far that scientific method becomes abstract and absolute." (J. R. Kantor, *The Logic of Modern Science* [Bloomington, Ind., 1953], 4.)

2.6:3a Cf. Aquinas, *Post. Analyt.*, Lib. I, Lect. I, No. 9.

2.6:4a Cf. *infra*, III.23.

2.6:5a A short bibliography of works on scientific method is given at the end of Filippo Selvaggi's article, "Scientifico, Metodo," in *Enciclopedia filosofica* V (Firenze, 1961), 1150. See also the bibliographies in R. Harré, *The Principles of Scientific Thinking* (Chicago: University of Chicago Press, 1970).

2.6:6a Cf. Margenau, *The Nature*, 450.

2.6:6b "Characteristically, the proponents of the United-Science movement picture science as a kind of surface, as a two-dimensional structure. They are fond of calling it a mosaic. ... There is a region below every problem which is illuminated and exposed to view by its very solution,

7. The fundamental error of the materialist theory of science lies in its failure to recognize the full proportions of reality. The same inability leads the materialist to suppress the formal aspect of science itself.[a] And, by the same token, the recognition of the formal aspect of science implies a recognition of the formal aspect of reality. The recognition of formality is basic to valid cognitional theory: it may be likened to the recognition of the meaning of quantity in the specialized science of mathematics. The mistaken supposition that the need of science to verify its conclusions in experience requires the reduction of all objects of thought to the level of sensory experience is analogous to the mistaken supposition of an erroneous mathematics that the presence of the qualities of points in lines and of planes in solids requires the reduction of all lines to points and of all solid figures to plane figures before they can be accepted as valid. Scientific reality is like a 'solid' figure that comes into its own as it emerges from the 'flat' figure of sense phenomena and begins to manifest the qualities proper to its intellectual state of being.

8. We call an 'insight' that act of recognition in the knowing subject by which his intelligence sees a new comprehensive meaning in the relevant data being considered. An insight is an act of intellectual intuition. While sensory intuition is the perception of sensory data as an organized whole, intellectual intuition is the apprehension of intelligibility in objects presented to the intellect. Intelligence, as the power of intellectual intuition, is also called reason. We may thus distinguish in human reason a transitional and an abiding element. The process of transition of human reason from one insight to another is called reasoning; it is carried forward by a native power of intelligence which is fairly constant in the normal person. The abiding element of reason is called understanding; it develops by the acquisition of insights and retains them as the disposition of the mind; it changes in the sense of development, but remains as a resource for future acts of reasoning. Reasoning is intelligence in concentrated action; understanding is intelligence at a given stage of development. Understanding is the pervasive aspect of human intelligence. It precedes, accompanies, and succeeds the reasoning process. Reasoning is the special instrument of the understanding by which it finds the proper conditions for an advance to a deeper understanding or for the application of understanding to a concrete situation.

9. The realm of science differs from the realm of imaginative thought both in its mode of thinking and in its results. Imaginative thought uses the creative imagination to focus the mind upon unreal images for the sake of producing sensuous or aesthetic pleasure. Scientific thought uses the power of reason to focus the mind upon real objects in order to produce understanding. Science is composed of *insight* on the part of the knowing subject, *meaning* on the part of the real objects that he knows, and *understanding* on the part of the intellect which provides his medium of thought. It is not a mere collection of unrelated facts verified by experience. It is structured knowledge, and the structure arises from the natural development of the mind itself. Material science is the collection of facts; formal science is the understanding of the facts in the intellect of the knower.[a]

10. What draws the mind of man from material to formal science is the power of intelligibility underlying real objects. This power, acting from outside the mind, produces intellectual vision of the kind of object that exists only in intelligence and that arises only from extramental intelligibility. This intellectual objectivity is the remote formal object *quod* of human intelligence,[a] and it is divided into degrees and aspects which we call meanings. A meaning is an intellectual object presented by the intelligibility of an extramental sourcé and formed into a recognized object by the selective power of intelligence. The act of recognizing a meaning on the part of the knowing subject is called an insight. Hence, a meaning is a unit of

and in this region new problems are always found. Investigation goes deeper and deeper into this third dimension, which the encyclopedists conveniently neglect." (Margenau, *The Nature*, 19.)

2.6:7a Cf. van Laer, *op. cit.*, 43-49.
2.6:9a Cf. *ibid.*, 43-44; Margenau, *The Nature*, 449-452.
2.6:10a Cf. van Laer, *op. cit.*, 2.

(remote) intellectual objectivity, and an insight is a unit of understanding. But understanding is not a merely transitory act. Insights register modifications of the intellect itself and represent steps in its growth. It is the abiding effect of insights upon the intellect that we call 'understanding.'

11. The recognition of the difference between *what* the intellect knows and *how* it knows what it knows divides the field of science into material and formal knowledge of reality. It also divides the field into the lower level of knowledge of the facts (*scientia*) and the higher level of understanding of the facts (*intellectus*).[a] It is understanding that advances science towards ever greater intelligibility and protects its conclusions from those forms of unscientific understanding called pseudoscience. The difference between science and pseudoscience lies materially in the truth or falsity of the respective conclusions, but formally in the medium of thought.[b] This medium first appears under the name of 'presuppositions.' A presupposition may be defined as "something required as an antecedent in logic or in fact" (*Webster*). Presuppositions are of various kinds on both the prescientific and the scientific levels. *On the prescientific level* they may be of such variant kinds as items of true antecedent knowledge, valid experience, illusion, bias, reasonable belief, unfounded belief, solid intuition, sheer fantasy, or wishful thinking. Presuppositions even on the prescientific level can be true as well as false and helpful as well as misleading.

The fact that thinking takes place on a prescientific level does not make it wrong. Such thinking preserves a right to respect unless it can be shown to be wrong either by ordinary common sense or by the more precise conclusions of a valid specialized science. *On the scientific level*, however, presuppositions may assume the form of items of true antecedent knowledge, self-evident facts or principles, formulated hypotheses, and other things of this nature, for a presupposition is but an element of thought antecedently required in order that a conclusion can be reached.[c]

12. There is a presupposition essential to all science in the sense that, being antecedently known, it makes scientific thinking possible. Since science is the knowledge of the real as such, the presupposition which constitutes the viewpoint of science as science is the concept of reality.[a] This concept is the medium of scientific thought as such, and it has both a subjective and an objective aspect according as it is referred to the subjective or the objective side of the primordial dichotomy of human intelligence. In its *subjective aspect* science is *real* knowledge of things; it is therefore not fanciful or illusory knowledge, but is rather a state of *insight* made possible by the development of one's understanding of things (*scientia qua scitur*).[b] In its *objective aspect* science is the knowledge of things *in their reality*; it is therefore the ordered totality of real things as they are known and understood by qualified

---

2.6:11a Cf. Aquinas, *Post. Analyt.*, Lib. I, Lect. XVII, No. 5; Lib. I, Lect. XLIV, No. 4. St. Thomas ponders the problem that understanding (*intellectus*) in one way is not a science at all and in another way is the very science of sciences (*supra*, 2.4:3a). The difficulty can be removed by the apt consideration that science is mediate knowledge of the real. Science in the full sense is insight into reality. Insight is a step on the graduated plane of understanding whose actuation depends upon the development of the understanding. We may define 'understanding' as the medium between the insight of a knowing subject and the meaning which he apprehends. Understanding arises in consciousness from the extramental medium which we call the intellect. Science in the full sense is not mere knowledge of reasoned conclusions; it is insight into the meaning of things. And just as insight is the 'life' of science, so also understanding is its 'soul.' We do not limit understanding to a source from which conclusions may be reasoned, for understanding not only initiates but also accompanies the process of science. (Cf. Aquinas, *Post. Analyt.*, Lib. II, Lect. XX, No. 15.) As the medium of science, it is structured and organic. It is the abiding factor of science as a mental phenomenon.

2.6:11b Cf. Aquinas, *Post. Analyt.*, Lib. I, Lect. XLI, Nos. 6-16.

2.6:11c Cf. *ibid.*, Lib. I, Lect. V, No. 6; Lect. VI, Nos. 2-5; Lect. XIX, No. 4; Lect. XLIII, Nos. 9-13; Lib. II, Lect. I, No. 1.

2.6:12a *Supra*, 2.4:3; 2.4:7; 2.6:6; and *infra*, 8.2:10.

2.6:12b Cf. van Laer, *op. cit.*, 4.

thinkers (*scientia quae scitur*).c Science, objectively taken, is both the knowledge of real things and the knowledge of the reality of things, and this distinction yields the phenomenal and the rational aspects of science, according to which the phenomenal aspect is a concentration upon the data of sense and the rational aspect is a correlative concentration upon the understanding of the data.d As science grows, it tends to render explicit the 'viewpoint,' or thought-medium, through which it functions by the recognition and formulation of 'postulates.' The application of these postulates to the subject-matter of science according to a methodology prescribed by the medium leads to conclusions, which in turn become 'principles' for further thinking along the same lines.

13. The term 'science,' taken in the broadest sense, defined above, includes the three stages which in scholastic terminology are called 'science' (*scientia*), 'understanding' (*intellectus*), and 'wisdom' (*sapientia*).a Science, in the framework of this latter distinction, is "a system of demonstrated conclusions," which appears as teaching in scientific expositions and textbooks and can be simply memorized to constitute knowledge about science.b Awareness of the scientific medium and thus of the reasons why the conclusions may be considered to be demonstrated constitutes the aspect of differentiated consciousness called *understanding*. And the functional awareness of the deepest and most comprehensive reasons bearing upon a given situation constitutes the maturity of understanding which is called wisdom.

14. When the knowing subject has an insight, he perceives a meaning that he had not perceived before. The insight consists in a grasp of some item of meaning in the thing he is considering, and the act of insight registers a growth in his understanding, at least within the field, or context, of his immediate consideration. The insight is possible for two similar reasons: the intelligibility residing within the thing considered has the power to raise the intelligence up to its recognition; the vitality residing within the intellect of the knowing subject has the power of development to this new state of understanding. Insights have a tendency to recur; the recurrence of insights is not a simple succession of the same identical experience, but rather a related succession of qualitatively different experiences. Inasmuch as the knowing subject adverts to his experience of the insights he is having, this experience becomes isolated in the form of objects not reducible in their proper characteristics to the sense experience from which they are distinguished. The knowing subject learns by experience that successions of insights are arranged along a line of development leading from the lower to the higher, where the higher is defined as the more intelligible and the less sensible, and that they are arranged in parallel fashion along a line of development leading from the less profound to the more profound, where the more profound is defined as the greater growth of the understanding and the less profound as the lesser maturity of the understanding. The path may thus be regarded as upward or downward as long as the direction is seen to be from imperfection toward perfection. The fact of such advance and growth is a datum of evident experience. But, when the process of growth is made the object of direct examination, it is possible to err in particular cases as a result of illusion, deception, self-deception (bias), or faulty reasoning.

15. Reality is intellectually divided into its matter and its form. With regard to the act of knowing, the form

---

2.6:12c Cf. van Laer, *op. cit.*, 5.
2.6:12d Cf. Aquinas, *Post. Analyt.*, Lib. I, Lect. XVIII, No. 9. The objective aspect of science is divided into the material object (sensory images), the remote formal object, and the proximate formal object (*supra*, 2.4:3a). To the subjective aspect of science pertains the *lumen quo*, or light medium, as it affects the capacity of the subject to perceive it. The objective and subjective aspects of the scientific medium are also referred to as the formal object

quod and the formal object *quo* (cf. van Laer, *op. cit.*, 43-49).
2.6:13a Cf. Aquinas, *Post. Analyt.*, Lib. I, Lect. XLIV, No. 11; Lib. II, Lect. XX, No. 15.
2.6:13b Cf. F. Wernz - P. Vidal, *Ius Canonicum*, I (Romae, 1952²), 78; J. F. McCarthy, *The Genius of Concord in Gratian's 'Decree'* (Rome: Pontificia Universitas Lateranensis - Institutum Utriusque Iuris, 1964), 7-9; or idem, in *Ephemerides Iuris Canonici*, 19 (1963) [105-151; 259-295], 111-113.

of any object is the identity-unit of intelligibility seen in it by the intellect; the matter is the excluded part of the dichotomy used by the intelligence in isolating the intelligibility to be recognized, and is, therefore, by very definition 'unintelligible,' at least within the context of the insight at hand. In terms of this definition, the 'matter' of reality is the merely sensory aspect of the phenomena being attended to, which aspect is left behind by intelligence as it isolates the intelligible form of the phenomena. The 'form' of reality is the dimension of meaning as it leads progressively away from the merely sensory aspect of phenomena, thereby also clarifying the notion of reality through the recognition that the 'reference to the data of experience' includes the reference to the data of *the experience of intellectual objects.*[a] Thus, the growth of the understanding, while it is a growth of the intellect of its subject in a subjective way that stands outside of the arena of consciousness, is also a growth of the intellect in an objective way that is recognized and known within the arena of consciousness. This objective mode of the understanding emerges in intellectual consciousness as the steps of progress along the intellectual dimension of reality. These steps are *in substantial things*, and, as units of meaning known by intelligence, they *are intellectual things*. The steps are *based upon sensible things* in the sense that the foot of the steps is resting squarely upon the data of sense phenomena and the steps are connected by the links of solid reasoning.[b]

16. Reality, then, is a kind of continuum whose solidity provides the firmness of scientific knowledge. At its base is the verity of the world known to sense experience. The line of its altitude is solidified by the verity of the reasoning process and divided into steps of graduated intelligibility by the separate and ordered identity of particular insights. The recognition of the third dimension of reality is the condition of advance to a full understanding of the meaning of science; it is the *pons asinorum* dividing the philosophers of truth from the manufacturers of pseudoscientific doctrines.[a].

17. The steps of formal reality lead upward and away from the phenomenal world standing at their base. The path of their progression stands within the real world, but not within the dimensions of the phenomenal world as such. The world of sense phenomena and the world of intellectual intelligibilities are not different worlds; they are different dimensions of the same world.[a] How can we be sure that the steps of the intellectual dimension lead 'upward'? It is a fact that an open mind cannot deny. As the intellect forms the dichotomy between sensory presentations and the meaning discovered 'within' the presentations, the knowing subject knows by an undeniable experience of insight into reality that the *meaning* of the sensory data is the 'higher' and more valuable of the two. It is the

---

2.6:15a Cf. Margenau (*The Nature*, 291), who regards it as evident that "the field of sense data, while coinciding partly with the real domain, does so more or less by pretension; it is unable to demonstrate its relevance by its own indigenous character." For this and other reasons, he says, "we are driven beyond the confines of Nature in our quest for physical reality."

2.6:15b "Among constructs, then, the problem seems to center. If external objects are constructs, they must hold the key to reality. ... Reality is conferred jointly by the process of fitting new parts into an already existing structure of ordered conceptions and by the process of empirical validation. Valid constructs, verifacts in short, are the elements of reality. Constructs are not valid because they refer to something real; on the contrary, they denote something real because they have been found valid. Hence the towering importance of epistemology." (Margenau, *The Nature*, 292.) Margenau here succeeds in locating reality within the mind, but he does not yet make the important distinction between reality as a set of *objects* embedded in the reality of the understanding and the constructs which human creativity may devise to link elements of reality together.

2.6:16a Cf. Margenau, *The Nature*, 449-450.

2.6:17a Cf. F. H. Bradley, *The Principles of Logic* (Oxford, 1883), 91-92. While Bradley describes the journey to the world of science as being away from the facts, we prefer to describe it as being away from the limitations of the purely material and sensuous to the world of intellectual perception.

intellectual meaning that is 'closer' to the intellect and provides its proper object. As the mind proceeds along the path of this intellectual dimension, it is conscious of a change in the kind of meaning that it recognizes: the meaning becomes more 'intellectual,' more 'intelligible,' and 'higher' in the sense of containing greater and fuller meaning in proportion to the sensory data with which it is mixed. The process is from almost full immersion in material things to almost full recognition of meaning transcending material things. The fact that this process is upward is verified in intellectual experience itself.

18. The recognition by the mind of reality in its intellectual dimension is the first step into 'interiority.' The steps of intelligibility lead somehow *into* the mind, not in the sense that they become subjective, for they stand within objectivity, but in the sense that they do require the instrumentality of a developing intellect in order to come into consciousness. The field of the intellectual dimension of reality lies between the knowing subject and the world of sensible objects. 'Interiority' is simply another name for the intellectual dimension of human consciousness; it should not be confused with idealist subjectivism. The meanings which constitute the 'world' of interiority are real objects lying between the knowing subject and his merely sensory experience.[a] The advance into interiority is in one way a movement toward the knowing subject and is in another way a movement away from him. It is important to recognize the reason for this contrary movement.

19. When Galileo Galilei had made the first telescope and turned it toward the heavens, he became the first of his contemporaries to realize that there are many more stars than people had supposed. His first little telescope brought into view the stars of the seventh magnitude, and by that alone increased the number of visible stars by about six times. When Galileo observed stars of the seventh magnitude, he was actually looking farther and deeper into space than any man had looked before. In this sense he was seeing what was 'behind' the visible heaven of the unaided eye, and the object of his gaze was farther away from himself than it would otherwise have been. But in another sense the object of his gaze was closer. The telescope was serving as an instrument to gather the light of those distant stars and reconstruct their image inside the telescope itself. What Galileo was looking at directly was not the stars themselves, but their image inside his telescope. By bringing the object of his vision closer to himself he was able to look farther away.

20. The intellect of man is an instrument that functions in similar fashion. The meaning of sensible phenomena, their intellectual dimension, lies behind them as known to the senses alone, and the intellect has to look beyond the presentations of sense in order to see them. But in order to bring these meanings into view, in order to make visible their intellectual reality, the intellect must be used as an instrument through which they may be viewed. The intellect must develop concepts that may be likened in their use to the lens or mirror of a telescope. The knowing subject is then in a position to catch sight of intellectual meanings through, or in, the instrument of his intellect. The objects of intellectual awareness, as objects, are by origin in extramental things; they are behind and beyond the presentations of sense experience. But these objects exist formally in the mind of the knowing subject; they are closer to him than are the presentations of sense; the higher he ascends along the steps of intelligibility and the more he penetrates into the depths of reality, the closer also to himself as the knowing subject will be the medium in which the object of his attention resides.[a]

21. The first steps up the incline of

2.6:18[a] "Among the limited set of constructs that can claim the rank of verifacts at any time one can discuss *systems*, *observables*, and *states*, but this is not an exhaustive classification. The property of physical reality attaches primarily to these, and we suggested that it be limited to them. ... Primarily, then, we regard as the content of the real world the verifacts called systems, observables, and states. ... (D)ata and constructs cannot both define a unique physical reality (for if they are indiscriminately joined together, we get all *experience* back again)." (Margenau, *The Nature*, 451-452.)

2.6:20[a] Cf. *infra*, 5.3:6[a].

science are restricted to the general area of common sense. The recognitions of the intellect isolate objects that are conceptual in nature but have the hazy outline of mere conceptions rather than the clear features of concepts. The first principles of reason appear, although they are not explicitly attended to. But the subject becomes aware of his mind and begins to use it correctly to arrive at truth verified both by experience and by reason. This is the exercise of common sense. At a certain point the path of ascent divides into two distinct areas of concentration, each area organized in terms of a special formality or approach that makes it a field of its own. The first area is that of reason as the complex of principles which provides the instrument for arriving at the recognition of truth. The second area is the world of sensible phenomena, within which the body of knowledge can be expanded. These two fields of thought may be given the names of 'philosophical science' and 'empirical science.'

22. We are speaking here of the development of science as a conscious experience of the contemporary individual thinker. As soon as the thinker has recognized the existence of principles of reason it becomes possible for him to investigate them. Such investigation usually is done through separate disciplines in the school he attends and in the books he reads. But the reason for the division can also be studied analytically. In the usual educational process the student is confronted with a division of studies into disciplines of science and disciplines of 'value.' The value courses regard such subjects as religion, sociology, history, while philosophy is postponed to the college level and even there remains a remote and abstract area of speculation. The focus of attention and of concrete relevance to the mental growth of the student is concentrated largely upon the study of the method and results of empirical science, cushioned by an acquaintance with the peripheral area of the 'humanities,' including literature and fine art.

23. Even so, the field of scientific attention will at a certain point divide into the separate disciplines of experimental science and mathematics. Experimental science is especially concerned with the confirmation of supposed facts by testing them over and over again in sense experience, so that the probability that the same experience will tend always to be the same becomes mathematically very large. This extroversion of experimental science towards an ever greater and more precise experience of the phenomena of the sensible world is known as induction. Mathematics is especially concerned with the deduction of evident conclusions according to the natural process of the intellect from a given set of premises, all included within the general notion of quantity. While mathematics can hypothetically proceed from any given set of such premises, it actually becomes a pseudoscience if it violates what is self-evident to the intellect or if its conclusions contradict what is evident to sensory experience. Mathematics, as a true science, must be constantly verified both by intellectual and by sensory experience, and it is the tremendous application to the advance of knowledge about the real world enjoyed by mathematics that has provided its greatest appeal in the community of scientific thinkers.

24. The division of scientific thought into a rational and an experimental factor is disciplinary rather than substantive, even though 'pure mathematics' prides itself in its total freedom from the practicality of 'applied mathematics.' The organization of distinct disciplines becomes possible, as a group of related concepts are united within a distinctive formality. The synthesis of the related concepts forms a 'theory' that is expressed in terms of related judgments and laws. There is a general concept giving formal unity to each such field and providing a boundary beyond which the field does not extend. But when a field is not thoroughly analytical it can be composed of a stylistic union of concepts into a functional unity that does not easily admit of full unification under a single formality. Such a union is descriptive, not essential.

25. As soon as the intellect of the thinker begins to synthesize individual items of real knowledge into more general conceptions, the field of his scientific knowledge is divided into a phenomenal and a theoretical aspect.[a] The phenomenal aspect is what is ex-

---

2.6:25[a] Cf. Fulton J. Sheen, *Philosophy of Science* (Milwaukee, 1934), 1.

cluded as the mind concentrates upon theory. The two fields are conceptually distinct, but they have a reciprocal relationship. While there is advantage to concentrating now on the one field, now on the other, it is a mistake to try to make them totally independent of one another. Both aspects belong to the one field of science, and to separate either aspect into a completely independent field simply means to falsify both.[b]

26. Thus, the scientific knowledge of the structure of the solar system as it is known today was made possible by a combination of observation and theory. Until the development of the telescope, the planets visible to the naked eye were known as 'wandering stars.' Galileo Galilei, followed by others, used the telescope to identify visually the planets as such, and the mathematical interpretation of their visual movements produced the geometrical description of the solar system. Astronomy is regarded as a specialized science heavily dependent upon sensory perception and mathematical calculation. The recording of data and the mathematical interpretation of data are separate activities, but they do not belong to separate sciences. They are different aspects of the same science, and the same is true of all the other specialized sciences.

27. The idea of 'empirical science' arises from the admission of mathematical deduction as the formal medium of a viewpoint otherwise technically limited to the inductive method.[a] The admission of mathematics permits intellectual development of science, since with mathematics are implicitly admitted the self-evident principles of reason.

28. For the sake of clarity, a parallel should here be drawn between the phenomenal and theoretical aspects of science, which have now been discussed, and material versus formal science. In defining the material and the formal, I have said that the material is what is left behind as the intellect grasps or sees an intelligible meaning in a sensory image.[a] Theory is formed by the unification of intelligible meanings in terms of more general meanings. In this sense, theory could be regarded as the form of science and what the theory is taken from as its matter. But this definition of matter and form admits two different and related notions of the matter of science. According to one notion, the matter of science is the phenomena from which it arises. According to the other notion, the matter of science is the 'subject-matter' that the science treats.

29. The recognition of 'material science' as the subject-matter of science and of 'formal science' as insight into the subject-matter is important for a full understanding of what science is. Scientific thinking leads to insights, expressed in concepts. The concepts are then linguistically expressed in terms, themselves joined in the form of propositions. The propositions are expressed as judgments verified by experience, and they constitute the sentences of scientific monographs and books. What is written in scientific works is generally regarded as the expression of knowledge about the real world; it can be learned by memory. Whoever possesses the teaching of scientific works in the form of memorized knowledge may be said to possess 'material science.' Only persons who have the insights by which the scientific conclusions were reached may be said to possess 'formal science.' The importance of knowing the difference between material and formal science lies in their respective reliability. The cogency of material science comes entirely from trust in the validity of the conclusions expressed therein. If the conclusions should be false because of a faulty use of reason or because of an unexpressed bias of the writer or for some other reason, the possessor of merely material science is unable to check the conclusions for himself. He is constrained to accept them at face value, unless either he can find some other authority to contradict them or they possess contradictions evident to the limited formal science of common sense.

2.6:25[b] "To deny the presence, indeed, the necessary presence, of metaphysical elements in any successful science is to be blind to the obvious, although to foster such blindness has become a highly sophisticated endeavor in our time." (Margenau, *The Nature*, 12.)

2.6:27[a] Thus Fulton Sheen (*Philosophy of Science*, 2) distinguishes three possible philosophies of science: the physical theory, the mathematical theory, and the metaphysical theory.

2.6:28[a] *Supra*, 2.6:15.

30. Reason is both the container and the vehicle of science. Statically considered, it is the container; dynamically considered, it is the vehicle. *Reason at rest* is initially the self-evident principles in terms of which the intellect is able to understand; subsequently it is the understanding of the knowing subject at every stage of growth in scientific knowledge. *Reason in action* is the transition of the intellect to greater knowledge and deeper understanding; it is the reasoning process itself. After the mind has become sufficiently differentiated to be able to examine its reason as an object, the organization of such disciplines as logic and epistemology becomes possible.

31. The rôle of intuition in the reasoning process is fundamental and pervasive. I have defined intuition as a direct and immediate perception of objects. Sense intuition is the direct and immediate perception of sensory phenomena; intellectual intuition is the direct and immediate perception of intelligible objects. Such perception is fundamental to the reasoning process, since it is both the awareness of the data of experience from which reasoning begins and the embodiment of the insights markng the formal advance of reason at every stage. The grasping of a concept is an intuition; the making of a judgment is an intuition. In fact, the whole process of intellectual development may be described as the deepening of intuition.ᵃ

32. Yet, a distinction is to be made between what is known as 'intuitive reasoning' and syllogistic reasoning. The intellect grows by a process of comparing and contrasting ideas, by the analysis and synthesis of objects. This process consists partly in a direct comparison of ideas and partly in an indirect comparison through the use of a third idea serving as a medium. The process of arriving at valid conclusions by the use of a middle term is often identified with the whole reasoning process, but it is more correctly named 'syllogistic reasoning.' Reasoning that takes place by direct comparison is extrasyllogistic and is properly called 'intuitive reasoning.'

33. The recognition that one thing is larger than another or heavier than another by a simple comparison of objects of experience in consciousness is an example of intuitive reasoning. Such comparisons are frequent, and they are even a major element in syllogistic reasoning, for the comparison of each extreme of the syllogism with the middle term is intuitive. Syllogistic reasoning, whether or not it is recognized formally as such in the consciousness of the thinker, has served as a powerful instrument of science, and its success has up to the present been so great that the importance of intuitive reasoning has been eclipsed to the point of neglect. But an adequate grasp of the meaning of science must take into account the important and related functions of both syllogistic and intuitive reasoning.

34. This is not to deny that the syllogism has had its rise and decline in the history of science as usually described. The discovery of thought-analysis in terms of syllogistic reasoning was of crucial importance to the development of Aristotelian thought, and it is heavily used in the writings of St. Thomas Aquinas. The attraction of syllogistic reasoning subsequently became so great that forms of scientific thought depending upon intuitive reasoning tended to fall into disuse. Then, with the rise of modern science of the empirical type, the value of intuition was brought back into focus, but unfortunately the focus was largely limited to the sensory level, the main exception being the use of middle terms on the level of mathematical abstraction.

35. The power and success of modern empirical science has been itself 'phenomenal' to the point of raising persistent doubts as to the validity and usefulness of any other type of science. These doubts have had deleterious effects. They have, for instance, led to the dissemination of pseudophilosophies aimed precisely at satisfying the natural desire of man to achieve a deeper understanding without actually giving him a deeper understanding. Rather than assist him to recognize the higher intelligibilities of the things with which he is surrounded, they have served to persuade him that there is

---

2.6:31ᵃ The fact of intellectual intuition is basic to the cognitive theory of Aristotle and Aquinas. While Margenau leans more towards empiricism as regards the notion of reality used in modern physics, he is open to the possibility of another kind of intuition. (Margenau, *The Nature*, 453.)

really no higher intelligibility to be found. To accomplish this they have utilized a method that might be styled 'inverse intelligibility,' which seeks the meaning of the more intelligible in terms of the less intelligible and of the more intellectual in terms of the more sensuous. The product of such pseudo-speculation has been a kind of 'tunnel vision' extraordinary for its narrowness as well as for its precision.

36. As personal scientific knowledge expands to an ever greater acquaintance with the reality of sensory experience, there is a natural tendency to expand also along the dimension of intellectual insight. The growing precision of detail begs to be comprehended in a fuller knowledge of the whole. As the astronomer extends his observations by the use of more powerful instruments, so that what before was a tiny segment of a field of vision becomes the entire field, he spontaneously attempts to fit his successive discoveries into a more consistent picture of the whole universe. But usually he limits the attempt to the universe of sensory phenomena. What has been lacking is the continuation of this work by the scientific comprehension of the whole of reality in terms of the intellectual dimension as well.

37. The acceptance of the intellectual dimension of reality is a presupposition of science. As a dimension, it necessarily has a static as well as a dynamic aspect. The prime expression of the static aspect of intellectual reality is the concept, with the understanding that it embodies. Whether or not there are universal essences corresponding to insights on a higher level than that of mathematics is a question that can be answered only by a careful study of abstraction on a higher level than that of mathematics. To pose the question in mathematical terms in order to arrive at the foreordained conclusion that there are no such essences is a mistake in scientific method.

38. The sciences are related to one another. They are, in fact, different parts of the one universal science, extended to embrace the whole of reality. The relationship of the sciences is graduated and progressive, so that higher science somehow cannot be arrived at without passing through the levels of the lower. Yet, there is the function of material science, which permits the intellect to increase its fund of knowledge as such (*scientia*) without in fact advancing in formal understanding (*intellectus*). The occurrence of a material aspect of science also makes it possible for fields of formal science to be situated on the same level of abstraction, so that only one field on a particular level need be passed through in advancing to a higher level. The relationship of scientific fields to one another and to science as a whole is studied in that branch of advanced science known as theory of science, or metaphysics.

ARTICLE 3
THE LOCUS OF HISTORY IN HUMAN CONSCIOUSNESS

3.1. The Locus of History in Prescientific Consciousness

1. Modern historians have long been aware of an abiding ambiguity in the conception of history.[a] "The word history is used in two senses. It may mean either the record of events or events themselves. Originally limited to inquiry and statement, it was only in comparatively modern times that the meaning of the word was extended to include the phenomena which form or might form their subject."[b] This ambiguity presents a narrower sense of history surrounded by a wider sense. The narrower sense of history as the record of the events is itself commonly divided into two definitions, according as it is 'objectively' or 'subjectively' considered. "In an objective sense history is, to use the words of James Harvey Robinson, 'all we know about everything man has ever done, or thought, or hoped, or felt.' Subjectively expressed, history may be regarded as a record of all that has occurred within the realm of human consciousness."[c] In the wider sense history embraces not only everything that man has ever known or been conscious of but also all of the phenomena of the world of physical nature. "It includes everything that undergoes change; and as modern science has shown that there is nothing absolutely static, therefore the whole universe and every part of it has a history. This idea of universal activity has in a sense made physics itself a branch of history."[d]

2. Most contemporary historians look upon history as concentrated in the area of the human past and as being an account of the events of which that past is constituted.[a] They consider

---

3.1:1[a] A brief specimen of contemporary discussion regarding theory of history is embodied in W. H. Walsh, *An Introduction to Philosophy of History* (London: Hutchinson and Co. Ltd., 1967³). (Hereinafter referred to as *IPH*.) See *infra*, Appendix II. A short bibliography is included. Longer treatments with supplementary bibliography, written from somewhat divergent viewpoints, may be examined in James Thomson Shotwell, *Introduction to the History of History* (Oxford, 1939), and in Harry Elmer Barnes, *History of Historical Writing* (Norman, Oklahoma, 1937). For a comparison of views held in the first quarter of the twentieth century, samples may be found in Ernst Bernheim, *Lehrbuch der historischen Methode und Geschichtsphilosophie* (Leipzig, 1908), and in H. E. Barnes, *New History and the Social Studies* (New York, 1925). *The Idea of History* by Robin G. Collingwood [Oxford-London, 1946 (hereinafter referred to as *IH*); see *infra*, Appendix III], published posthumously, is now considered to be a classic on the notion of history from the idealist point of view.

3.1:1[b] J. T. Shotwell - E. F. Jacob, "History," in *Encyclopaedia Britannica*, Vol. XI (London, 1961), 594: "It was perhaps by a somewhat careless transference of ideas that this extension was brought about. Now indeed it is the commoner meaning. The 'history of England' is used without reference to any literary narrative. Kings and statesmen are termed the 'makers of history' and sometimes it is said that the historian only records the history which they make. History in this connection is obviously not the record, but the thing to be recorded." Cf. *infra*, IV. 16.

3.1:1[c] H. E. Barnes, "History, Its Rise and Development: A Survey of the Progress of Historical Writing from Its Origins to the Present Day," in *The Encyclopedia Americana* (international ed.: New York, 1967), 205.

3.1:1[d] Shotwell - Jacob, "History," *loc. cit. supra* (3.1:1[b]), 594.

3.1:2[a] The sketch of the modern historian's conception of history contained in this paragraph is not based upon a statistical investigation of the opinions of historians. It consists simply of observations gathered from the description given by

their study to be primarily a cognitive activity, whose object, the past, is to be investigated and reconstructed with reference to its own objectivity, not with reference to the caprice, bias, or other subjective motive of the historian.[b] They recognize that, because history deals with the past, memory impressions are an indispensable element of its raw material.[c] It is now a matter of convention that the historian is directly concerned with the study of human events against the background of physical nature.[d] The ambiguity of history has tended to divide all historians into two schools of thought regarding their approach to human events: the 'positivist' school, which stresses the massive accumulation of 'facts' with the greatest precision, striving (naively) to eliminate all philosophical interpretation of these facts; and the 'idealist' school, which assumes that past human activities and experiences are a matter of human intuition, not of scientific measurement.[e] Nevertheless, the generally acknowledged aim of all historians is to reconstruct the past in terms of a meaningful narrative, thus implicitly admitting some set of presuppositions giving unity and system to their results, although there is no consciously accepted set of principles agreed upon by historians which would permit them all to draw the same conclusion from the same piece of evidence.[f]

3. The presence of a double sense of history among professional historians, not in the form of an understood distinction but as a troublesome ambiguity, stems from the fact that modern historians have not sufficiently succeeded in defining their subject with respect to itself or to other branches of knowledge.[a] The abhorrence of most working historians for 'idle speculation' about their subject has enabled them to develop methods of high precision in the ascertaining of 'facts,' but the relative lack of concomitant effort to analyze their own thinking has precluded the elimination of fundamental ambiguities that have sown equivocation in the interpretation by various historians of the 'facts' that they have ascertained.[b]

4. This ambiguous field of history, as visualized by contemporary historians, is surrounded by an even wider historical background that is not called 'historical' according to the modern denotation of the word, but which contains forms of thought resembling in some way the modern idea of history. Because the ways in which these forms of thought coincide with modern history or diverge from it have never been precisely isolated and defined, it has been relatively easy for modern sophists to stretch the basic ambiguity of history into false conclusions about these other forms as well.[a]

5. A glance at any large dictionary will reveal that the word 'history' has at least eight definitions, some very similar in meaning, some widely divergent, but juxtaposed and intertwined in a manner suggesting the absence of analytical distinction of their order and logical relationship to one another. The following definitions given by *Webster* might be compared and contrasted: 1) "a narrative of events connected with a real or imaginary object, person, or career"; 2) "a systematic written account comprising a chronological record of events (as affecting a city, state, nation, institution, science, or art) and usually including a philosophical explanation of the cause and origin of such events — usually distinguished from *annals* and *chronicles*"; 3) "the branch of knowledge that records and explains past events as the sequence of human activities; the study of the character and significance of events"; 4) "a treatise presenting systematically related natural phenomena (as of geography, animals, or plants)"; 5) "a drama based on real events"; 6) "the events which form the subject matter of a history"; 7) "a series of events clustering about some center of interest (as a nation, a department of culture, a natural epoch or evolution, a living being or a species) upon the character and significance of which these events cast light; the character and significance of

---

W. H. Walsh in *IPH* and is therefore a direct testimony of Walsh's opinion alone. It is Walsh's express opinion that is here the direct object of our analysis.
3.1:2[b] *IPH*, 16, 21.
3.1:2[c] *IPH*, 30.
3.1:2[d] *IPH*, 31-32.
3.1:2[e] *IPH*, 42-47.
3.1:2[f] *IPH*, 35.
3.1:3[a] *Infra*, 3.1:9[c].
3.1:3[b] *Infra*, III.21.
3.1:4[a] *Infra*, IV.16.

such a center of interest"; 8) "past events, especially those events involving or concerned with mankind."[a]

6. The diversity of these definitions is indicated by the fact that they may signify a plain narrative or a philosophical narrative of events; they may signify a narrative of real events or of imaginary events regarding political, social, scientific, or artistic developments; they may signify human events or events of physical and biological nature. Modern historians would exclude some of these definitions from the field of history, properly so called, yet they are all somehow credentialed by the fundamental ambiguity of history. In fact, if the history in some way accepted as genuine by professional historians includes "everything that undergoes change," then all of the forms described by *Webster* are history. And if history is restricted in the narrower sense to "all we know about everything man has ever done, or thought, or hoped, or felt," then once again all of the forms described by *Webster* are history. Yet, ambiguity reigns, for historians tend to restrict their subject to events touching upon human life, as distinguished from events in physical and biological nature, and they tend to deny that "a narrative of events connected with an imaginary object, person, or career" is history at all. Narratives of imaginary events are part of what we know about everything that man has thought, they have occurred within the realm of human consciousness, and nevertheless they are not admitted as history.

7. The notion of 'total history' as a conception of knowledge embracing everything that undergoes change, everything that man has ever done, or thought, or hoped, or felt, everything that has ever occurred within the realm of human consciousness, is entertained at least hypothetically by some historians and flatly rejected by most empirical scientists, because it seems to flatter the pursuit of the former and dissolve the autonomous field of the latter.[a] The forms of change included under the notion are many and diverse, yet possessing certain common characteristics assigning them to the same general class or category. If this most general category should not be called history, then what should be its name?

8. A cursory review of some of the forms of thought contained within total history will indicate their variety. The kind of events that historians have considered most central and important has differed with different times and cultures. In modern times the study of history has often centered upon political and military events, but this 'classical' approach has now widened into a focus upon other types. By the early twentieth century, secular historians were described in the synthesis of Harry Elmer Barnes as divided into eight principal schools of interpretation, each seeking the meaning of history in an area contrasting with the older theory of political causation.[a] The eight newer approaches included the 'great man,' the economic, the geographical, the spiritual or idealistic, the scientific, the anthropological, the sociological, and the intellectual or 'collective psychological' theories. The content of history that emerges is different for every school. But these forms are not all exclusive of each other. There is political history, economic history, environmental history, cultural history. There is ideal history, exemplified in the deeds of great persons. There is national history, institutional history, private history. There is history of science, history of art, history of music, of poetry and prose, history of philosophy, history of drama, history of creative imagination, history of thought. There is history of the very flow of consciousness, whether personal, interpersonal, or collective. And there are other forms as well.[b]

9. The development of each individual mind has its history; every thought-form present within society has its history.[a] There is, then, also the history of history as a thought-form present in society.[b] It is not our purpose

---

3.1:5[a] Sub verbo, *Webster's Third New International Dictionary* (1965). Cf. the parallel definitions of history in Funk and Wagnall's *Standard Dictionary of the English Language* (1960).
3.1:7[a] *IPH*, 15.
3.1:8[a] These eight schools of historical interpretation are described at length in H. E. Barnes, *New History and the Social Studies*.
3.1:8[b] *Infra*, 6.3.
3.1:9[a] *Infra*, III.15.
3.1:9[b] Cf. Shotwell, *The History of History*.

to review the history of history,[c] but we do wish to note that what may seem to set history apart from other recognized categories of knowledge is the incorporation of its objects in a past made present by imagination.[d] It is the construction of its objects by the imagination that seems to make history what it is, inasmuch as, if the objects were present to direct sensory perception, they would not belong to history at all.

10. But by this criterion the rôle of imagination in history remains ambiguous. There is certainly a different use of the imagination in the cases of political history, biography, and autobiography; the mixture of imagination is not the same in the history of mining and the history of painting. Nor can the biography of a general or the description of a convention be equated with an historical drama, an historical novel, or an epic poem. Much less can the account of a famous deed be equated with a fable, myth, or fairy tale, with a macabre mystery, or with an adventure of a legendary character. All of these forms are narratives of the past constructed with the help of the imagination, and yet there is a fundamental difference running through them.[a]

11. One may appeal to the presence in some of 'creative' imagination, but all historical imagination is creative in some way. One may speak of true history as constructed from memory, rather than from fantasy, but fantasy itself depends upon memory — the memory of prior sensory perceptions. Within all of the forms of history mentioned above both art and imagination play a functional part. The difference that makes the dichotomy of all of these forms is rather the conformity or lack of conformity with *reality*. One part of the dichotomy includes only the real portrayal of real events. The other part retains all of the unreal events and the unreal portrayal of real events. All of the forms are history; that part of the forms wholly contained in reality is historical science.[a]

12. There is a mode of consciousness embracing all of the forms of 'history' contained within the notion of 'total history.' This mode of consciousness is not simply concerned with feeling and appetite; it is not the self-awareness of the knowing subject as such. Rather, it is concerned with objects, and one common feature that unites these objects in a single very general class is the awareness that they are situated in the past. Historical consciousness, in the sense of the awareness of history in any of its forms, is the consciousness of the past. The objects with which history is concerned were at some time and in some way present, but when they were present they did not belong to the consciousness of history. Anyone who has the awareness of history in any of its meanings knows that he is looking at things past, although he may not have reflected on what it means to be past. It is the consciousness of the past that is the distinguishing feature of all history, and that is why history in its broadest denotation may be defined as "the consciousness of the past."[a]

13. Since the consciousness of the past is a consciousness of objects, history in its widest sense may also be defined as "*knowledge* of the past," where knowledge is defined as the consciousness of objects, real or unreal,[a] and the past is taken reduplicatively. History is the knowledge of the past as such.[b]

3.2. *The Locus of Historical Science in Scientific Consciousness*

1. Historical science arises when the mind begins to concentrate upon the real objects of history.[a] It is the gath-

---

3.1:9[c] The divisions of history have been confused with one another due to the lack of a clear general definition of history. It would be an unnecessary digression to discuss here the notions of history which have been advanced. An illustrative collection of definitions is presented in F. J. Teggart, *Prolegomena to History* (1916), Part III, Sect. 1. The alternative definitions which have been proposed represent varying degrees of historical acumen and intuitive understanding, but none of them is an exact definition.
3.1:9[d] Cf. *IPH*, 30.
3.1:10[a] *Infra*, IV.7.
3.1:11[a] *Infra*, IV.6; IV.12.
3.1:12[a] *Infra*, 6.3:1.
3.1:13[a] *Supra*, 2.3:19.
3.1:13[b] For an answer to Walsh (*IPH*, 30-31), who finds this definition indefensible, see *infra*, II.10.
3.2:1[a] *Supra*, 2.4:2; 2.4:3.

ering of historical objects within the notion of reality that constitutes the historically scientific mind.[b] As it begins to operate, it undertakes a systematic distinction between those forms of historical consciousness regarding 'real events' and those forms regarding 'unreal events.'[c] The history of unreal events is excluded from its direct field of attention, although it can also attend to 'imaginary' forms of history indirectly, that is, under the aspect of reality.[d] This simple attention to real events, the disposition to verify events that are narrated or thought by reference to one's notion of reality, the body of related happenings that results from this gathering together of 'real' events into a special area of consciousness, is already historical science, but it is science on the level of common sense alone.[e] By applying the "right reason of sound judgment, not involving technical concepts"[f] to narratives of the past, a thinker can produce a certain degree of scientifically valid history, provided also that his thinking is free of bias and other subjective influences.[g] But the depth of meaning is limited.

2. The field of history contained within the limits of the real is enormous, even when the events which constitute its object are reviewed by common sense alone. There are, for instance, cosmic events, such as solar phenomena, the appearance of supernovae, and other changes in the galaxy. There are geological events, such as the birth of mountains and oceans, changes in the course of rivers, earthquakes, floods, and volcanic eruptions. There are minute events, such as the birth of a mosquito or the appearance of a blade of grass. There are momentary events, such as the blink of an eye or the switch of a cow's tail. Historical common sense does not attach equal importance to all of the events providing possible objects of attention. It tends by virtue of the logic of its thinking to narrow the field of its attention to events that are 'significant.' It seeks to bring together events that are meaningful, and it begins to organize them in terms of general conceptions. Thus, it distinguishes between 'natural history' and 'human history,' between 'public history' and 'personal history,' between political, economic, and cultural history. It creates units of time, such as hours, days, weeks, months, and years, and it progresses to eras, ages, and epochs. It divides human history into ancient, medieval, and modern, and goes on to create numberless other conceptions: the British Empire, the Iron Curtain, the Victorian Age, the Romantic Movement, the Permissive Era, etc. Common sense makes and uses such terms, but it does not consciously ask itself such questions as how or why it makes them, what is their validity, or how historical knowledge can be justified in relation to the other sciences.[a]

3. There is a natural tendency to specialize historical science beyond the level of common sense.[a] Since all science begins from the notion of reality and advances by a growing awareness of the reality which is its medium,[b] it follows that historical science will grow only by an increasing attention to its own medium and that this special medium is located within the more general medium of reality.[c] It does not therefore follow that the growth of historical science is exclusively epistemological or idealistic in that the questions it asks would necessarily be in regard to its own validity and its relation to the other sciences.[d] Its growth covers the whole field of attention to its proper medium within reality, and this may be attended to directly without explicit reflection upon its underlying principles of knowledge.[e] However, it is important in this discussion to treat the matter also in its epistemological implications, precisely because it is in this area that the greatest confusion seems to have arisen. For there are features of historical knowledge that are specific to history, and there are other features that are common to history and to other branches of knowledge; there are features that are specific to historical knowl-

---

3.2:1b *Infra*, IV.19.
3.2:1c *Infra*, IV.12.
3.2:1d *Infra*, IV, 17.
3.2:1e *Supra*, 2.5:3-4.
3.2:1f *Supra*, 2.5:2.
3.2:1g *Infra*, II.35-II.36.

3.2:2a *Supra*, 2.6:21.
3.2:3a *Supra*, 2.6:36.
3.2:3b *Supra*, 2.6:6; 2.6:12.
3.2:3c *Infra*, III.10.
3.2:3d *Supra*, 2.6:18.
3.2:3e Cf. *supra*, 2.6:21; 2.6:24.

edge as restricted to the consciousness of the real, and there are features that are common to all of the forms of historical consciousness, real or unreal. The inability to distinguish the specific from the common features leads to an imprecise conception of the field of historical science and opens the door to a sophistic use of the common features in order to introduce certain elements specific to historical science into imaginary forms of history.[f]

4. It is presently a matter of convention among working historians to leave to the investigation of other fields the study of some kinds of past events. Chemical reactions, for example, are events left almost entirely to the examination of natural scientists. Certain classes of events are yielded to astronomers, geologists, biologists, and others to the point of virtual exclusivity, while historians devote their interest to human events.[a] Now, this division of labor is of both practical and theoretical necessity, but the lack of a general theory uniting the various approaches to events in more general underlying concepts has not been without its inconvenience. It has led no small percentage of 'scientists' to deny any serious and scientific validity to the explanations of historians. It has led historians to ignore problems emerging in the field of 'natural science' precisely for the reason that scientists are studying events without the benefit of principles applying to events as such. The result of this lacuna in scientific theory has often been either a 'scientism' professionally blind to the historical reality of things,[b] or a subjective idealism professionally blind to the more comprehensive dimensions of true science.[c] When historians advert to the ambiguity of their subject, when they distinguish between the 'facts' and the meaning of the facts, they are actually searching for the advance of a science, namely, the science of history as a discipline having its own identity within the one total science of known reality.[d]

5. The science of history advances beyond the limits of common sense by becoming consciously aware of its own medium of thought.[a] The awareness of this medium as reality is implicit in common sense; the ingress into historical science as such requires the explicit awareness of the specific medium of historical science as such. This awareness will be vague and confused at first, but as historical science grows in the mind of the knowing subject, that is, within his objectivity, the specific medium of history will become more clearly and sharply defined.[b]

6. The common-sense level of historical science is confused about the nature of historical reality. It is driven to devise methods of determining with ever greater precision what actually took place in the past. This determination is seen as verification in the vestiges of the past remaining as sensory phenomena. But there are unanswered questions of imposing magnitude. The vestiges are not the objects of history; the events of the past have to be reconstructed with the help of imagination. And the meaning that historians claim to see in the events of the past is not verifiable on a sensory level.[a] When the historian becomes conscious of the intellectual dimension of the reality he has in view,[b] and when he begins to see the specifically historical aspect of this intellectual dimension, then may he truly be said to have become a scientist of history, or a historian on the scientific level.[c]

7. The common-sense historian asks intelligent questions about the nature of his subject, but he does not know how these questions are answered. The

---

3.2:3[f] Cf. *infra*, III.6.

3.2:4[a] Walsh affirms that history "as normally understood" takes no cognizance of great segments of the past, such as all those ages that preceded the appearance of man on earth (*IPH*, 31). This common understanding of 'history' follows the now conventional division of the past between 'historians' and 'natural scientists.' However, there are obvious features of the past that transcend this conventional division, and it is the explicit purpose of the present discussion to draw attention to some of these features.

3.2:4[b] *Infra*, 5.3:5.
3.2:4[c] *Infra*, II.53-II.56.
3.2:4[d] Cf. *infra*, III.21.
3.2:5[a] Cf. *infra*, III.22.
3.2:5[b] Cf. *infra*, II.10; II.11; II.17; II.23; II.24.
3.2:6[a] *Infra*, IV.6.
3.2:6[b] *Supra*, 2.6:17.
3.2:6[c] *Infra*, II.25; II.34; II.37; II.54; III.13.

historical scientist knows why these questions are asked and how they are answered. He is able to pose them analytically and in order.[a] Since reality has several dimensions, of which the three described so far are its 'two-dimensional' configuration and its 'three-dimensional' configuration, according as it is considered with or without the intellectual component,[b] a scientific approach to questions regarding the historical medium must take into consideration the division of reality into its partial and fuller dimensions. The search for the past 'as it actually took place' pertains especially to the 'two-dimensional' aspect of historical reality; the search for the meaning of the past pertains to its three-dimensional aspect.[c] Both of these aspects are related. It is, in fact, by discovering the relationship between the two that one can achieve an insight into the precise kind of understanding that the mind of man can derive from the past.

8. Historical science, to be sure, distinguishes between levels of historical meaning on the basis of temporal perspective. Events recorded within a short perspective of time are regarded as material for the composition of more advanced historical science. A diary is an historical form recording personally or institutionally related events within the perspective of twenty-four-hour periods. Journalism is a form of historical writing attentive to an exact recording of events from a viewpoint of popular interest within the perspective of the current day, or, at least, within the outlook of a short period of time. Annals are records organized with relationship to yearly periods of time. Chronicles usually describe periods of several years with emphasis upon the narration of events on an assumed 'eyewitness' level and with little attempt to explain the facts and make them more intelligible. Public registers, police blotters, archives, and other sources of this kind record events in a certain chronological order, but with no attempt to organize them into a consistent narrative. All of these short-term historical forms are usually distinguished by historians from what they consider to be history proper, that is, historical explanation.

9. If history is the knowledge of the past as such, then historical science is the knowledge of the real past as such. It seeks to comprehend the reality of the real past. The explanation it gives has to do with historical reality; the meaning it discovers has to be divided into units of historical intelligibility traceable to the capability of past events to be understood. Historical science discovers and understands the meaning inherent in nature under the aspect of the past. It asks how there could be a real intelligibility in a past that no longer exists, and it finds the answer to this question.

10. Science is a being existing actually in the mind of the knowing subject, potentially in extramental things; the object of science exists formally in the mind of the knower, materially in things.[a] Just as the physicist or the chemist recognizes the objects of his science by abstracting from sensory phenomena the intelligibility that he finds within them, so also does the historian. The sensory phenomena which provide the material object of historical science, and from which the historian must abstract the intelligibility he finds therein, are the recorded or reconstructed events of the past. The formal object of historical science, the proper object containing the meaning or intelligibility of history, resides actually in the mind of the historian and reductively in the sensory phenomena from which it was derived.[b] Hence, it is fruitless and methodologically erroneous for an historian to force his attention exclusively upon the uncovering and verification of the sensory aspect of the phenomena to be found in the historical field, without attending to the formal conceptions of his own mind.[c]

3.2:7a *Infra*, III.12.
3.2:7b *Supra*, 2.6:6; 2.6:15.
3.2:7c *Infra*, II.9; II.42.

3.2:10a *Supra*, 2.4:8; 2.6:11; 2.6:15; 2.6:17.
3.2:10b *Infra*, III.16.
3.2:10c *Infra*, II.18.

Article 4

THE RELATIONSHIP OF PAST AND PRESENT IN HISTORICAL SCIENCE

4.1. The Object of Historical Science

1. The material object of a science consists in the phantasms in which it finds relevant meaning. These phantasms present themselves in the field of view proper to the respective science. The formal object of a science is the kind, quality, or level of intelligibility that the science has discovered or seeks to discover within the field of its material object. The functions of the material and formal objects of a science are reciprocal, the field of objects being determined by the formality of the viewpoint and the viewpoint being modified by the yield of insights afforded by the objects.[a]

2. History has a lot to do with the past and with the present. Its concern with the past is perhaps the most obvious thing about it, and its relationship to the present is undeniable, even if often aenigmatic.[a] Many puzzling questions about the nature of historical science can be resolved by the use of the principle that the material object of history is the phantasms representing past events in the mind of the historian and the formal object of history is the meaning of past events contained in the phantasms.[b]

3. The present, in its most general sense, is what is at hand, in view, actually stimulating our sensory or our intellectual perception; its opposite is the not-present. The not-present may be divided into the inexistent and the absent, according as the object which it represents in no way has existed, does exist, or will exist. The past is a species of the absent. It may be defined at least descriptively as "the remembered absent." This definition of the past distinguishes it from the absent in general by its collocation in the consciousness of the remembered.

It excludes the present, because the present is being actually experienced, and its excludes the future, because the future is absent but not remembered. Thus, past and present are not the original parts of the same dichotomy. The present is distinguished from everything that is not being actually experienced as the perception of an extramental source. The past is immediately distinguished from everything that is absent and not made present by the intervention of memory, including the future by direct opposition. Thus it results that the opposition between past and present is not the same as the opposition between past and future, so that two different kinds of functional relationship emerge.

4. The phantasms which form the material object of history are all contained within the remembered absent; they are all constituted by the evocation of sensory perceptions that are not actually present to perception except in the form of records.[a] The material objects of history no longer exist. But the same may be said of other fields of knowledge. The phantasm evoked by a mathematician when he thinks of a circle or a triangle is not actually present to his external senses; it is a reconstruction of similar perceptions that he has once had, not exactly the same, but sufficiently similar to give him the material with which to construct his image. The insight which the mathematician has into the meaning of a circle or a triangle is not, of course, identified with the phantasm, but he cannot have the insight without a certain minimum of sensory imagery providing its material base. What gives the kind of meaning that the historian finds in the remembered absent is the formality of his approach, that is, the sort of questions

---

4.1: 1a *Supra*, 2.6:10; 2.6:12.
4.1: 2a Cf. *infra*, III.1.

4.1: 2b *Supra*, 3.2:10; *infra*, II.39.
4.1: 4a *Infra*, II.33.

he is inclined to ask about the remembered past, and this formality is not located in the past or in the future; it is located in the present.[b]

5. The abstract past is static and colorless; it is the emptiness of a field after everything that it contained has moved out of it. The concrete past is dynamic and colorful; it contains the images of everything that ever was. The concrete past arises as one adverts to the vital fact that the remembered absent of the past is absent only in the sense that it is not present to the experience of the external senses, but it is present to an indefinite degree by virtue of memory. The notion of the concrete past resides in some variegated phantasm of dynamic imagery; there is movement, action, life. The phantasm of the past is 'made present' in the medium of the sensitive imagination, and the knowing subject abstracts from this phantasm the notion of the past as an intellectual object. Out of the past in general emerges the idea of the historical. This notion of the past is implicit and vague. It contains movement as a meaningful whole, and this movement is easily identified as 'development.' Events and episodes appear as related to one another in a manner that is at first difficult to analyze, but the relationship is seen to have a special form of its own. The past is seen as extended in time, although the rôle that time plays is not initially clear. Things like music and rhyme are also beings whose meaning is extended in time, but the relationship of their parts is not historical, since the distinctive element is lacking.[a]

4.2. *The Present of Historical Science*

1. Phantasms of past sensory experience become present to the knowing subject through the evocations of the sense imagination. Man is conscious of a certain not totally controllable power to develop these images in one of two ways. He can develop them by reference to the aesthetic or sensuous pleasure they afford to his sensitive appetites and to the idealism of his aspirations. The continuity which comes forth from such a development has the form of a day-dream, a story, a pleasing succession of imaginary events in which he himself may or may not play an imaginary part. But, in any case, the meaningfulness of the meditation is directly tied to the feeling of pleasure that it evokes. Man has a certain freedom, on the other hand, to develop the images of his imaginative memory by seeking the intelligibility which they contain through a realistic development, out of which may come the science of history.[a] This involves putting aside merely indolent imagery and putting one's imaginative memory to a more inquisitive use. It involves tying the imagination to the bounds of known reality and making it work assiduously within those limits.[b]

2. While the past is by definition not present, any insight into the meaning of the past is necessarily present.[a] Such insight is an immediate intellectual experience of the knowing subject, and it produces a growth in his understanding which is also present at the moment of the insight and which abides as the condition of subsequent growth.[b] Every individual mind has its own history made up of conscious events joined in developmental relationship, and in this sense every phase of consciousness is history, while every personal advance in knowledge of any kind is historical science. But to phrase the matter in these terms is to state it paradoxically, because historical science is a field of its own, distinct from other sciences, and it is this distinctive character that we are here seeking. The point at hand is that there is a general present common to history and to the other sciences. This is the presence of one's own mental activity. Within the mind of every thinker both static and dynamic elements are present. The naive thinker imagines that he simply gazes at meanings across an empty field, but the scientist recognizes intervening formalities. The historian makes use of his general mental equipment, above all of the understanding which he has derived from every kind of scientific growth and which constitutes his generic present, but his special medium is the historical present.

4.1:4[b] *Infra*, II.10.
4.1:5[a] *Infra*, II.32; II.39.
4.2:1[a] *Supra*, 2.6:12.

4.2:1[b] *Supra*, 2.4:2.
4.2:2[a] *Infra*, II.39.
4.2:2[b] *Supra*, 2.6:8.

3. The term 'historical present' is used in grammar to designate the employment of the present tense where the past is more literally intended. Thus if one says, "The horse crosses the river," when the actual meaning is that the horse crossed the river, he is using the historical present. This grammatical usage indicates a certain complexity in the notion of the present, but it does not fully express it.

4. In historical science, the term 'historical present' is applied to that area of the general present that is strictly historical in character. It is concrete, mobile, and tied to a vision of the past.[a] It is concrete, because from its specific character are excluded such present things as timeless truths of the universal and necessary order. It is mobile, because its ultimate meaning can be recognized only in terms of the contemporary situation of the viewer. It is tied to a vision of the past, because it is the point of reference from which the vision of the past springs. Historical vision is necessarily retrospective, and it is the retrospective gaze from the present at the remembered absent that brings the past into existence. The past is a being of the mind with a fundament in reality, the reality of the things that were and are no more, but also and formally the reality of the *meaning* which abides after the temporal things in which it remotely inheres are gone.

5. Conscious attention to history of any sort requires an appropriate succession of images; concentration upon historical science requires in addition that the images be rooted in reality. This means that the historical scientist will consciously seek a set of images having reference to reality and will constantly verify the resulting images in reality. Very often the images he chooses to consider are contained in a record or account. When the historian brings before himself a narrative, document, record, or similar source, the container of the source becomes present to his external senses, but if he simply recalls the source as having been so present at an earlier time it becomes present only through his imaginative memory. If the source is a book, to use an easy example, the senses will experience its physical characteristics, including the visual impressions of the words written thereon. Simultaneously the mind of the historian will concentrate upon the book and will focus its attention first upon the phantasms evoked by the words. Within the phantasms he will grasp the meaning contained, and this meaning will be the remote formal object of his consideration. If the meaning intended by the writer is historical, the formal object of the reader will be historical, that is, the kind of intelligibility that he derives from an understanding of what is written will have the form of the time-related and developmental meaning that we have identified as characteristic of historical consciousness. The historical scientist will, in addition, seek to understand the meaning of what is written as related to reality. He will verify the facts described by testing them against all of the evidence he can accumulate. He will look for consistency and inconsistency, harmony and contradiction in the account that is before him, as well as for clues concerning the intention of the writer. If the account begins with a phrase like "once upon a time," to quote a common instance, it affords strong internal evidence that the writer of the account did not intend to adhere to reality.

6. The historical scientist will also seek the fullness of meaning that the images narrated may contain, whether or not this fullness was consciously intended by the writer of the account. In fact, the ultimate aim of the historical scientist is not simply to grasp the meaning as related to the intention of the writer or of the persons described in the written account, but rather to grasp the meaning as related to his own understanding.[a] The narrative under examination is a formal medium: the historian is not looking directly at the events narrated; he is viewing them in the medium of the written account. But this medium is only the proposed formal object of his viewpoint; its stability is totally dependent upon the results of his verification of it in reality. The actual formal object is the understanding of his own mind as it concentrates upon the concrete object of the account against the background of everything that he himself knows and understands about everything.

4.2:4a *Infra*, II.9; II.26.

4.2:6a *Supra*, 2.2:2b; *infra*, 8.2:3ff.

7. Controversy over the meaning of historical science is partly due to confusion over the aim of historical research. Many historians have assumed that the ultimate aim is the attainment of the true historical facts by the process of verification in evidence present to the external senses. That this aim is important should be clear, since historical science is exclusively concerned with the real, and the real is what is verified in experience. But that this is not the ultimate aim should be equally clear, since historical science is concerned with meaning, and meaning as such is not verifiable in sensory experience.[a] The element lacking in the confused view of historical science is precisely the function of verification in intellectual experience. Once this fact has been recognized, it becomes obvious that the ultimate aim of historical science is the ever fuller and deeper understanding of events which themselves have been verified as far as possible in sensory experience.[b]

8. The historical understanding to be derived from the contemplation of past events is different from the understanding to be gained from the approach of other sciences. It resides in a different area of intellectual consciousness. Empirical science comprehends changing phenomena in terms of necessity and universality, at least to the extent that its approach is mathematical. Historical science comprehends changing phenomena in terms of the concrete and particular as regards the distinctiveness of its approach, utilizing necessary and universal concepts as secondary equipment.

9. The outcome intrinsic to events as described in a narrative of the past and from the past may be called the relative present. The outcome of any event or of any series of events is the historical present in the relative sense, for the outcome is the intrinsic intelligibility of the events leading up to it. Every relative historical present is itself contained in the absolute historical present of the contemporary thinker. The knowing subject has his historical present, and he cannot escape entirely from that present when he reviews events of the past. He can modify his absolute present so as to make it more clear; he can purify it of bias, error, and ignorance; he can make it a more effective point of view by recognizing it as an object and attending explicitly to it. But he cannot escape from this present, for it is the medium of his own historical understanding.

10. How does the student of history turn his absolute historical present into an object of attention? He first adverts to the fact that the process of history as an intelligible whole is continuous and extends from the past right up to the moment in which he exists. The whole of phenomenal reality is developing simultaneously in such wise that a cross-section can theoretically be made through its entirety at any given moment. Each successive moment adds new genetic intelligibility to the development of the phenomenal world as well as to the world of human events. Therefore, the fuller intelligibility of an historical account is somehow to be found in its reference to the further knowledge of results extending up to the moment in which the student of history actually is living.[a]

11. The remote formal object of historical science is genetic meaning. History presents the past under the aspect of genetic change. History is therefore the awareness of the past from the viewpoint of genetic change. This viewpoint resides in the mind of the historian; it is an intellectual object (a proximate formal object) visible to the knowing subject of any historical knowledge, but explicitly adverted to only by those who have reached the stage of historical science. The viewpoint of genetic change is the special present of the historian as such.[a]

12. The genetic viewpoint is visible in the proposed formal object of historical science. The writer of an historical account has recognized meaning in a sequence of events. If he had not recognized meaning, there would be no described sequence and there would be no described events. The meaning is associated with a progress of the events from a beginning towards a conclusion, and the conclusion of the narrative, or of any unified part of the narrative, expresses the genetic viewpoint of the writer. He had to know the outcome in some way in

---

4.2:7[a] *Supra*, 2.6:9.
4.2:7[b] *Supra*, 2.6:17.

4.2:10[a] *Infra*, II.26.
4.2:11[a] *Supra*, 2.3:11; *infra*, II.61.

order to be able to write the account; if he had not known something about the outcome, it would not be an historical account. This is the difference between historical consciousness and the mere consciousness of writing down data with no anticipation of what they signify outside of themselves. But the genetic viewpoint is not wholly contained in the account under examination. Its fullness and ultimate meaning is contained rather in the historical present of the one examining the account. Even the outcome of what is described in the narrative from the past is dependent in its fullest sense upon the perception of meaning in terms of the present of the contemporary thinker.

13. The consciousness of the immediacy of the present as the most recent form of the dynamic continuity of the historical process leads to the development of historical perspective. The most elementary meaning of events is contained in a very limited abstraction from the sensory phenomena in which they are recognized. But the fuller meaning sought by the scientist of history requires a greater abstraction from the strictly material circumstances of the events. This abstraction is not 'total' in the sense of an advance to 'universal' concepts, at least, it is not primarily so. It is rather a 'formal' abstraction of the intellect by which it moves to deeper understanding in its formal viewpoint and to higher meaning in its formal object.[a] It involves the abstraction of concrete intuition.[b]

14. Thus, historians tend to distinguish chronology from mere records, the former having some arrangement according to time relationship. Journals, annals, and chronicles are ascending types of historical account in which a gradual abstraction from the most material immediacy of temporal perspective is illustrated. But history does not emerge on the higher level until simple chronology has been transcended and the reasons behind historical events begin to appear. The account of events in terms of their historical reasons is known as historical explanation. When historians talk about history, they are usually referring to historical explanation.

15. While historical explanation is essentially dependent upon a precise knowledge of past events in their concrete circumstances, it cannot exist in valid form without a certain abstraction from the material circumstances of the absolute present constituting its proximate viewpoint. This fact may seem paradoxical until one reflects that depth of understanding is itself a certain abstraction or pulling away from the material limitations of shallow understanding. This is true for insight into knowledge of every kind. In the case of historical insight, it involves a general capacity of the intellect to recognize more general meanings in the process of concrete development. Why 'depth' of understanding should be equated with 'generality' of the viewpoint may not be immediately obvious, but I am not speaking here of generality in the sense of universality.[a] Somehow it should be obvious to everyone that being able to see the 'big picture' is an advance over being able to see only fragments and isolated details. As one does come to see the bigger picture and the details in their relation to it, he automatically knows by an insight that is itself an immediate conscious experience that the depth of his understanding of the objects concerned has increased.[b]

16. The scientific observer of the historical past looks into his historical present and, by the use of his imaginative memory, views events of the past as though they were in some way present. The mode of presence of these events, as has already been pointed out, is incorporation into the material present of the imagery evoked as well as into the formal present of intellectual understanding. The past becomes an intellectual object as the result of a dichotomy separating the imagined present from the present of actual sensory experience; it becomes a scientific object through its identification as remembered or reconstructed, not simply imagined or made up. The 'real past' as the material object of historical science has verified membership in reality. The past as such has intelligibility of its own, but its meaning can be perceived only in the light of the present. We say that the present is the principle, the origin, of his-

---

4.2:13[a] *Supra*, 2.4:9.
4.2:13[b] *Infra*, 5.2:17; IV.13.

4.2:15[a] *Infra*, II.62.
4.2:15[b] *Supra*, 2.6:14.

tory in the sense that the conclusions which constitute historical knowledge spring therefrom and can in no other way exist. The past of historical science is the visibility of objects as they are observed through the present.[a]

17. The present can be considered as an object of cognitive awareness. Before science has arisen at all in the mind, the present is not recognized as an objective medium standing between the material object and the knowing subject himself. On the prescientific level the present is identified in consciousness with one's own subjectivity, so that the viewer of the past seems to be looking straight at it from the vantage point of his unclouded subjectivity. Historical science arises as the thinker begins to see that he is actually looking at the past in terms of certain conceptions constituting his viewpoint, and this science increases as he brings these conceptions into view in such wise as to isolate the objective element from the part that remains subjective and unintelligible.[a] In the process of purifying the objective medium from subjective elements he eliminates bias, emotion, ignorance, and other factors distorting his viewpoint and falsifying his judgment.

18. The succession of events on the phenomenal level can rather easily be recorded and chronologically interpreted right up to the moment of the writing. This is already history of the elementary type. The amount of historical interpretation required for the recognition of the cronological significance of a phenomenal event just past is available from ordinary abstraction. The events transpiring in the course of a forest fire or in the flooding of a river, for example, can be immediately interpreted on a chronological level, because, as phenomenal events, they are all far enough below the viewpoint of the adult mind as to be immediately in perspective. In this case the succession is evident on the sensory level, and the presence of common sense is sufficient to make the interpretation. The same may be said for the chronological interpretation of ordinary human events, although a little more abstraction is needed as well as skill in distinguishing the essentials and depicting them fairly. And, when it comes to human events, viewpoints begin to intervene, showing an implicit need for technical training to achieve full objectivity.

19. The difference of viewpoints regarding even short-term events stems from a difference of background and education. Different people witnessing the same event may notice different things about the event and describe it differently afterwards. Thus, after witnessing the same automobile accident, a mechanic might describe it in one way, a lawyer in another, a physician in another, and a housewife in still another, in accordance with the use of differing specializations of the minds concerned. The housewife and the physician might not realize the legal implications of the accident; the mechanic and the lawyer might not notice the effect of the crash upon the persons involved. But the viewpoints are not mutually exclusive. What is important to note in the present context is that development of the mind in any field increases its ability to recognize meanings in events and thus to influence the chronology of the events. But development of the mind in other fields is not the specific kind of abstraction required for the recognition of historical meaning on a level transcending chronology. It is simply the general abstraction that is necessary and sufficient for the proper recognition and chronological recording of certain aspects of events not evident to the untrained observer.

20. The problem is different for historical explanation. The student of history very soon realizes that human events have a particular relevance for him, and as he moves above the level of common sense he finds that many types of human events need to move a little distance into the past before they can be adequately interpreted. In fact, in a certain sense, some events, such as a political movement, have to be completed before a proper historical interpretation can be given of them. The historian will thus distinguish between what is past and what is contemporary, including in the contemporary what may perhaps have happened months or years ago. The contemporary is the historian's present, the medium in which he thinks. He selects from the present the conceptions that are true and objective, he

---

4.2:16[a] *Infra,* II.9.

4.2:17[a] *Supra,* 2.6:20.

takes note of the latest state of affairs, he tries not to make doubtful issues the basis of his judgment, but judge he must concerning the past. The historical scientist casts his gaze far enough into the past to be able to read meanings in things in terms of relative presents, of outcomes long past themselves. To a certain point the influence of his absolute present will be negative; the awareness of it will simply serve to keep alien ideas out of the picture. But as the level of the science increases, more and more of the absolute present must be brought into use. This is the point at which historical explanations become falsified if they are not expressed by a mind that is fully honest and fully informed, not merely about the past, but also and especially about its present. Judgments about the course and meaning of historical development in the larger sense are inevitable. Those who have refused to make them have simply defaulted from the historical scene, leaving the control of the field in the hands of those who did not. Most of the general judgments about the past have been defective, not because they were unwarranted, but because they were unscientific. They have proceeded from oversights and biases that should have been eliminated, but which escaped the process of scientific analysis.

21. The absolute present of historical science is the counterpart in the historical scientist's mind of the full intelligibility of the past. It comes into view as a usable medium to the extent that the historical scientist achieves abstraction that is itself historical in character. There must be a concomitant development of the mind in a general scientific way, but there is also something specific. The specific medium is associated with the rising of the mind to a recognition of more elevated aspects of the historical process itself. It is therefore a developed understanding of the historical process as a concrete whole. It is contained in intellectual ideas that are general but not 'universal' in the Aristotelian sense.

4.3. *The Past of Historical Science*

1. The historical past is the remembered absent as extended in time.[a] The absolute past is the remembered absent with regard to the absolute present of the observer. The distance of the absolute past is measured from the absolute present of the observer. Just as the absolute past is what once was and has already passed out of existence at the moment of observation, so the absolute future consists in what is not yet at the time of observation but will be in the course of development. Strictly speaking, there is no vision of the absolute future, at least with reference to history. It cannot be seen in its concrete character for the simple reason that every unit of history is unique. Certain general features of the probable future can be anticipated by those who are capable, and man has a certain amount of control over the future that he himself builds, but the vision of historical science is entirely of the past. The historian does not include the future within the field of his science. His science is applicable both to the present and to the future, but not in the sense that the historian can claim a vision of the absolute future.

2. The present may be absolute or relative. He who observes the past stands in his own present and views the events of the past as though they were present. Events can be perceived only inasmuch as they are rendered present to the senses. They must become relatively present. The past as such is an extended being whose dimension is time. It is composed of many former presents arranged in time-sequence. Any point of time in the past can be located by reference to the absolute present, which is the point in time at which the one observing the past stands.

3. The absolute past may be divided by an indefinite number of relative presents into an equal number of relative pasts and relative futures that are visible to the historian. The division can be made in terms of the corporeal time registered by the calendar, so that up to a given moment on a given day in the absolute past all of the phenomena in the world are taken to be relatively past and all of the intervening phenomena up to the moment of the absolute present are taken to be relatively future. The relative present is in this sense a point having no dimension or content of its

---

4.3:1[a] *Infra*, II.32.

own, but simply constituting the principle of the division into past and future. But the division can also be made in terms of a 'genetic time' that is related to corporeal time but is not limited by it. Such a division distinguishes a recognized genetic development into past and future in terms of a present having both dimension and content. Its dimension is of a different order than that of the development divided; it consists in the intelligibility of an outcome giving fuller meaning to the events leading up to it.

4. The dimension of historical science arises from the distinction between the past as past and the past as present. The intellectual medium of historical science is present in the understanding of the historian; in this medium is reflected the meaning of past objects not present to the mind of the historian. The origin of historical science as *historical* is the distinction between past and *future*; out of this distinction arises the notion of genetic relationship. The origin of historical science as *science* is the perception of past objects in the medium of reality; historical science grows by the progressive discovery of the meaning of past reality in the presence of the understanding.[a] We may therefore say that the *remote* objects of historical science are formally past in the sense that they are knowable only in the intellectual medium of the past, and the *proximate* object of historical science is formally past, not in its being as an object, but only in the sense that the *content* of the concept which is the object is 'the past.'

5. The medium in which the past as such is perceived is the present. When phenomenal reality is viewed in its historical context, it manifests itself as evolving from the past through the present and into the future. Since the future lies outside of the historical perspective, the absolute present is seen to be the most recent phase of the historical development. Through his knowledge of the absolute present, the observer maintains a certain awareness of the direction which the developments of the past have assumed and of their results. It is this awareness of directions and results which gives intelligibility to the past as such.[a]

6. The observer need not attend directly to his awareness of the absolute present. He can 'transport himself' back into the past in such wise as to imagine himself re-experiencing events long transpired. The successfulness of such a venture depends upon the accuracy of the historical reconstruction presented to his senses. Hence the vital importance of complete data and precise instruments. One who is reading a good account of the storming of the Bastille can, in his mind's eye, see almost as much as one who was actually there. Better still, one who is watching a moving picture of an event, e.g., a football match, sees just as much as one who was actually there. In the latter case we have an example of a perfect reproduction of an event, effected by exact instrumentation.

7. In one sense, the reproduction of a past event with all possible exactness is the historical ideal. In this sense, the moving-picture camera is the ideal historian. But in a more profound sense the serviceability of such devices is extremely limited. Neither he who observes the original football match nor he who views the film of it becomes thereby an historian. The first observes only the present; the second observes the past as present. But the historian seeks the past as past. The difference lies in the perception of the historical medium. If a person watches a film of a prize fight having taken place some time before and has heard nothing about the fight or its outcome, his viewing the fight is not retrospective in the historical sense. But if he knows from the beginning how the fight turned out, he may recognize features that he could not otherwise have noticed, such as signs of fatigue in the contestant who lost. If he has additional knowledge of the sequel, such as that blows received by one of the contestants developed into serious injury, this will also influence his view of the fight. The more penetrating view, based upon knowledge of results, is historical insight.

8. There is a natural mode of understanding what is past. Words spoken five minutes ago, a sentence not yet completed, belong, humanly speaking, to the present, not to the past. We call such events 'contemporary.' They remain contemporary until significant

---

4.3:4a *Infra*, III.17.

4.3:5a Cf. *infra*, II.44.

developments have taken place. The lapse of time required varies with the type of event and with the circumstances. A football match played last week is contemporary, except for those who are minutely interested in the events which succeeded the match. An address given yesterday is contemporary. It becomes historical when it enters the historical perspective. Such an historical perspective implies an awareness of a new present, composed of subsequent developments, in which the old event can be seen. Sufficient time must have elapsed for the mind of the observer to draw away a little from the immediacy of the event. This means that new events must have come to the fore. One can speak of the writings of Voltaire 'in the light of' the French Revolution. The event of the French Revolution set the writings of Voltaire in a certain historical perspective.

9. The absolute present may be simple or complex. When past events are being viewed in the light of what is purely contemporary, the medium is the simple present. But past events can also be illuminated by the light of other past events. One can, for example, study the events of 18th-century France in the light of the French Revolution. In this case the French Revolution is taken as a present into which the previous events evolve. The observer stands in a double present: his own absolute present and the assumed present of the French Revolution. Both presents influence his judgment of the events. The measure of historical science is the precision with which the two presents are related to one another.

Article 5

THE FORM OF HISTORICAL UNDERSTANDING

5.1. Historical Causality

1. A cause is that on which a thing depends for its being.[a] So there are as many kinds of causes as there are kinds of influences upon the being of things. Historians are concerned with the understanding and explanation of those causes which uphold the becoming of individual things; they study the causes of events. Historians ask themselves why events were what they were and as they were, and it is the discovery of these causes that produces historical science.

2. The word 'cause' suggests to most people the idea of the 'efficient' cause, the extrinsic agent bringing about the effect.[a] Historians in their search for causes have tended to seek the efficient causes of events, and their search has centered upon the human agents of human events. But there are other influences upon the being of things. The intrinsic cause by which a thing is constituted in its species was named by Aristotle the 'formal' cause.[b] It consists in the essence of substances and accidents and is recognized as the intellect distinguishes the intelligible part of an identity from its container. The identity of the remainder is what Aristotle called the 'material' cause. The efficient cause tells us why it is in terms of an external agent. The other extrinsic cause, named by Aristotle the 'final' cause, tells us why a thing is in terms of its own nature. We mean by 'nature' the permanent basis of change within an identity, and the intelligibility of the change is called its finality.[c] If these four causes are compared and contrasted in relation to historical events, it is the final cause that will emerge as the central causality of historical concern.[d]

3. The final cause of a thing may be considered more ultimately as entirely extrinsic to the thing itself. For instance, the 'purpose' of an individual insect might be ascribed as the propagation of offspring and the preservation of the species. Or the final cause and purpose of a thing may be considered more proximately as its own development. Thus, the finality of an individual insect is its own growth and satisfaction. In this case the finality is extrinsic to the identity of the individual but intrinsic to the individual itself. But history is concerned with a relationship that may be called 'genetic causality.'[a] It is a relationship of development between two states of a thing, standing outside of the static conception of the essence of the thing and of the identity of the thing as a concrete individual being. Genetic causality is concerned with change in concrete individuals as such. The cause of this boy is the man he grew into. The cause of this seed is the flower it grew into. Everything that develops or evolves has a history. The understanding of this development is historical science.

4. Historical science is the knowledge of beginnings in their outcomes, of imperfect things in the perfection that they attained.[a] The formal intelligibility of historical science derives from the knowledge that the historian has of the outcome of the development. Genetic meaning is discernible in the

---

5.1:1a Cf. T. Aquinas, *In duodecim libros Metaphysicorum Aristotelis expositio* (Turin: Marietti, 1964), Lib. V, Lect. I, No. 751.
5.1:2a *Ibid.*, Lect. II, No. 765.
5.1:2b *Ibid.*, No. 764.
5.1:2c For Aristotle the material cause is the intrinsic thing out of which something is made, the final cause is that for the sake of which something exists, and 'nature' is that from which comes the beginning of movement of a thing acting because it is what it is. (*Ibid.*, Nos. 763 and 771; Lect. V, No. 810.)
5.1:2d *Infra*, IV.14.
5.1:3a *Infra*, II.43.
5.1:4a *Infra*, IV.7.

units of concrete development which we call 'events.' Each event is an intelligible result of a development from an intelligible beginning in the same thing or situation. Historiography is the description of change in terms of genetic meaning. As an expression of understanding it is dependent upon the mind and the insight of the writer, but the material objects of its description are extramental realities. The historian does not create the genetic meaning of events; he simply recognizes and describes it.[b]

5.2. *The Form of Historical Reasoning*

1. Reasoning has been defined above as the intellect in operation.[a] The process of modification of the intellect can be considered either vertically or horizontally. The vertical development is the process of deepening the understanding, the acquisition of deeper insight into things.[b] The horizontal development is the application of acquired understanding to the acquisition of greater knowledge about things without a concomitant deepening of the understanding of things.[c] The distinctive kind of reasoning by which historical conclusions are reached is both vertical and horizontal. It is vertical, for the historian does in fact arrive at a deeper grasp of reality as a result of his study. It is also horizontal, for the historian can use the understanding he has acquired to draw conclusions that add to his knowledge of historical facts without the incorporation of such acts of reason into the deepening of the understanding as such.[d]

2. Historical reasoning as a distinct identity differs in an intelligible way from the reasoning of classical science. Thinkers about theory of history have often noted that classical science reasons in terms of general concepts and arrives at a body of general truths, while historical science is concerned with concrete and individual truths. Under analysis this means that classical science reasons in terms of universals, or 'open classes' of things that are potentially infinite in their extension to individuals coming within them, while historical science reasons distinctively in terms of individuals or at least of classes that are not 'open.'[a]

3. There has been a tendency of some thinkers to conclude that, because historical science does not deal universally with things but only particularly and concretely, it cannot be science at all. But this is an unsound conclusion.[a] The fact that history arrives at knowledge of the real makes it scientific on the level of common sense at least.[b] And the fact that its mode of reasoning as historical is not unfolded in terms of universal essences does not thereby exclude it from the level of specialized science, for it has never been shown that all technical understanding of the real is limited to operation upon open classes of things.[c]

4. On the contrary, there is an obvious case in favor of historical science. The philosophers of old recognized that man as man can grasp the individual and concrete, but they did little to develop this area of knowledge.[a] Modern empiricists and positivists have shifted to an emphasis upon the concrete features of things while adhering to mathematical and self-evident principles as a paradoxical component of their science.[b] Thus we have the occurrence of two mutually exclusive notions of science. The Aristotelian notion and others of its type exclude from science all conclusions that are not reasonably deduced from self-evident and necessary principles. The notion of empirical science, on the other hand, excludes from the realm of the scientific all conclusions that are not results of the inductive method of reasoning in terms of classes of things that are not strictly universal but have enormous mathematical probability of recurrence. Does it not seem possible that both views have validity in their own right and that both notions are therefore species of the broad and comprehensive concept of science in general?[c]

5. But the question before us is more particular, for history does not

---

5.1:4[b] *Infra,* II.55; II.60.
5.2:1[a] *Supra,* 2.6:8; 2.6:30.
5.2:1[b] *Supra,* 2.6:17.
5.2:1[c] *Supra,* 2.6:36; 2.6:38.
5.2:1[d] *Infra,* 6.1:3.
5.2:2[a] *Infra,* II.9ff.

5.2:3[a] *Infra,* IV.18; IV.19.
5.2:3[b] *Supra,* 2.6:21.
5.2:3[c] *Infra,* II.62.
5.2:4[a] *Supra,* 2.3:18[a].
5.2:4[b] *Supra,* 2.4:5.
5.2:4[c] *Supra,* 2.4:3; 2.4:4.

seem to fall under either of the notions of science, whether deductive or empirical. Deductive science depends upon syllogistic reasoning from universally valid truths. Thus, if the essence of man is taken to be 'rational animal,' then the statement, 'all men are rational,' is a universal truth applicable to any individual of the class. If this truth is taken as a major premise and under it is subsumed a truth verified in concrete experience, such as, 'Edwin is a man,' a deductive conclusion can be drawn, to wit, 'Edwin is rational.' The Aristotelian and the empirical scientist will view this argument differently. The Aristotelian will find it shaky in its minor premise, which is derived from inductive experience and surrounded by the possibility of error. He will admit inductive premises, but he will not ascribe to them the degree of validity that he ascribes to universal essences perceived by the intellect alone and the truths that are deduced from them. But the empirical scientist will find the major premise questionable. He doubts whether it is valid to define man as rational animal and then make this definition the basis of a scientific argument.

6. The focus of classical philosophy upon universal essences and necessary laws places knowledge of individual truths in the margin of the barely intelligible. The focus of empirical science upon the concrete reality of what is verified in sensory experience admits mathematical concepts as the paradoxical dimension of its own formal understanding and excludes the intelligibility of metaphysical concepts from the field of science. What empirical science has done is to develop an area of intelligibility that was overlooked by the deductive philosophy that preceded it. It did not thereby eliminate the field of intelligibility already uncovered.

7. It is possible that history deals with a field of intelligibility that is not recognized either by classical philosophy of the Aristotelian type or by empirical science. The possibility of specialized science is not divided completely between essential and empirical science, as though the former had a monopoly on deductive science and the latter on inductive science. In fact, both deduction and induction are used in either case, the difference being the emphasis given by essential science to deduction and by empirical science to induction. The area between the two is occupied by a not fully investigated field regarding the intelligibility of individuals as such.[a]

8. The undeveloped character of this 'middle' field of intelligibility can be ascribed partly to the excessive emphasis of ancient and medieval science upon essential forms and deductive reasoning and partly to the excessive emphasis of modern empirical science upon material forms and mathematical reasoning. Both the earlier and the later approaches to science have leaned heavily upon mathematical reasoning and syllogistic thinking to the detriment of other possible ways of deepening the understanding of reality and arriving at truth.

9. The form of reasoning specific to historical science is for this reason still waiting to be analyzed in detail. But something can be said of it by way of indication. The unique character of historical reasoning is evident both in the kind of concepts it forms and in what it does with them.[a]

10. Classical science of the ancient and medieval type tends to restrict the dignity of 'concepts' to the recognition by intellectual insight of the intelligibility of classes of things and the impression of this recognition upon the intellect of the knowing subject as an element of understanding. These concepts represent insights into the universal and unchanging meaning within the identity of a thing, in such wise that the identity expressed by the concept is not the identity of the thing as an individual, but rather the identity of the class represented by the individual. Because this earlier science has been systematized on the metaphysical level, the essential forms said to correspond in the individual things to the essences grasped by the human intellect are also defined on the metaphysical level, with the result that they are often badly understood by persons who are not thinking on the metaphysical level; indeed, such persons are tempted to exclaim that there are really no essential forms at all.

11. Classical science of the modern empirical type tends to restrict the dignity of concepts to the mathematical

---

5.2:7a *Infra*, II.16.

5.2:9a *Infra*, II.53.

or quasi-mathematical definition of what is seen to be essential in the data of sense as regarded from the mathematical or quasi-mathematical point of view. To the extent that the intelligibility uncovered by modern empirical science is mathematical, its essential concepts, just as in the case of the earlier science, represent insights into the universal and unchanging meaning of the identity of things, but the 'things' whose identity is grasped are mathematical conjugates. Modern empirical science of the statistical type has gotten away from its bond to the universal and the necessary and extended itself into the realm of reasoning about the particular and the contingent as such, but only because the development of mathematics along statistical lines has provided an intelligibility also in this direction. The fact is that whether empirical science is dealing with the certain or with the merely probable, the medium of its reasoning is largely mathematical.

12. The syllogistic reasoning of the earlier science contains a hidden dependence upon mathematics. In the case of the syllogism cited above, the conclusion, 'Edwin is rational,' is seen to be valid for the reason that, if all men are rational, then this one man is rational. It is evident that one is part of all, and this evidence is mathematical. Similarly, a syllogism demonstrating that children are rational and firemen are rational and so forth would be seen to be valid, for the reason that some is part of all, and this fact is also mathematically evident. It can be said with considerable reason that the evident character of syllogistic comparisons of one, some, and all are so simple and native to thought that they antecede mathematics as a specific kind of thought and belong to the native character of thought itself. I should not like to contest such a claim, but I would like to point out that it is right down at this simple level of human thinking that historical thought seems to branch off from the approach of classical science, and it may be the very simplicity of the difference that has caused it to be missed.[a]

13. The historian does not find the characteristic meaning of his science in what is static and unchanging in a substance or an object of sensory experience. Yet, he cannot find meaning in what has no discernibly abiding element. There is no meaning in chaos or in pure flux. The historian finds meaning by isolating and grasping two or more states or stages in the development of the same individual thing.[a] These two stages have a genetic relationship. The second, or later, of the two stages is an event; it has 'come forth' from the first. The entire thing underlying the development is changed by the development, not in its essential identity as a class, nor in its identity as an individual, but in the status of its individuality. When a man grows old, his whole concrete being grows old, and the historian can distinguish stages in this development. He may discover that the growing old of this man developed into corporal weakness and mental wisdom. His insight into the man as a youth and in the prime of his life is illuminated by his knowledge of the man in his old age. Adulthood and old age are recognizable events in the lives of individual persons that cannot be deduced from the universal concept of man as a rational animal. Each individual has historical intelligibility of his own.[b]

14. In order to be able to do his reasoning, the historian must have general terms identifying historical objects within the development of concrete substances. Infancy, childhood, youth, manhood, and old age are general terms used to identify stages in the development of a man. They do not represent the essence of the man, but they do represent the 'essence' of stages in his development. In other words, they represent what is recognized as essential in a stage of development and what can serve as a criterion of comparison with the individual case. This kind of generalizing is not in its own essence different from the conceptualizing of either classical philosophy or that of empirical science. The difference lies in the recognition of what is essential according to the point of view of the examiner. The Aristotelian calls essential what abides every change in the individual, and this is metaphysical. The empirical scientist recognizes developments, but he defines them mathematically and tries to ex-

---

5.2:11a *Supra*, 2.6:27.
5.2:12a *Infra*, II.56.

5.2:13a *Infra*, III.18.
5.2:13b *Infra*, II.58.

press them in general laws of the classical or the statistical type. The historian calls essential what is more proximate to the individual development as such, and he finds relevant those classes of things that afford intelligibility to the individual development as such.[a]

15. This fact has led many observers to conclude that history is therefore just common sense and is not worthy of being called science. But underlying this conclusion is the assumption that the degree of understanding, or the level of science, is commensurate with the degree of universality.[a] It is an assumption that even some empiricists make, in spite of themselves, for they are wedded to the concreteness of sensory experience, but they are also wedded to the universality of mathematical conclusions, or, in the case of statistics, to the colossal breadth of its probability. There is no reason why the historian cannot at times make use of the same degree of mathematical probability, since he does use the proficiency of the other sciences in a manner ancillary to his own, but he can also find sound intelligibility in a type of abstraction that is not strictly mathematical, and the level of science is measured not by the breadth of its total abstraction, but by the depth of its formal abstraction.[b] Hence, if history is found to possess a depth of formal abstraction going beyond the level of common sense, and if the understanding that it acquires through this abstraction is of real things, then history is a specialized science on a par with classical philosophy and modern empirical knowledge.[c]

16. The value of historical abstraction lies especially in its proximity to the concrete situation. Abstraction of the classical type is mere distraction if it is not anchored to an appreciation of the relation of the concept to the whole of one's understanding.[a]

17. The kind of reasoning involved in strictly historical thinking may be described as 'intuitive' rather than syllogistic, if the term is properly understood. All insight in any science is intuitive. The mind must grasp meaning as an immediate experience; it cannot arrive at meaning on the mere level of process. However, the manner of setting up the mental situation in which the mind can grasp a meaning and thus come to an understanding can differ. Classical science of the ancient and modern types characteristically sets up the situation of an insight through the interposition of universal middle terms. The general form of the reasoning depends upon the evident principle that two things equal to the same thing are equal to each other. When the conclusion, 'Edwin is a man,' was drawn above, it was based upon such a comparison. The middle term in that case was 'man,' and it was reasoned, in effect, that since rational is equal to man and Edwin is equal to man, then Edwin is equal to rational. This equality was not mathematical, and the argument would not have been valid if 'man' had not been universal in the major premise and limited in the conclusion, but the term 'man' was able to serve unequivocally as a kind of imperfect identity linking the two extremes. The validity of the reasoning was based also on the evident intuition that 'one' comes under 'all.'

18. Historical reasoning is based upon the same principle that two things equal to the same thing are equal to each other. Historical meaning is sighted in the medium of a concrete concept, and it adheres to the developmental thing represented by that concept. In historical reasoning, the subject of an historical conclusion is joined to the predicate because of the intuitively recognized identity of both the subject and the predicate with the thing known in the concrete concept. Intelligence compares two stages of the same concrete thing and sees meaning in the earlier stage by its underlying identity with the later stage.[a] It understands the beginning by knowing the outcome. Thus, for example, the historical meaning of a maternal conception can be illuminated by the fact that twins were delivered.

5.2:14a *Infra*, II.11; II.61; III.17.
5.2:15a *Infra*, III.24.
5.2:15b *Supra*, 2.4:9; 2.5:3.
5.2:15c *Infra*, II.62.
5.2:16a *Infra*, 5.3:6.
5.2:18a *Infra*, 5.2:21. Thus a past thing is *real* by reason of its identity with itself as a present thing, and the 'is' of the scientific historical conclusion affirms the extramental reality of the meaning contained in the *becoming* of a concrete thing as such.

The birth of the twins affects the way in which the conception is thereafter viewed. The insight is called immediate in the sense that it does not result from mediate, or syllogistic, reasoning. But it produces understanding all the same.

19. A classic misunderstanding of genetic science has arisen from the teaching of Aristotle on genetic causality. In a well-known passage Aristotle observes: "From experience, from every coming to rest (ἐκ παντὸς ἠρεμήσαντος) of the universal in the soul, of one alongside the many, (arises) that which, being one and the same in all, is the starting-point (ἀρχή) of art and science. If it regards generation, it is art; if it regards being, it is science."[a] What he means is that a passing array of individual sensory experiences of the same kind of object, by a kind of concretion of separate units, becomes unified as a stabilized object of its own in consciousness and that from this unity arise art and science. He states that the perception of generation is art and not science. For Aristotle art is the right reason, the right method of doing and making things. He holds that science regards necessary things and therefore the immutable nature of things, while the being of things produced by human art is contingent, as is obvious from the fact that things begin to be. He maintains that scientific knowledge concerns eternal and therefore ungenerated and incorruptible things; contingent things cannot be scientifically known, since, when they are not actually being experienced, we do not know whether they exist or not, as we do not know whether Socrates is sitting unless we are presently experiencing that he is sitting. St. Thomas, explaining this doctrine, notes that the knowledge of contingent things can be useful; the knowledge of contingent things in their particularity pertains to practical science, but not to speculative science.[b]

20. The restriction of the object of human understanding to the immutable essences of things is misleading. Aristotle admits that the universal overview of individual sensory objects is the starting-point both of art and of science. Such an overview can be gathered from a series of objects in temporal sequence as well as from a series in atemporal sequence, as is obvious from listening to a melody or to human speech. He then goes on to restrict the meaning of the contingent to its practicality as a product of human art, and this is not a precise division, since the whole process of Nature manifests a contingent as well as a necessary aspect, as is obvious (according to the same line of reasoning) from the fact that natural things come to be. It follows that, while there is by very definition no mutability in 'nature' taken as the immutable essence of a thing, there is massive mutability in 'Nature' taken as the world in which we live. There can also be real and scientific knowledge of contingent things. While I cannot know that Socrates *is sitting* unless I directly or through another person am actually experiencing that he is sitting, I can know without present sensory experience that Socrates *sat*. When an event has taken place, from the historical point of view it becomes eternal, immutable, and necessary, and thus it has all the essential requirements of valid science.

21. Historical science deals with concrete universals. Aristotle, in maintaining that science is only of universals, ascribes the origin of universals to a kind of concretion of sensory objects. I have attributed the origin of universals to the distinction by the intellect of the form of sensory objects from their identity.[a] Once the form has been distinguished, it can be applied to other identities, and therefore every individual sensory object becomes a germinal universal the moment its identity is recognized. The historian deals characteristically with intellectual objects that by virtue of their recognized identity stand in the order of the universal and the abstract, but are not extended to the point of Aristotelian universals. The concepts of the historian as such are germinally open classes of things considered in their historical situation. Thus the

---

5.2:19a My own translation. Cf. Aristotle, *Posterior Analytics* (Greek text with parallel ET by Hugh Tredennick [London: Loeb Classical Library, 1966]), 258. Cf. also Aquinas, *In Post. Analyt.*, Lib. II,

Lect. XX (*textus Aristotelis*, No. 437).
5.2:19b Aquinas, *In decem libros Ethicorum Aristotelis ad Nicomachum expositio* (Rome: Marietti, 1964), Lib. VI, Lect. III.
5.2:21a *Supra*, 2.3:18a.

concept of 'Scotland' is germinally open in the sense that the historian can conceive of an unlimited number of Scotlands, and it is closed in the sense that the historian's consideration is limited to this concrete thing which is Scotland. But when the historian examines seventeenth-century Scotland in the light of twentieth-century Scotland, he is in fact comparing two Scotlands to this underlying identity which is the concept of Scotland. A development is recognized in terms of universals of the classical scientific type (communication, telephone, etc.), and historical meaning emerges from a comparison of propositions pertaining to two stages of the development with their identity in the developing thing itself. Developing Scotland is a concrete universal whose concept is applicable to any number of stages of the development, and in that sense to any number of Scotlands. It is the identity of stages in the same developing thing that supplies for the middle term of the Aristotelian syllogism and thus makes historical reasoning possible.

22. We conclude that the prime object of human intelligence is the universe of reality — an object whose universality is primarily concrete. If intelligence is limited to the perception of universals, we are constrained by reason to admit that the universals of classical science and the universals of historical science are species of a more general universal which by objectivity recognized as such.[a] While, in the order of total abstraction, historical universals are considered less abstract than are the universals of classical science, nevertheless, in the order of formal abstraction, historical universals can be and often are more abstract and more 'intellectual' than are the universals of classical science.

5.3. *The Value of Historical Science*

1. When the characteristic reasoning of history is viewed on its simplest level, it looks dull and pedestrian. The same is true of classical science, ancient or modern. But as with the generalizing types of science, so in the case of historical science, life, momentum, and interest appear as the depth and complexity increase. The distinctive value of developed historical science, considered in relation to the condition of classical science and of the educated man at the present time, may be summarized under the following headings: A) Historical science can deal capably with human problems that are inaccessible to classical science of either type. B) Historical science has the potentiality of reprocessing the 'empirical residue' of classical science and of reclaiming important values that are needed for full human development. C) Historical science can bring classical science and valid theory of every kind closer to man by organizing under historical understanding the methods and conclusions of those areas of thought that are not properly historical in themselves. D) Historical science can serve as a portal of entry into an appreciation of those forms of reasoning that transcend the formalities of classical science and are not properly historical in themselves, but are similar to historical science on the primordial level of reasoning.

2. In the first place, the chronic problem of classical science has always been its treatment of the individual as an anomaly. It accepts the individual, but does not quite know what do do with it. Classical philosophy has tended to develop the generalizing capacity of the mind in terms of universal concepts while simply assuming that the crucial application of its general conclusions to concrete situations would be rational as a matter of course. In the absence of a science of the concrete, an almost superstitious veneration of the practical man has often reigned alongside of the exercise of reason in those classical philosophers who have had the humility to recognize their plight. Empirical science has fared better on the level of material phenomena, not because of a suitable provision for the understanding of the concrete within the formality of its own reasoning, but because of the healthy tradition of technology that has grown and prospered in the company of empirical science, despite the patronizing attitude that many students of 'pure science' have shown towards their more practical colleagues. But empirical science has felt the same embarrassing ineptitude in dealing with the problems of the concrete human situation.

3. What historical science could do

5.2:22a *Supra*, 2.3:18a.

to remedy this unfortunate weakness of classical science is understood to be supplemental, not substitutional. The realms of understanding opened up by classical philosophical science are breathtaking. Historical science accepts their discovery as real historical fact. The power, success, and utility of empirical science have been phenomenal. Historical science admits their importance and builds upon it. The aim of historical science is the development of another field of intelligibility willing and able to coexist peacefully in the human mind with classical science of either type. Perhaps what is unique about historical science in this regard is its proper mission of reconciling classical philosophy and empirical science with one another by determining where the field of each is located in intellectual consciousness and how the boundary is drawn between them and historical science itself. The result, then, of a full historical development within intellectual consciousness could be a new and greater appreciation of science of every kind and a new enthusiasm for the incorporation into human understanding of areas of reasoning long forgotten or never comprehended that are not strictly historical but lie on the other side of the portal of historical effort.

4. In the second place, the desired development of historical science can be visualized as a methodical movement back into the antecedents of the present time in order to discover the meaning that is lying within the field of the past. This historical movement is essentially retrospective. It is known to most classical scientists in a vague and prescientific way that the present focus of science is the result of a long historical process in which earlier ways of looking at the world have been either greatly refined or discarded altogether. If we take the microscope as an illustration of this process, we might note that the discovery of many facts regarding the microscopic world has required the perfection of lenses capable of bringing into view objects that defy observation by the unaided eye. The field of vision in the microscope is extremely small and precise. Now, a microscopist can easily take his eye away from the microscope and focus it on the larger world around him, and the awareness of the 'context' of the microscopic world keeps his view of it more balanced. Similarly, empirical science has 'evolved' out of the viewpoints of the ancient and the medieval classical thinkers by a process according to which the field of vision has become remarkably more precise and no less remarkably more restricted. But the empirical scientist is for the most part at a loss to change the focus of his vision as a scientist to a focus on the larger world surrounding it in a way that is both serious and meaningful. The means of doing so are simply not at hand unless an instrument is provided along the lines of the historical science presently being discussed. It is not primarily a question of a shift of consciousness from the 'scientific' focus to the focus of common sense. It is rather a matter of shifting intellectual consciousness from a focus on the empirically scientific to the scientific meaning of the cosmos. Therefore, the solution does not lie in simply taking one's eye away from the 'microscope' for a period of diversion. It lies in the conscious ability to proceed back into time and gather from around that empirical focus the world of intelligible meaning that it left behind as it refined itself. This is the 'empirical residue' considered on the historical level, and this is the key to the fuller understanding of man in his human situation.

5. In the third place, what is most conspicuously lacking in the 'worldview,' or central outlook, of the average empirical scientist is an integrated picture of the whole of reality. Those scientists who attempt no serious explanation or knowledge of the rest of reality transcending the fragment of their specialty are simply disoriented and partially mature men. The solutions adopted by the others in order to adjust themselves to human life as it is are largely makeshift and inconsistent. Many have simply resigned themselves to being professional scientists and amateur human beings. Some try to fill the gap by being affectionate spouses and parents, although their love is not fitted into their rational outlook. Others are religious, but seldom in the sense that their religious convictions are scientific. It is more a matter of an emotional outlet. Others are dedicated to hobbies, sports, etc. Now, these avocations can

be good and healthy in themselves, but they leave a gnawing dissatisfaction wherever minds that are proficient in one realm of thought have to squeeze themselves into semi-rational contexts in order to provide some balance for their own rationality.

6. By scientifically reprocessing the 'empirical residue,' historical science can bring into view the rationality of the world surrounding the focus of empirical science. It can place empirical science into the context of the larger world. With the help of a tradition of thought extending back over more than three thousand years, it can find the thread of unity tying the sciences together into a human view of the entire cosmos.[a]

7. In other words, it is the proper work of historical science to trace the family tree of science itself. The amount of study already done on the history of science is sufficient to indicate that common sense has given birth at widely separated dates to several sister sciences, each of which has tended to produce her own children. One of the children of common sense is classical philosophy of the essentialist type, and she has several descendents. Another child of common sense is classical science of the empirical type, and she has several descendents. A third child of common sense is 'intuitive' reasoning of the type illustrated by historical science, and she seems also to have several descendents. It is this third daughter-science of common sense that constitutes the direct object of the present discussion, but with indirect attention to the wider background. We are thus attempting directly to clarify what historical science is and how it operates. Indirectly, we are attempting two additional things: a) to open the door of investigation to other possible sciences of the same intuitive type as historical science; b) to fit the three above-mentioned daughters of common sense into the larger picture of science as a whole as seen from the historical point of view.

8. In the fourth place, before the contemporary thinker has undertaken an historical investigation of the thinking of the past, he is necessarily un-

---

5.3:6[a] The human view of the entire cosmos pertains to the field of formal science. It is true that, since all intellectual abstraction is both 'total' and 'formal' (2.4:9[b]), the distinction between the total abstraction of classical science and the formal abstraction of historical science is basically conventional: total abstraction, in the sense that it is also a development of the understanding of its possessor, is formal abstraction; and formal abstraction, in the sense that it is a recognition of its own intellectual medium, is total abstraction. Thus, all scientific concepts are in the third dimension of intellectual perception, as opposed to the 'flat field' of sensory presentations (2.6:6). Nevertheless, because classical science does not in its formal approach make use of the most universal concepts of human understanding (2.3:18[a]), its characteristic abstraction has a two-dimensional configuration, as opposed to the three-dimensional structure of formal science (7.8:3).

The concepts of classical science are formed, as it were, by the intersection of pairs of infinite lines (proximate genera and specific differences) in a purview that does not technically include the whole concept of reality (5.2:22). Using the analogy of the telescope, we may say that all scientific understanding is like the perception of distant things as reflected in the mirror of sense perception and focused in the eyepieces of intellectual concepts. The field of classical science receives a dimension of depth from the functional variation of its 'eyepieces' (open classes of things), but the depth is limited, and attention is concentrated on the field as flat. Historical science uses the formal concepts of reality and other concrete concepts to bring the images in its field closer to the eye of human intelligence (like the use of shorter eyepieces on the telescope) and thus is able to look more deeply into objectivity as present to the viewer (4.2:13). In this sense the viewpoint of historical science adds a dimension of depth to the relatively flat field of classical science.

The concrete concepts of historical science (5.2:21) are, therefore, an intellectual medium that reflects the meaning of realities invisible in the focus of classical science. Paradoxically, this includes the 'terrestrial' meaning of one's own self as a psychological subject, and of one's own past, present, and future, in its relation to reality as a concrete whole. But it more directly regards those objective 'galaxies' of reality that shine with massive intelligibility in the distant field lying beyond the range of classical science.

aware of the amount of achieved understanding that lies buried under the 'modern' focus. Ignorance of several forms of thought that are now out of vogue and out of the focus of attention, but still operative at least negatively in the problematic of contemporary thought, has made contemporary thinking shallow. It is often for the contemporary non-historical thinker not so much a matter of what he has figured out as of what he has chosen to ignore. And he can appeal to the agreed focus of attention to justify his choice. There is a kind of 'gentlemen's agreement' among many contemporary thinkers that the forgotten forms of thought are to be considered non-existent.

9. Historical science is the suitable instrument for studying the development of the 'modern outlook' and placing it in a more intellectual context. It can reconsider what is valid and what is invalid in steps of development now taken for granted, but not rationally so. It is able to find at each stage of the development those elements that were rejected by the thinkers of the time but not with sufficient reason. It can determine the presuppositions of the present outlook and examine them scientifically, bringing to light also their historical meaning. By this process what is now assumed and taken for granted by the modern thinker can be elevated to the level of scientific conviction. Part of the 'empirical residue' to be examined historically is made up of the now rejected classical philosophy. Another part is contained in the 'intuitive' reasoning of thinkers who were following other methods.

Article 6

THE FIRST DICHOTOMIES OF HISTORICAL SCIENCE

6.1. Historical Method

1. A method is a set form of procedure. A scientific method is the form of procedure by which the science is acquired; it is determined objectively in the sense that its central and essential features are inescapable, and subjectively in the sense that there are also optional features of personal or conventional nature depending upon the particular genius or ideosyncrasies of the respective users of the method. Any method proceeding from intellectual consciousness will assume one or another of three general forms: it will be either a simple art or the science of an art or the art of a science. If the method is a simple art, it pertains to the prescientific area of consciousness. To the extent that it is not a reasoned adherence to the awareness of the real, it is accounted as simple and prescientific art, no matter the amount of unscientific theory it can claim to justify its results. If the method is the science of an art, it is scientific in its form, but artful in its purpose. For example, medicine is substantially the art of making people healthy and keeping them that way, but it is based upon theory of the real, so that a medical student is expected, not only to learn the art of his pursuit, but also the theory, or 'science,' of the art. Finally, if the method is the art of a science, it is artful in form, but scientific in purpose. For example, there is an art to learning mathematics, to performing an experiment, to writing a scientific description. These three forms of method arising from intellectual consciousness are similar and related; in any discussion of method they must be carefully distinguished.

2. Historical method is the art of a science. It is the skill, partly to be discovered, partly to be devised, by means of which the student arrives at a possession of historical science, that is, of historical knowledge both remembered and understood. The implementation of historical method has two main directions of concentration: a) the verification of data constituting the material object of historical science; b) the verification of the reasoning according to which historical conclusions about the material object have been derived. In recent centuries some ingenious methods have been worked out for testing the veracity of statements and signs in books, documents, and monuments from the past. Overall method has been advanced by these developments on the level of the verification of data. But historians have not given equal attention to the perfecting of special methods for the verification of historical reasoning about the data admitted as true, and this neglect has lowered the scientific value of the historical conclusions drawn. It is not simply a question of recourse to logic, for the science of logic, as it is commonly understood, is a function of classical science. In the present instance, it is a matter of rendering precise and explicit the valid operation of historical reasoning as distinct from the reasoning of classical science.[a] The historical scientist must become familiar with the operation of human reason as it derives understanding from the genetic relationship of concrete, individual objects.[b] The recognition of this characteristic act of reasoning requires the experience of insights in a sufficient number of occurrences. An analysis of the act of historical reasoning requires detailed treatment. Within the focus of the mind concentrating upon the past as its field of objectivity, the historical scientist must investigate the intuitive aspect of his science and its complex relationship with forms of mediate inference,[c] he must become familiar with his specifically historical

---

6.1:2[a] *Supra*, 5.2.
6.1:2[b] *Supra*, 5.1:3; 5.1:4.
6.1:2[c] *Supra*, 5.2:17; 5.2:18; 5.2:21; and *infra*, III.18.

present as the specific medium of his science, and its complex relationship with his general present and with the series of relative presents in the past he is contemplating.[d] But this investigation is not a meditation upon the mood of his consciousness; it is rather a realistic inquiry into facts of concrete import.

3. We may thus distinguish three major phases in the implementation of historical method. The first phase is reconstruction of the historical facts in the form of verified events. The second phase is the historical understanding of the reconstructed historical facts. These two phases are reciprocal and mutually interdependent. The historian who has not developed his capacity of historical understanding will tend to give an incorrect and distorted reconstruction of the facts on the basis of his unscientific understanding. The historian who does not strive to work with verified facts will not achieve much understanding of historical reality. A third phase in the implementation of historical method is the exploitation of the knowledge and understanding acquired through historical science by its use in solving problems outside of the field of the past. Historical science is not only speculative; it has a practical application. The proper employment of historical insight in the assessment of present situations is a special skill of considerable importance. History is not a vision of the future, but the ability of a good historian to anticipate the future in many situations is significantly greater than that of the man lacking historical insight. The historian must maintain a clear distinction between the three phases of his method and not attempt to use the third beyond its rational limit.[a]

6.2. The Art of Historical Science

1. Historiography is the art by which historical science is expressed.[a] There is an essential identity between written history and the objective aspect of historical science, because written history embodies in narrative form the demonstrated conclusions of the science. But the identity is not complete, since the faithfulness with which the science is reproduced in the narrative depends also upon the artistic ability of the writer. This does not mean that the historiographer is entitled to falsify reality in any way. He is obliged to reproduce the truth. It means, however, that the perfection with which the historian depicts the past depends not only upon the exactness with which he possesses the knowledge, but also upon his ability to express with precision what he knows.

2. One who collects and retells the narratives of historians is himself an historian of a sort. All historians begin in this fashion. Children learn history by memorizing it. But history is a discipline of the mind which requires more solid food as it begins to grow. To deny that the pictorial character of chronology is history at all would only falsify the issue. It is the beginning of history. The chronicle bears to history the same relationship as arithmetic bears to mathematics. The chronicle is the first form of history.

3. The historiographer is a scientist before he is an artist. His written expression of historical truth is only the finished product of his work. His research does not proceed according to the direction of time, but backward into time. He seeks the beginnings of conclusions already known to him. His knowledge of conclusions determines his selectivity both in research and in writing. He wants to know the processes according to which things have grown to be what they are. These processes are brought to light in terms of series of events. Events are also the exemplary cause, or model, of history. As a chronicler, the historiographer seeks to reproduce the events as exactly as possible and to set them in chronological order. As an historical scientist, he seeks to set forth the historical meaning of these events.

4. The intelligible pattern of past events does not spring ultimately from the mind of the historian. The historian does not create history; he simply recognizes it. Since history is an art as well as a science, there is a creative element involved in the *writing* of history, but the object of history is intelligibility extrinsic to the mind which

---

6.1:2[d] *Supra,* 4.2.
6.1:3[a] Cf. *infra,* II.5.
6.2:1[a] Cf. N. Petruzellis, "Storia e storiografia," in *Enciclopedia filosofica* (2d rev. ed., Florence, 1967), Vol. VI, 192-203, with copious bibliography (201-203).

sees it. The historian does not read meaning into history, for meaning is not something which the historian adds to the data he has compiled. Meaning lies potentially in the events of the past, waiting to be penetrated by the mind of the historian who interprets those events. There is no such thing as a fact without a meaning; there is no such thing as a recorded fact without an interpretation. Inasmuch as history is an understanding of the genetic meaning of events, it depends upon a modification of the historian's intellect. In this sense history is relative to the human mind concerned. But the events themselves are not relative to the mind of the historian; neither are the intelligible meanings which they contain. Historical meaning, rightly recognized, pertains to objective reality.

5. The primary purpose of any work of historical science is the exposition of historical truth. The thing which establishes a work of historical science in its literary genre is the apt expression of historical truth. This is the *finis operis* of scientific history. As long as the work remains truly scientific, other extrinsic or practical aims of the historian will not change the intrinsic finality of the scientific work as such.[a] The fact that the historian must use selectivity both in his choice of sources and in his narration of facts does not deprive him of the chance to be perfectly objective. Since sciences differ in the formality of their approach, certain objects will easily fall within the purview of a particular science, while others will remain on the periphery. Some events are obviously good material for the historian, while others are unlikely. The historian must ignore meaningless details if he is going to construct a meaningful history. A subsequent historian may find great meaning in details which the earlier writer has disregarded. Such a finding of new meaning implies either new evidence found in the meantime or the acquisition of a more solid scientific medium by the later researcher.

6.3. *The Broadest Division of History*

1. The matter of history is things past. The form of history is the past as such.[a] Since the material object of the past as such has no real existence outside of knowing minds, it is the knowledge of the past as such that constitutes *actual history*, while the past as present, or the present as it turns into the past, is properly called *potential history*.[b] The past of history is extended in the dimension of time. Actual history is the knowledge of things according to the dimension of time; it is the knowledge of events in their sequence of development, proceeding descriptively from the more remote to the more recent past and intuitively from the more recent to the more remote past.[c]

2. History depends upon the use of imagination. *Real history* uses the imaginative memory in such wise as to adhere as closely as possible to the real development of real things as it really took place. *Fictional history* uses the imaginative memory in such wise as to depart wittingly or unwittingly from the real development of real things as it really took place.[a]

3. *Historical science* is the knowledge of the real past.[a] Formally, the science of history is the knowledge of the real past as really past. The formal being of historical science as science is the intelligibility of the real, and this formality is generic, applying equally to all science. The formal being of historical science as historical is the intelligibility of the really past as past, and this formality is specific to historical science. Real history is practically to be identified with historical science except that the science implies a studious interest in the reality as such.

4. Historical understanding may be defined statically as the ability to perceive the genetic meaning of things. It may be defined dynamically as the movement of the intellect in terms of genetic meaning, and this movement may be theoretical or practical. If the movement of the intellect is theoretical in the sense of the recognition of greater intelligibility, historical science is the development of the intellect itself in the area of genetic meaning. If the movement of the intellect is practical, historical science is the discourse of the mind as it concentrates

---

6.2:5a *Infra,* II.25; II.42; III.11; IV.5.
6.3:1a *Supra,* 4.1:2.
6.3:1b *Supra,* 3.2:10; *infra,* II.10.

6.3:1c *Supra,* 6.2:3.
6.3:2a *Supra,* 3.1:10; 3.1:11; 3.2:1.
6.3:3a *Supra,* 3.2:9.

upon genetic meaning in order to arrive at a practical purpose.[a]

5. Merely chronological history (*chronology*) is limited to the recording of events in their material circumstances and succession. Explanatory history (*historical explanation*) includes an explicit explanation of the events recorded. The difference between the two is one of degree, for there is always some historical interpretation in even the barest chronology. In fact, the very recognition of events and their being set in chronological order requires some awareness of their genetic relationship, and the expression of genetic relationships is the essence of historical explanation.

6.3:4[a] Historical consciousness tends to organize historical knowledge into different disciplines; the organization tends to become conventional.

There is a major division between natural history and human history. The investigation of natural history is now left largely in the hands of 'natural scientists,' while professional 'historians' concentrate upon the human past. This division is to a certain extent necessary and to a certain extent unfortunate and unnecessary (*infra*, II.10). In fact, natural scientists have tended not to bring to their study a fully developed ability to recognize historical meaning in the history of the physical universe, especially as regards genetic intelligibility. A 'headless history' has resulted, i.e., history without its finality.

The division of human history into the historic and the prehistoric is illogical in its terminology and is seldom used any more. But there is a disciplinary division between the areas of paleontology, archaeology, and anthropology on the one hand and civilized history on the other.

The division of Western history into ancient, medieval, and modern is superficial at best. Cf. E. Kevane, "Augustine and Isocrates," in *American Ecclesiastical Review*, 149 (1963) [301-321], 316-321; N. Berdyaev, *The End of Our Time*, trans. by D. Atwater from the Russian (New York: Sheed and Ward, 1933), 9-120.

ARTICLE 7

THE LOCUS OF HISTORICAL THEOLOGY IN HUMAN CONSCIOUSNESS

7.1. Sacred Theology According to St. Thomas

1. Thomas Aquinas avers that sacred doctrine is a science in its own right, truly scientific and truly distinct from philosophy and natural science, for sciences are distinguished according to the differences in their formal viewpoint or approach (*diversa ratio cognoscibilis*) and the formal viewpoint of sacred doctrine is God as He has revealed Himself.[a] He notes that God is the subject of sacred science, in the sense that the subject is to a science as the object is to a vital power (for instance, as color is to sight) and the object of a power is whatever comes within the characteristic of that power (*sub cuius ratione omnia referuntur*), but in sacred doctrine all things are treated under the viewpoint of God (*sub ratione Dei*) inasmuch as they are God Himself or at least they are seen in their relationship to God. He adds the further reason that, since the whole science is contained virtually in its principles, the subject of a science is the same as the subject of the principles of the science; but the principles of sacred science are the articles of faith, which is about God. Thus, everything in Sacred Scripture is contained under God, not as parts, species, or accidents, but as in some way ordered to Him. Aquinas admits that the theologian does not know what God is, but he uses the effects of God either in nature or in grace as a substitute for a definition and thus demonstrates things about the divine Cause in terms of his effects.[b]

2. St. Thomas points out that sacred doctrine is a unified science, for the unity of a science stems from the unity of its formal object or approach, which is the 'viewpoint' of God as He has revealed it.[a] Sacred science develops from principles known, not by the natural light of human understanding, but by the light of the knowledge possessed by God Himself.[b] Sacred Scripture, he says, is unified by the unity of its formal object, for everything it contains is presented from the viewpoint of its relationship to what is divinely revealed.[c]

7.2. The Act of Faith According to St. Thomas

1. In the explanation of Thomas Aquinas, to believe means "to think with assent" (*cum assensione cogitare*). The verb 'to think' can be taken in any of three principal ways: a) most generally, it may refer to any act of consideration on the part of the intellect; b) more properly, it refers to an act of consideration by the intellect together with questioning and implies development towards the certitude of intellectual vision; c) the act of the cogitative power. The second way contains the whole distinctive character of faith in its definition as "thought with assent." [a] St. Thomas, following Aristotle, maintains that the movement of consciousness consisting in deliberation about universal meanings (*intentiones universales*) belongs to the intellectual realm, while deliberation about particular meanings (*intentiones particulares*) belongs to the sensitive realm of consciousness. This deliberation about particular meanings on the lower level of sensitive consciousness is the act of the cogitative power. St. Thomas sees this act as peculiar to man in this that, while there seems to be no difference in the way that sensible forms present themselves to

---

7.1:1a *S.Th.*, I, q. 1, art. 2.
7.1:1b *S.Th.*, I, q. 1, art. 7.
7.1:2a *S.Th.*, I, q. 1, art. 3.

7.1:2b *S.Th.*, I, q. 1, art. 2.
7.1:2c *S.Th.*, I, q. 1, art. 3.
7.2:1a *S.Th.*, II-II, q. 2, art. 1.

lower animals and to man, yet there is a difference in the way that meanings are recognized, because the lower animals see meanings (*intentiones*) only by instinct, but man by some sort of comparison (*per collationem quandam*). These particular meanings are stored in the memory, and the very notion of the past is one of them. The faculty of recognizing particular meanings is called the 'estimative power' in lower animals, but in man it is called the 'cogitative power' or 'particular reason' (*ratio particularis*), for it compares particular meanings as intellectual reason compares universal meanings. Man, like the animals below him, has a sense memory by which he can immediately remember past things; in addition he has the power of 'reminiscence' by which he can seek syllogistically, as it were, into the memory of past things according to their individual meanings.[b]

2. The act of faith, in the exposition of Aquinas, is midway between the thinking that leads the will towards consent to believe and the thinking that leads the intellect towards understanding what one already believes.[a] The assent of faith is the act of the intellect according as it is determined by the will to make the affirmation. Assent seems to be an act of the will, but in this case it is taken to be an act of the intellect, since the reason for the assent is the testimony of God.[b] The object of faith is prime truth. While prime truth constitutes the formal character of the object of faith as the intellectual medium on account of which one assents to what is believable, it also serves as the object, or goal, of the will of the believer. The object of the will is the good, but in this case the good of the will is prime truth, so that the will, in moving towards its end, is moving the intellect towards the prime truth which is the proper end of the intellect.[c]

3. St. Thomas teaches that only rational creatures can have an immediate ordering to God, for lower creatures than man can at best know singular things, but rational creatures can know the universal character of the good and of being (*rationem universalem boni et entis*) and are thus ordained to the universal principle of being. But man cannot attain his destiny of knowing God as the consummation of goodness and of being without believing in God. Even for the acquisition of scientific knowledge on the natural level a certain amount of human faith is necessary, for the pupil must believe his teacher in the earlier stages of the learning process in order that he may move on to a perfect possession of the science. All the more necessary is faith for the acquisition of that perfect supernatural science which consists in the beatific vision.[a]

7.3. *The Gift of Understanding According to St. Thomas*

1. In the exposition of St. Thomas, faith is a virtue perfecting the human intellect, and therefore perfecting the human understanding. It is impossible to exercise the virtue of faith by believing in something that is false.[a] What is believed by faith cannot at the same time be scientifically known (*scitum*), because scientific knowledge is derived from principles that are seen in the sense that they are *per se* known, and one cannot believe in what is in sight. Nevertheless, theology is a science, because it reasons from principles of faith, namely, from the authoritative words of Sacred Scripture, and it thus proves for believers from the principles of faith, just as science proves for everyone from naturally known principles.[b]

2. St. Thomas observes that faith is more certain than the gifts of the Holy Spirit, for it is their fundament. Of the seven gifts of the Holy Spirit, wisdom, understanding, and knowledge (*scientia*) regard necessary things and are therefore more certain than the others. Even as natural virtues wisdom, understanding, and knowledge (*scientia*) are less certain than faith, although from the viewpoint of the knowing subject (*ex parte subiecti*) they may seem to be more certain; such an impression of the greater validity of natural insight and knowledge

---

7.2:1[b] *S.Th.*, I, q. 78, art. 4.
7.2:2[a] Aquinas, T., *Scriptum super Sententiis magistri Petri Lombardi* (Lethielleux edition: Paris, 1929-1947), *In III Sent.*, dist. 23, q. 2, art. 2 (Vol. III, p. 726).
7.2:2[b] *S.Th.*, II-II, q. 2, art. 1.
7.2:2[c] *S.Th.*, II-II, q. 4, art. 2.
7.2:3[a] *S.Th.*, II-II, q. 2, art. 3.
7.3:1[a] *S.Th.*, II-II, q. 1, art. 3.
7.3:1[b] *S.Th.*, II-II, q. 1, art. 5; I, q. 1, art. 2.

is subjective (*quoad nos*) and relative (*secundum quid*). But the gifts of knowledge, understanding, and wisdom surpass faith in the order of manifestation.[a]

3. Understanding in general denotes a certain intimate knowledge (*intus legere*) penetrating beyond exterior and sensible qualities to the essence of things. The object of understanding is what a thing is (*quod quid est*). The gift of supernatural understanding is a supernatural light enabling its possessor to penetrate beyond the limits of natural understanding. The gift of understanding introduces the believer to a glimpse by supernatural light of the spiritual things hidden within the words and figures of Sacred Scripture. The gift of understanding lights up these spiritual objects in a manner similar to that in which natural essences are illuminated by the first principles of the natural understanding. The gift of understanding is called a 'divine instinct' (*instinctus divinus*) in the sense that this gift is instilled by God over and above what can arise from the natural dynamism of human reason. Natural understanding grows through the process of discursive reasoning, which begins from natural understanding and terminates in deeper understanding. But the understanding of revealed truths is not reached by discursive development of the natural understanding.[a]

4. The gift of understanding does not afford *perfect* insight into the truth of the propositions that pertain directly and essentially (*directe et per se*) to the object of faith. Thus, it is not possible for the believer in this life to 'understand' that the One God is Three Persons or how the Second Person of the Blessed Trinity could become incarnate. For St. Thomas, perfect insight is the comprehension of the essence of a thing. This he excludes from what is directly and *per se* the object of faith, but he admits the occurrence of imperfect understanding, which he describes as insight into the fact that external appearances do not contradict the truths believed by faith.[a]

5. St. Thomas observes that the gift of understanding pertains both to speculative and to practical insight; it is first and principally in the speculative intelligence, and from there it is extended to practical understanding.[a] It is distinct from the other gifts of the Holy Spirit, because three of the gifts pertain to the appetitive drives, namely, piety, fortitude, and fear of the Lord. The other four pertain to supernatural cognition based upon faith. But understanding is different from wisdom, counsel, and knowledge (*scientia*), for it regards apprehension and they regard judgment. Right judgment about divine things pertains to wisdom; right judgment about created things pertains to knowledge; application to singular things pertains to counsel.[b]

6. To the gift of understanding corresponds the sixth beatitude, "Blessed are the clean of heart, for they shall see God" (Matt 5:8). There is a dispositive cleanness of heart by which the appetites are purified of inordinate affections, and this is effected through the appetitive virtues. For the vision of God there is required also a complementary (*quasi completiva*) cleanness according to which the mind is not contaminated with phantasms and errors impeding spiritual thought about God. By the imperfect vision of God which is the gift of understanding in the pilgrim state we can know what He is not; the more fully we realize how much God exceeds whatever our understanding can grasp concerning Him, the more perfectly do we know Him. The beginning of understanding which is founded on faith in the truths that God has revealed about Himself consists more in the realization of what God is not than of what He is.[a]

7.4. *The Gift of Knowledge According to St. Thomas*

1. The gift of knowledge (*donum scientiae*) is the possession of correct judgment concerning what is to be believed and what is not to be believed. This certainty of judgment is not said to be arrived at by a process of discursive reasoning but rather in a simple way that more closely resembles the divine simplicity.[a] According

---

7.3:2[a]   *S.Th.*, II-II, q. 4, art. 8.
7.3:3[a]   *S.Th.*, II-II, q. 8, art. 1; I-II, q. 68, art. 1.
7.3:4[a]   *S.Th.*, II-II, q. 8, art. 2.

7.3:5[a]   *S.Th.*, II-II, q. 8, art. 3.
7.3:5[b]   *S.Th.*, II-II, q. 8, art. 6.
7.3:6[a]   *S.Th.*, II-II, q. 8, art. 7.
7.4:1[a]   *S.Th.*, II-II, q. 9, art. 1.

to the cognitional theory of Aquinas, any cognitive disposition (*quilibet cognoscitivus habitus*) formally regards its own medium and materially regards what is known through its medium. The gift of knowledge is distinguished from the gift of wisdom in this that the gift of knowledge pertains more to knowing God through the medium of created things, while the gift of wisdom pertains more to knowing God in terms of divine things. On the natural level cognition of things not surpassing man is called knowledge (*scientia*), and this term implies certitude of judgment reached through the medium of secondary causes. On the supernatural level to know what is to be believed pertains to the gift of knowledge, and this gift regards the knowledge of human and created things. Faith, it is admitted, is about divine and eternal things, but faith itself is a temporal thing in the consciousness of the believer (*in animo credentis*). To know the things believed according to themselves pertains to the gift of wisdom, and this gift implies union with the things known. It corresponds to charity, which joins the mind of man to God. Hence, the gifts of knowledge and wisdom are both science in the more general sense, but, because wisdom is the science which reaches the certitude of its judgments in terms of the highest and most ultimate cause, it is given the honor of a special name.[b]

2. The gift of knowledge is primarily speculative, but practical by extension. The reason for its extension to the practical is that the prime truth which is its speculative object is also the ultimate end of human activity.[a] To the gift of knowledge corresponds the beatitude, "Blessed are those who weep, for they shall be comforted" (Matt 5:5). The weeping is necessary, because those who seek creatures as their goal will lose the true goal of their lives. Creatures are to be regarded as good for the knowing subject only to the extent that they refer him to God. Weeping over one's own past mistakes pertains to the gift of knowledge. The joy which comes from creatures as referred to God corresponds to the gift of wisdom.[b]

7.5. *Revealed Truth*

1. To begin my reflections on the notion of historical theology I call to mind the fact that scientific truth is at one and the same time correspondent and coherent. It is correspondent for the reason that it is a formulation of a mode of meaningful existence known by compelling evidence to be self-standing apart from the subjectivity of the knowing subject.[a] It is also coherent for the reason that it fits together into an intelligible and a reasonable pattern not contradictory to the laws of human understanding.[b] All expressions of thought which correspond in this manner to reality as self-standing outside of the knowing subject and cohere with everything else known about it are themselves 'objectifications' of scientific consciousness. They are always able to be tested against the reality to which they correspond, and, if their coherence with the rest of the scientifically known is not perfectly seen by the knower, he must at least have reasonable grounds for presuming that more complete knowledge will fit them coherently into the harmonious unity of meaning.[c]

2. Sacred theology is based upon the extramental existence of revealed things. Corresponding to the extramental meaning of revealed things is the power of supernatural faith in the knowing subject. The power of faith is not another faculty added to the intellect of man. It is rather a new ability to see, instilled by an act of God into the intellect of man. The existence of revealed truth cannot be shown except by indirect arguments to a person who does not have the power of faith. Since faith is a power of intellectual sight different from the power of 'natural reason' but, like natural reason, analogous to the power of physical sight, the existence of the object of faith is evident only to those who have the power of faith, just as the existence of natural meaning is evident only to those who have natural intelligence and the existence of color is evident only to those who have the power of physical sight. On each of these three levels, the supernatural, the rational, and the physical, it is

---

7.4:1b *S.Th.*, II-II, q. 9, art. 2.
7.4:2a *S.Th.*, II-II, q. 9, art. 3.
7.4:2b *S.Th.*, II-II, q. 9, art. 4.

7.5:1a *Supra*, 2.4:8.
7.5:1b *Supra*, 2.6:12; 2.6:36; 2.6:38.
7.5:1c *Supra*, 2.3:2; 2.4:1; 2.6:9; 2.6:15b.

possible in some instances for one who does not have the power to respect its existence in those who do have it, but an intuitive awareness of the existence of the object is not attainable by those who do not have the power to which it corresponds. Faith is not a mode of consciousness opposed to the contemplation of the real; the object of faith forms rather a continuum with the reality of the rationally conscious subject as such; faith is a mode of knowing of which the believing subject can be realistically aware at the same time that he is rationally conscious.[a] The realm of faith and the realm of natural knowledge do indeed regard distinct groups of objects visible in the consciousness of the same knowing subject, and the focus of attention will therefore always be concentrated more on the one or on the other. This unity of consciousness in the believing subject is the origin of sacred theology, for sacred theology is faith seeking comprehension, and comprehension takes place by a certain expansion of the act of faith into the area occupied by natural understanding.[b]

3. If man by the use of his intellect and will is to attain an end or fulfillment which is above his nature and absolutely unknowable on the level of his own nature, then what is absolutely necessary is the revelation by God of the existence and direction of this goal.[a] The intervention of the charismatically guided magisterium of the Church is relatively necessary to the extent that individual men could not by their individual effort determine what has in fact been revealed or how it is to be understood. That is why revelation has occurred within the matrix of the Church and exists as a social as well as a literary fact. The intervention of professional theologians can be useful in providing expertise both to individuals and to the official Church in determining the locus and meaning of revelation. And, since the attainment of the goal of man requires the use of intellect and will by each individual to be saved, that personal theology which consists in the comprehension by the individual of the existence and meaning of his goal beyond the confines of his natural existence is also absolutely necessary. Personal comprehension has the same degree of necessity as has divine revelation, for individual understanding is the formal actuation of a revelation that would otherwise remain meaningless.[b] Revelation is not sacramental in the sense of having a sanctifying power outside of the order of consciousness. It is psychological and essentially ordered to comprehension within human consciousness.[c]

7.6. *Faith and Reason*

1. The act of faith is transient; the virtue of faith is its abiding element. Charity is the supreme virtue; it is higher than faith. Love is entitative in the sense that it transcends the cognitive order; it is the love of charity that gives life and value to faith. But the cognitive must be primary in the cognitive order, and the present discussion as a discussion obviously pertains to the cognitive order. Cognitive objects must therefore be the first discussed, and appetitive objects in relation to them. Thus, in asking ourselves what is meant by faith and by reason, by knowledge, understanding, and wis-

---

7.5:2a Reality is a continuum whose several dimensions can be simultaneously perceived by a single knowing subject (*supra*, 2.6:6; 2.6:15; 2.6:16). Following the system of Henry Margenau (*supra*, 2.6:6a), I have called the dimension of intellectual perception the 'third dimension' of reality. It should be clear, however, that the field of sensory perception may also be regarded as having the three dimensions of length, breadth, and thickness, and in this frame of reference the level of intellectuality is correctly designated the 'fourth dimension' of reality. The infused theological virtue of faith extends the dimension of intellectual percep-

tion by inserting a proximate formal medium (*lumen quo*) in which the person knowing by faith sees the meaning (remote formal object) of things whose reality is too distant from natural man to be visible in his unaided understanding.

7.5:2b The act of faith, as an act of supernatural understanding, is primarily directed at revealed objects. The infused intellectual medium (*lumen fidei*) stands closer to the knowing subject as such than do most or all of the concepts of natural understanding.

7.5:3a Cf. *S.Th.*, I, q. 1, art. 1.
7.5:3b *Infra*, I.23.
7.5:3c Cf. Jn 1:1-14.

dom, we do not wish to lose sight of the fact that primarily we are trying to know and to *understand* what these objects *mean*.[a] In doing so we do not proceed from the assumption that to discover and to know these things is more important than to love what they signify. On the contrary, we seek to know and to understand the objects of faith in order that we may love what they signify.

2. Faith is an elevation of the consciousness of the believer to the perception of supernatural objects. The essence of supernatural consciousness lies in the awareness of the supernatural intellectual object of that consciousness and in the appetitive experience of embracing that object as an object of the will. This supernatural consciousness is cognition to the extent that it is an experience of cognitive objects. It is scientific knowledge (*scientia*) to the extent that it is an affirmation of the real existence of these objects.[a] It is understanding to the extent that the supernatural meaning of these objects is impressed in an abiding way upon the cognitive disposition of the knowing subject. It is wisdom to the extent that it is knowledge (*scientia*) synthesized within the recognized awareness of the supernatural medium itself.[b]

3. When reason is contrasted with faith, the precise opposition is between natural understanding together with its natural development on the one hand and supernatural understanding together with its characteristic supernatural development on the other. Any such supernatural development is a gift of grace, but it takes place as a conscious activity of the believing subject. Hence, reason as opposed to faith is not to be identified either with the act of understanding or with the movement by which understanding develops, and this use of imprecise terminology has led to much confusion. It is, therefore, preferable to contrast natural understanding with supernatural understanding, the light of natural intelligence with the light of supernatural intelligence, and natural reasoning with supernatural reasoning.[a]

7.7. The Object of Theological Understanding

1. The problem of the material object of faith consists in a certain seeming contradiction: the reality to which faith conforms is God Himself, and God cannot be a material object.[a] Here a distinction must be made between two kinds of reality. God is ultimate reality in the sense of intelligibility and formal actuality. Something else must be ultimate reality in the sense of the unshakable fundament against which scientific judgment is always tested as the material criterion. This ultimate material reality for faith is the observable signs and symbols containing divine revelation, the Scripture that cannot be broken, the interventions of God that cannot be explained away. Hence, the material object of faith is the inspired word of Sacred Scripture together with the supernatural manifestations of God in the past and in the present in a way that can be validly ascertained by the believer. This material object of faith forms the bedrock of belief and the immovable foundation of the wisdom that is constructed upon belief.[b]

2. The words (and other signs) of divine revelation emerge on the scene of reality as an abiding addition to the things otherwise comprising the world. These words are not necessarily formed from miraculous elements. They are taken from elements already existing, and they are usually pronounced through the instrumentality of human individuals. Contained in books and scrolls, they are open to the scrutiny also of non-believers; they are like other writings. But they differ from other writings in this that they are the material substratum of a realm of meaning which transcends the meaning of the natural world and invites its reader (or hearer) to rise in his consciousness to a kind of understanding that stands on a level of objectivity surpassing all of the concepts of natural comprehension.[a]

3. The supernatural meaning of the revealed utterances stands 'within' them and 'on the other side' of them, towards God.[a] It is primarily the

---

7.6:1a *Infra*, I.8.
7.6:2a *Supra*, 2.4:3.
7.6:2b *Supra*, 2.6:13.
7.6:3a *Supra*, 2.6:8.

7.7:1a *Infra*, I.21.
7.7:1b *Infra*, I.37.
7.7:2a *Infra*, I.3; I.11.
7.7:3a *Infra*, 8.4:10-12; *supra*, 2.6:15-20.

meaning intended by God as He has revealed something of what He knows about Himself. It is an invitation to journey towards God by ascending along a new dimension of meaning and thus sharing in the understanding and love of God as He really is.[b] Such a journey leads to the personal fulfillment of the pilgrim who undertakes it within the expanses of his consciousness.[c] The deepening of supernatural understanding is the 'automatic' result of the gradual ascent of the incline of meaning 'behind' the revealed utterances.[d] Graduated meaning is contained 'objectively' and extramentally in the revealed sentences; the comprehension of this meaning is an ascent into objectivity; it is the discovery of self-standing supernatural reality; it is the inchoative stage of the vision of God.[e]

4. On the level of faith *as such* (and prior to the *understanding* which is given in faith) the believer sees the simple credibility of what is materially signified by the signs and symbols of revelation (apart from any formally supernatural understanding of the meaning contained in the material expression), and by the act of faith as such he *affirms the reality* of these supernatural phenomena.[a] In order to make this act of faith, he must use his natural understanding positively and negatively. He uses his natural understanding positively to apply the natural concept of reality as a formal medium to the material expression of the intervention of God as an intervention of God and makes the judgment that this intervention is *real*. He uses his natural understanding negatively to exclude doubts as to the reality of the intervention of God. This positive and negative use of the natural understanding resolves itself into special problems, questions, and situations and thus develops into theology as a 'rational' science.

5. The act of faith *as an act of understanding* is spontaneous in itself. It is spontaneous insight into the reality of the object of faith made possible by the illumination of the obscure presence of God in the mind of the believer. The discursive level of sacred theology develops as the faith of the believer is questioned by his natural intellectual consciousness.[a] This questioning arises as the implications of faith are diffused into the natural intellectual life of the believer.[b] But in order even to be able to see the sheer credibility of the object of faith a certain minimum understanding of that object is necessary, for otherwise no conflict with the biases of natural man could be recognized. There is recognition of the meaning of the articles of faith in the sense that they are seen to be propositions not demonstrable

---

7.7:3b *Supra*, 7.5:2a.
7.7:3c *Supra*, 2.2:8.
7.7:3d *Supra*, 2.6:14.
7.7:3e *Supra*, 2.6:18.
7.7:4a *Infra*, I.26.
7.7:5a *Infra*, I.10; I.26. What we usually mean by an 'act of faith' is a judgment that an object presented for belief is real, and this judgment uses the concept of reality. But any act of recognition of a revealed object is an act of faith, and the spontaneity of this act may conceivably transcend much or all of the reflection involved in the natural affirmation of reality.
7.7:5b René Latourelle observes that the act of faith involves both an adherence to what one has been told and a tending towards vision. "Il y a donc dans la foi, comme acte et comme état de connaissance, dans la mesure où elle s'oppose à l'acte et à l'état de la vision, un dynamisme qui suscite la quête de l'esprit. Un début de recherche intellectuelle est toujours présent dans la foi. Inchoativement,

croire implique déjà la théologie." *Théologie - science du salut* (Bruges-Paris: Desclée de Brouwer, 1968), 40; ET: *Theology: Science of Salvation* (New York: Alba House, 1969), 30. Latourelle notes that the theologian applies himself to understand the mystery contained in the words of revelation; it is from faith that theology derives its value as a science of the real (*"comme science du réel"*), for by his faith the theologian holds as absolutely certain a set of truths that he cannot know either from his experience or from his reflection and "his faith assures him that the propositions denote the divine reality itself." *Ibid.*, 41 (ET, 31). But faith is an intellectual experience of the divine reality itself, and I have undertaken in the present writing to show that theological understanding is an intellectual reflection upon the divine reality, not only as the assumed extramental source of what we believe but also as the cognitive medium in the light of which we have already begun to understand the divine revealed reality.

to the natural understanding (and this negative sort of meaning is recognized even by non-believers), and there is a minimum recognition of the presence of God in the form of a principle of understanding which is not within the pattern of the principles of natural understanding and yet is understanding in its own right. Thus, the awareness that belief is meaningful and reasonable (and not absurd) is the formal element of the act of understanding — the possession of the gift of supernatural understanding in its minimum degree. It is the minimum of understanding on the supernatural level which must always be present in the life of faith and extends itself as a supervenient formal medium over the entire intellectual consciousness of the believer. That is what is meant by saying that the conscious life of the believer unfolds *sub luce fidei* (under the light of faith).[c]

6. The cognitive formal object of faith is God *as God*. The act of faith can thus exist only in an intellectual creature. The appetitive object of faith informed by charity is God *as God*. Charity, too, can exist only in an intellectual being. When we say that both faith and charity are directed towards God as God, we are excluding the alternative that they could be directed merely towards effects or manifestations of God that are not God Himself. A non-intellectual creature can be moved towards effects of God, but it cannot move towards God Himself. A non-intellectual creature can know on its own level some of the effects of God, but it cannot know God Himself. Faith is directed towards the knowledge of God Himself; it is the beginning of the vision of God. Because it is the beginning of vision it must somehow be vision, even though obscure and incomplete.[a] Entitatively, faith is the beginning of the vision of God in the sense that it is like an embryo whose mature state will be the vision of God. But cognitively as well faith is already the beginning of the vision of God. It is called faith on the level of natural consciousness, within which the supernatural light is not yet strong enough to give the impression or realization of seeing God as He is. In that sense the believer uses his free will to walk forward in darkness, trusting in the truth of what God has revealed. This walking forward takes place through acts of charity, which are *non-intellectual* in the sense that the movement of the will takes place in a kind of darkness as far as natural consciousness is concerned, but which are *intellectual* in the sense that the same movement is guided by an obscure supernatural light illuminating the object towards which the believer is tending. This obscure supernatural light is the light of faith, and it is necessary for the act of charity, whether as an abiding act or as successive transient acts.

7. If the concept of science is enlarged to include perceptions of the meaning of reality that transcend the limits of deduction from analytical propositions, it becomes possible to consider God as God to be the formal medium of theological science. The understanding of the science is identified with an inchoative perception of God, which is not the same as absolute perception of God and which is compatible with absolute non-perception of the essence of God in the way required for the concept of science elaborated by S. Thomas.[a] It is clear that the grasp of meaning through the medium of the obscure presence of God is incomparably inferior to the grasp of meaning through the comprehension of the essence of God. It is also clear that the grasp of meaning through the obscure presence of God is immensely superior to the grasp of meaning through the unaided natural understanding, and it is this realm of superior meaning that is occupied by the gift of wisdom, extended also to the science of theology.

7.7:5[c] *Infra*, I.3.
7.7:6[a] Cf. *S.Th.*, II-II, q. 8, art. 2.
7.7:7[a] It is the opinion of St. Thomas that wisdom, understanding, and knowledge, *inasmuch as they are supernatural gifts*, derive their entire certitude from the certitude of faith, but, *inasmuch as they are intellectual virtues*, they are based upon the natural light of reason, which lacks the divine assurance of faith (*S.Th.*, II-II, q. 4, art. 8, *ad tertium*). But this does not exclude the occurrence of supernatural understanding: *supernatural* in the sense that the object of the understanding is the revealed God; and *understanding* in the sense that it is seated within the subject-object relationship of natural intelligence.

8. God as God is not a universal. He is the universal cause, but this is universality in the concrete sense. When we speak of the 'universe,' we are speaking of something concrete, not of something abstract in the sense of an open class of things. God is not an open class; He is not a genus or a species; He is not an abstraction. Worthy of our study and consideration is the possibility that the vision of God will be a concrete vision of pure intelligibility, arrived at, not primarily by an ascent of our universalizing power to a grasp of the broadest genera, but by an ascent of the intellect along the dimension of formal abstraction to the state of intellectuality in which we are able to grasp pure intelligibility.

9. The ultimate objects of faith and wisdom are concrete. Actually, even the ultimate objects of natural understanding and natural wisdom are concrete, for universals are but vehicles through which the mind reaches the concrete meaning in things.[a] On the level of faith the supernaturalized intellect does not stop at the locus of universals; it goes beyond them to the objects of meaning which they represent. The object of faith is not the essences of the characteristics of God as open classes (beauty, wisdom, power, eternity, etc.) but God Himself as concretely known in these qualities. Similarly, more proximate objects of faith are Jesus Christ (not the humanity of Christ, not the abstract truth of Christ), the Real Presence of Christ in the Holy Eucharist, the presence of grace in one's own soul, the character of the priestood in a validly ordained priest, and other things of this sort. In brief, the primary object of faith is three concrete divine Persons in God, not the abstract concepts of their relations to one another.

10. The light of faith can be understood in either an incomplete or a complete manner. The incomplete description of the rôle of faith in theology would attribute to the light of faith a merely congruent function in the deriving of theological conclusions. The articles of faith could thus be pictured as comparable to writing on the ceiling of natural consciousness which the believer accepts by his act of faith. His mind cannot penetrate to a perception of the intrinsic truth of what is written; he simply accepts it out of trust in the veracity of God, and he is moved to make this act of trust by an interior grace which elevates his will to embrace the goodness of God revealing these things about Himself. The acceptance of the articles of faith in the incomplete explanation of the rôle of faith would have a merely congruent effect upon the conscious life of the believer in the sense that they would motivate him to perform certain voluntary acts that he would otherwise not perform and to omit certain voluntary acts that he might otherwise have performed. Thus, for instance, since it is an article of faith that Baptism is necessary for salvation, he will seek to be baptized and to persuade others to be baptized. Again, since it is an article of faith that the receiving of the Body and Blood of Christ is necessary for the nourishing of the life of grace, he will seek to receive the Holy Eucharist as a conscious act within the reach of natural reflection. Similarly, he will be motivated to avoid acts that are prohibited by divine law.

11. According to the incomplete notion of faith, the rôle of the gifts of the Holy Spirit is complementary to natural consciousness. The gift of knowledge (*scientia*) is merely knowing what the articles of faith are and how they have been expressed in divine revelation. The gift of understanding is the ability to derive conclusions from the articles of divine revelation through the reasoning process of natural consciousness. The gift of wisdom is judgment according to the highest cause and is therefore the highest level of theology.[a] These three gifts are seen to elevate natural consciousness in the sense that the divinely revealed truths which are known, applied to problems within natural consciousness, and reduced to conclusions synthesized in terms of the highest cause, have a regulatory and ordering influence upon the entire rationally conscious life of the believer. They extend his spectrum of consciousness and give it a certain maturity. Implied in this explanation is also an extraconscious congruency of the life of faith. The virtue of faith informed by the virtue of charity unlocks the door which will

---

7.7:9[a] *Supra*, 5.2:22.

7.7:11[a] Cf. *S.Th.*, II-II, q. 45, art. 1.

be opened after physical death to the vision of God. Human acts performed through the virtues of faith and charity accumulate merit which is registered entitatively on the soul of the believer but is not a phenomenon within the consciousness of the believer. This merit will become a visible factor after the passage through corporal death to particular judgment and eternal life.

12. For a complete recognition of the rôle of faith in theology, faith as belief must be distinguished from faith as understanding. Faith as *belief* is knowledge (material science) derived from adherence to God as He is believed and actually unseen. Faith as *understanding* is vision by the light of the presence of God as He is obscurely seen and only virtually unseen. The light of the presence of God is the primary formal object of theology, defined as the comprehension of faith. Because this light is obscure and mysterious, the comprehension is also obscure and mysterious. The light of natural understanding is the secondary formal object of theology; its use as a medium of thought produces the insights and conclusions of theology as a secondary science, but always with ultimate reference to the primary formal medium (the light of the presence of God), so that insights and conclusions are also drawn, at least indirectly, in terms of the primary formal medium in an obscure and mysterious way.

13. The concurrence and reciprocal interaction of the primary and secondary formal objects of faith produce theological comprehension on a higher and a lower level. On the higher level there is insight and imperfect comprehension regarding the meaning of reality in terms of the recognized immediate presence of God. The resulting 'mystical theology' is organized, not according to the order of natural reason, but in relation to God as the Supreme Intelligible. On the lower level there is insight and imperfect comprehension of the meaning of reality with remote reference to God as God but with proximate reference to the categories of natural understanding, so as to produce theology of the rational kind. Such theology cannot be ultimately reduced to human reason, since it takes its premises from revealed truth and is covered remotely by the light of supernatural understanding, but its proximate formal principles are subject to the analysis of discursive reason.

14. The pure form of mystical theology is called infused contemplation. It is a form of understanding according to which the contemplative knows that he is perceiving the presence of God and experiences intellectually certain effects of this perception in a meaningful way that transcends the principles of natural understanding (although some of the principles of natural understanding are virtually operative). What constitutes the state of infused contemplation is the awareness on the part of the knowing subject that in an obscure and mysterious way he is perceiving reality in God in such wise that as a knowing subject he is conscious and yet the actual dominance of his natural consciousness has receded to a level of virtuality. The insights and vital acts which take place in the state of infused contemplation can only with great difficulty be described in human terms.

15. All primary theology is in some way mystical. That developed stage of primary theology which is usually called the state of infused contemplation is reached at the moment or during the intervals in which reflective awareness in the light of the presence of God becomes the dominant form of consciousness. At this point the contemplation of God as God becomes a distinct and autonomous form of consciousness. But there is also a less developed stage of primary theology in which awareness by the light of the presence of God in the sanctified subject is spontaneous rather than reflective and is covered by the stronger presence of the light of natural understanding, so as to remain subconscious rather than noticeable. This direct and non-recognizable perception of the supernatural presence of God, hidden subconsciously within the observable areas of natural consciousness, seems to be operative in everyone possessed of living faith (the state of sanctifying grace), above all as the believer by the exercise of charity regulates his conscious life in keeping with the demands of faith, but even, perhaps, when the believer is asleep, unconscious, or otherwise deprived of the full use of his conscious powers. Mystical understanding has an autonomous existence in the human intellect distinct from

rational consciousness as such. Its growth is indirectly dependent upon the use of natural understanding, but the mysterious content of this understanding is infused by God, not acquired by human industry.

16. The virtue of faith and the gifts of the Holy Spirit are themselves seated in the intellect of the believer. By the 'intellect of the believer' I mean the entitative foundation upon which arises the primordial dichotomy between the knowing subject and the object of his knowledge. This dichotomy is the first natural expression of intellectual consciousness; it is not the intellect itself. It is not, therefore, inconceivable from the mere fact that supernatural vision is an elevation of the natural intellect that this vision is prior to the primordial dichotomy and need not be expressed in accordance with its laws. Still, it is difficult to draw meaning from the valid analogy of sight with regard to supernatural understanding without presupposing the existence of something like the primordial dichotomy on the supernatural level as well.[a]

17. Supernatural understanding and natural understanding constitute different realms within a single consciousness. The two realms of meaning consist in the perception of reality by the one knowing subject through two different media of understanding. It is necessary that the immediate focus of attention be upon one medium or the other. If the immediate focus is upon the supernatural medium, the knowing subject is in the state of infused contemplation. If the immediate focus is upon the natural medium but with remote reference to the supernatural medium, the knowing subject is in the state of natural contemplation. In the second case, granted that the knowing subject has the virtue of living faith, the minimum of infused contemplation required for the virtue of living faith and perhaps a good deal more is present and operative within the consciousness of the knowing subject, but as subconscious rather than as conscious. The believing subject cannot advert to the presence of infused contemplation except to the degree that he has entered into the state of infused contemplation.

18. Nevertheless, in the state of natural contemplation of the believer possessed of living faith, there is a reciprocal action between the two levels of consciousness. The infused contemplation is subconscious in the sense that it is not recognized in the reflection of the knowing subject; it is supraconscious in the sense that it is a higher consciousness than that with which he is familiar. The exercise of prayer, penance, devout recollection, and other things of this sort is an activity related to the higher state of consciousness, disposing the believer to experience a development of his supernatural vision and of the quality of his supernatural love for God. As faith deepens and charity increases, the conscious awareness of supernatural objects also tends to increase.

19. The natural consciousness of the person sanctified by living faith has value purely in relation to the supernatural consciousness hidden within him. There is thus a strict obligation to regulate one's conscious life in keeping with the finality of the higher consciousness by the studious attempt to discover God with one's mind and to love Him with the entire force of one's will. The interaction of the two levels of consciousness excludes the idea that they represent two entirely different universes of discourse or two worlds so different that there can be no communion between them. The truth is that the intellectual consciousness springing from nature must seek to conform itself and develop itself according to the higher norm of faith, while the higher consciousness of supernatural understanding must tend to penetrate and transform every facet of natural intellectual consciousness. These two tendencies of conformity and transformation are the true basis of the dynamism which underlies the process of sacred theology.

7.8. *The Theology of Concrete Meanings*

1. St. Thomas did not devote much time to the search for the intellectuality of concrete meanings. He took as his working principle the proposition that human understanding is of universals, and he stayed in large part within the limits of this principle. But, in speaking of what he calls the 'cogitative power', he admits that the 'particular reason' that is exercised by it

---

7.7:16[a] *Infra*, I.14; I.19; I.20; I.21; I.25.

is human and higher than the merely instinctive discourse of brute animals.[a]

2. The historical theology emerging in our era takes its point of departure from the recognition that concrete meaning is an intellectual object.[a] The explicitation of a theory of concrete understanding can provide the basis for the resolving of many questions and doubts arising from the failure of universalizing understanding and deductive reasoning to provide intelligent answers in areas standing outside of the limits of that approach.[b] The aim of historical theology is to provide a theoretical basis for the overcoming of these particular doubts and problems and for advance along the dimension of supernatural understanding. The development of historical theology is not seen as a substitute for the universalizing speculation of the past. It is rather a formulation of theory that is complementary to the universalizing speculation of the past and serves to enhance its value by drawing attention to its concrete relevance in the present era.[c] Thus, the thrust of historical theology is positive and expansive, realistic and spiritually sensitive.

3. Formal abstraction corresponds to what has been called the 'third dimension' of meaning, while total abstraction has the two-dimensional character of an infinitely extendible plane.[a] By a parallel use of terminology we may say that 'formal' theory has the third dimension of intellectuality for its object, just as classical theory has in view the first two dimensions. Historical theology has for its proper and precise object that portion of formal meaning that is historical in character. This means that the realm of formal meaning is broader than the field of historical theology and that, consequently, other fields of formal meaning can be distinguished from it. But to the extent that the general realm of formal meaning remains unclaimed by other special disciplines it is maintained and developed as a matter of convenience by historical theology.[b]

4. According to formal theory, the supernatural gift of science has for its minimum content the articles of faith. This content can be immensely developed. In fact, the gift of science is the entire network of conclusions resulting from the gift of understanding in the person possessing it.[a] The gift of science is the material substratum of the gift of understanding: material not only in the sense that it contains the material object of faith but also in the sense that it is the verbal and non-vital expression of a living understanding. This scientific matter can be formulated, written down, and communicated to other persons. But it cannot be formally understood by another person who does not have to a sufficient degree in his own intellect the gift of understanding.[b] A person moves to understanding of the verbal expression of the gift of science in a discursive manner in the sense that he must look successively at the conclusions of this science through the degree of supernatural understanding that he already possesses so as not to obstruct the propensity of his supernaturalized intellect to rise to a new understanding. Such a development of understanding cannot take place without the assistance of interior grace. It is and remains a gift.

5. The gift of wisdom according to formal theory is the increase of supernatural understanding as viewed on the level of supernatural consciousness. Theology in the formal sense is the increase of supernatural understanding within the realm of consciousness; it takes its meaning and value from its essential tie to wisdom. The gift of wisdom is concrete to the full extent: not only as regards its object but also as regards the full commitment of its possessor. It cannot consist in the idle consideration of an intellect separating itself from what it is thinking about; it is rather the focus upon the meaning of the object being considered in its full impact upon the subject doing the considering.

6. Formal meaning exists within the concrete identity of intellectual objects, but its comprehension is authentic only to the extent that it is subjectively identified with the insight of wisdom.[a] Alleged meanings that contradict the intuition of wisdom are not true mean-

---

7.8:1[a] Aquinas, *De Anima*, Lib. II, Lect. XIII, Nos. 395-398. Cf. *supra*, 2.3:18[a]; 7.2:1.
7.8:2[a] Cf. *supra*, 5.2:21.
7.8:2[b] Cf. *supra*, 5.3:1.
7.8:2[c] Cf. *supra*, 5.3:5; 5.3:6; 5.3:7.

7.8:3[a] *Supra*, 5.3:6[a].
7.8:3[b] Cf. *supra*, 5.3:8; 5.3:9.
7.8:4[a] *Supra*, 2.6:9.
7.8:4[b] *Supra*, 2.6:29.
7.8:6[a] *Supra*, 2.6:13.

ings at all. In the real world with which every human being is faced, any outlook or attitude violating the insight of infused contemplation as rooted in supernatural faith cannot itself be a true insight; it is merely a bias slanted against the healthy intellectuality of its possessor. Theories and conclusions, accepted on the level of natural consciousness, that contradict or obstruct the finality of supernatural charity, based upon the insight of infused contemplation, are but distorted reflections of the valid science to which the human intellect is called. Historical theology, as a valid scientific expression of formal meaning authentically understood, must therefore take its origin from formal insights validated on the level of wisdom and extended from there into the sphere of natural consciousness.

7. The twofold task of nascent historical theology thus comes to the fore. On the side of positive understanding it must develop the intellectuality of the believer in areas unreachable by the methods of universalizing theology. On the side of negative understanding, it must refute and eliminate the pseudo-theologies that have arisen in the area of concrete meaning and flourished on fallacies because of the absence of proper methods to discover and refute them. It is the rôle of negative insight to see the falsity of these pseudo-theologies and of the pseudo-sciences in which they inhere, but the process of exposing their errors leads to valid principles and true insights, and in this sense even the negative function of historical theology becomes ultimately positive.

8. The formal origin of historical theology in supernatural wisdom sets up the condition that this branch of theology can advance only in the presence of cleanness of heart. Its system of conclusions, like those of any other science, can be materially understood and mechanically shuffled about in the absence of a formal understanding of them, but the formal understanding requires cleanness of heart — greater in measure as the understanding is deeper — and thus the advance into formal historical understanding must be accompanied by the purification of intention. There is no formal appreciation of the meaning of objects that is not echoed by a proportionate humility regarding the meaning of oneself. There is no formal step forward towards a greater understanding and love of God that does not have a resonance in a step backward from personal pride of place to a position of greater repentance for the unworthiness of one's own record and background.

9. Because faith is located in scientific consciousness, the sound theology derived from faith must also be located in scientific consciousness. Sacred theology is, in fact, merely the system of conclusions derived by the interposition of more restricted intellectual media between the believer as a knowing subject and the general medium of supernatural understanding. The general medium of supernatural understanding is the power of radical intelligence specified by the supernatural light of the presence of God. The locus of sacred theology is primarily in supernatural consciousness and secondarily in natural consciousness, and this order is correct even though the science is often presented by theologians from the viewpoint of natural consciousness. To the extent that expositions are not conformed to the higher insight of infused contemplation, their conclusions are distorted or false.

10. Since historical theology is a branch of sacred theology, it is located primarily in supernatural consciousness and secondarily in natural consciousness. Its conclusions are ordered to the insight of wisdom in the full sense of infused contemplation. The knowing subject of historical theology contemplates the object of faith in the supernatural presence of God in such wise as to derive conclusions related to wisdom. Historical theology differs from propositional theology because of the difference in the kind of secondary intellectual media of which it characteristically makes use. Propositional theology uses universal concepts as the intelligible medium of its understanding and its reasoning process. Historical theology uses concrete concepts as its intelligible medium, and its reasoning terminates in concrete meanings.

11. Historical theology is a state of knowledge located in that area of human consciousness which is historical, scientific, and supernatural. The specific locus of historical theology in human consciousness is the awareness of that scientific objectivity which is brought into view when the historical medium as such is interposed between

the knowing subject of intellectual consciousness and the object of his faith. We may therefore succinctly summarize the results of our search for the science of historical theology in the following formulae:

Science is the knowledge of reality as such.

History is the knowledge of the past as such.

Historical science is the knowledge of past reality as such.

Theological science is the knowledge of revealed reality as such.

The science of historical theology is the knowledge of past revealed reality as such.

Scientific historical consciousness, elevated by the light of supernatural understanding, is the proper theater of historical theologizing, and it is in this area that historical theologians must concentrate their attention in order to produce the scientific results proper to their field.

ARTICLE 8

A REPLY TO BULTMANN'S ARGUMENTS FOR THE NEED OF DEMYTHOLOGIZING

8.0. *The Need of Demythologizing*

1. *In 1941 Rudolf Bultmann projected the 'demythologizing' of the New Testament.*[a] *Section 8.0 of the present article is merely a summary of the need for demythologizing as expressed by Bultmann himself. There is no attempt in this first section to interpret or evaluate the statements of Bultmann except for the elementary interpretation involved in paraphrasing and selection.*

2. The New Testament preaches the event of redemption in a manner which presupposes (*entspricht*) a mythical view of the world. The world-picture of the New Testament is a mythical one.[a] Preternatural forces exert their influence upon the natural course of events, and miracles often occur. Man himself is by no means the master of his own life. Supernatural forces intervene within the very thought and volition of man to tempt him with evil suggestions or to inspire him with good thoughts and heavenly visions. God may open the mind of man to hear his word and endow him with the supernatural power of the 'Spirit,' or he may 'harden the heart' of a man.[b]

3. History is depicted as being both set in motion and guided in its course by supernatural 'powers' (*übernatürlichen Mächte*); it does not unfold in a continuous and unbroken sequence. History is conceived as being divided into two major epochs: the aeon of the present world controlled by the enslaving influence of evil 'powers' — Satan, sin, and death; and the aeon of the coming era of eternal blessedness to be inaugurated by a cosmic catastrophe followed by the raising of all men from the dead, the last judgment of God, and the entry of the justified into eternal life, the unworthy being banished to eternal damnation in hell.[a]

4. To the extent that the kerygma of the New Testament is enwrapped in the language of ancient mythology it is unacceptable to modern man, because modern man is convinced that the mythical view of the world is permanently obsolete (*vergangen*). It would be both unreasonable (*sinnlos*) and impossible for Christian preaching to expect the man of today to accept such a view. It would be unreasonable, because the mythical worldpicture as such is by no means specifically Christian. Mythology is nothing else but the world-picture of a prescientific era. It would also be impossible, because a man cannot force himself to adopt an obsolete picture of the world. Our world-picture is already determined for us by our situation in history. Naturally our picture can be modified and we ourselves can help to change it. But such a modification can be effected only on the basis of a new set of facts so compelling that it excludes the picture held up to that moment.[a]

5. It could happen that, in the first fervor of a new discovery of science, truths contained in ancient myths be overlooked. But the New Testament mythology as such can never be re-

---

8.0:1a R. Bultmann, "Neues Testament und Mythologie," in *Offenbarung und Heilsgeschehen: Beiträge zur evangelischen Theologie* VII/2 (Munich, 1941). Reprinted in *Kerygma und Mythos: Ein theologisches Gespräch* I (Herausgegeben von Hans Werner Bartsch, Hamburg-Volksdorf, 1948 [hereinafter referred to as *KuM* I]), 15-48; ET: "New Testament and Mythology," in *Kerygma and Myth: A Theological Debate* I, trans. by R. H. Fuller (London, 1953) (hereinafter referred to as *KaM* I), 1-44. See *supra*, 1.1:1b.

8.0:2a "Das Weltbild des Neuen Testaments ist ein mythisches."
8.0:2b *KuM* I, 15 (*KaM* I, 1).
8.0:3a *KuM* I, 15 (*KaM* I, 1-2).
8.0:4a *KuM* I, 16-17 (*KaM* I, 3).

vived, because all of our thinking today is shaped irrevocably by science.[a] To ask modern man blindly to accept the New Testament mythology would be unreasonable, and to insist upon its acceptance as an act of faith would be a reducing of faith to works. The fulfilling of such a claim would involve a *sacrificium intellectus* leaving the believer strangely divided and insincere.[b]

6. Science and technology have advanced so much over the past few centuries that no mature individual can any longer take seriously the New Testament view.[a] Since it is no longer possible to believe in the three-storied universe and its mythical components taken for granted by the creeds we have inherited, we cannot honestly recite them without first stripping their truth-content of the mythological framework in which it is enclosed. No mature thinker can any longer suppose that God dwells in a local heaven. "There is no longer any heaven in the traditional sense of the word. The same applies to hell in the sense of a mythical underworld beneath our feet." Hence, man can no longer believe in the legends of the 'descent of Christ' into hell or of his 'ascent' into heaven. Nor can we any longer place our hope in a return of the Son of Man coming on the clouds of heaven.[b]

7. It is no longer possible for an informed person to believe in the existence of 'spirits,' whether good or evil.[a] We now know that the planets are not moved about in the heavens by angelic beings. We know also that disease and infirmity arise from natural causes, not from the influence of demons or evil spells. The miracles of the New Testament can no longer be regarded as authentic, and "to defend their historicity (*Historizität*) by recourse to nervous disorders or hypnotic effects only serves to underline the fact."[b] Even for those phenomena, physiological and psychological, for which we cannot find immediately the cause, we know that eventually the cause will be scientifically discovered.[c]

8. Nor can we any longer maintain the mythical eschatology of the New Testament. It has been exploded by the fact that the parousia of Christ never took place as the New Testament had predicted and that history did not come to a cosmic end. Everybody knows that history has continued on its course and will continue until perhaps some natural catastrophe brings it to an end.[a]

8.1. *The Question of Sacred Authority*

1. While doctrinal authority cannot be invoked in disputation with a writer who denies its validity, Bultmann's difficulties can be resolved only by a person whose mind is enlightened by the teaching of sacred authority. On the peripheral level of material science,[a] authority does not pertain to the substance of the defense against

---

8.0:5[a] *Ibid.* Bultmann reaffirms this in "A Reply to the Theses of J. Schniewind," *KuM* I, 136 (*KaM* I, 120). Cf. *supra*, 1.2:2; *infra*, 8.3:12[f].

8.0:5[b] *KuM* I, 17 (*KaM* I, 3-4).

8.0:6[a] "*Welterfahrung und Weltbemächtigung* sind in Wissenschaft und Technik so weit entwickelt, dass kein Mensch im Ernst am neutestamentlichen Weltbild festhalten kann und festhält." *KuM* I, 17 (*KaM* I, 4).

8.0:6[b] "Kein erwachsener Mensch stellt sich Gott als einen oben im Himmel vorhandenes Wesen vor; ja, den 'Himmel' im alten Sinne gibt es für uns gar nicht mehr. Und ebensowenig gibt es die Hölle, die mythische Unterwelt unterhalb des Bodens, auf dem unsere Füsse stehen. Erledigt sind damit die Geschichten von der Himmel- und Höllenfahrt Christi; erledigt ist die Erwartung des mit den Wolken des Himmels kommenden 'Menschensohnes' und des Entrafftwerdens der Gläubigen in die Luft, ihm entgegen (1. Thess. 4,15 ff.)." *KuM* I, 17 (*KaM* I, 4).

8.0:7[a] "Erledigt ist durch die Kenntnis der Kräfte und Gesetze der Natur *der Geister- und Dämonenglaube.*" *KuM* I, 17 (*KaM* I, 4).

8.0:7[b] "*Die Wunder des Neuen Testaments* sind damit als Wunder erledigt, und wer ihre Historizität durch Rekurs auf Nervenstörungen, auf hypnotische Einflüsse, auf Suggestion und dergl. retten will, der bestätigt das nur." *KuM* I, 18 (*KaM* I, 5).

8.0:7[c] *KuM* I, 17-18 (*KaM* I, 4-5).

8.0:8[a] "*Die mythische Eschatologie* ist im Grunde durch die einfache Tatsache erledigt, dass Christi Parusie nicht, wie das Neue Testament erwartet, alsbald stattgefunden hat, sondern dass die Weltgeschichte weiterlief und — wie jeder Zurechnungsfähige überzeugt ist — weiterlaufen wird." *KuM* I, 18 (*KaM* I, 5).

8.1:1[a] *Supra*, 2.6:29.

Bultmann's charges, but an effective response can issue only out of the light of wisdom informed by sacred teaching.

2. The principal purpose of a response to Rudolf Bultmann is its positive formative effect upon the mind of the one who comprehends the response. The principal negative effect of a failure to respond adequately to Bultmann's charges is the corrosive effect of these charges in the minds of his readers. Bultmann's readers cannot afford to toy with his ideas in a manner inexpressive of authentic Christian belief. They must continually have recourse to the sources of belief,[a] finding therein the creative answers to the devastating implications of Bultmannian thought. In that way, a study of Bultmann's writings leads to a reinforcing of genuine faith in the minds of his readers and to the building up of that Christian hope and fortitude which pave the road of charity to the gates of heaven.

8.2. The Question of Science

1. Bultmann's initial argument [a] may be expressed as follows:

The mythical world-picture is obsolete. But the New Testament preaching presupposes a mythical world-picture. Therefore the world-picture presupposed by the New Testament preaching is obsolete.

In corroboration of the minor premise Bultmann points to a number of mythical elements observable in the text of the New Testament preaching: the three-storied universe, the inclusion of angels and miracles in the process of world events, direct interventions of God and supernatural forces in the mind and heart of man. These elements are held to be obsolete because all of the thinking of modern man is irrevocably shaped by science and by the place which modern man occupies in history.

2. Among the questions provoked by Bultmann's argument are the following: What is the place and function of a world-picture in human consciousness? What is the place and function of presuppositions in human consciousness? What is thinking that is irrevocably shaped by science? What is modern man? What scientific proof is there that the elements in the New Testament identified by Bultmann as being mythical are really mythical?

3. Since we have chosen to examine Bultmann's argument from the viewpoint of science, we must question its validity on the level of science. Science is knowledge of the real.[a] It is mediate knowledge of objects given in experience, whose specific thought-medium is the concept of reality.[b] What characterizes *science* as a viewpoint is its awareness that reality is the medium and criterion of its knowledge; what characterizes *scientific thinking* as such is its being based upon objects given in experience and its consistent adherence to the principles of reason contained in the human intellect.[c] Those among Bultmann's readers who are thinking scientifically have the awareness that they are not looking directly at the New Testament across a kind of vacuum, as one might look at the moon or the stars.[d] They realize that between them as knowing subjects and the text of the New Testament stand two by no means perfectly transparent media: the mind of Bultmann, expressed in his exposition, and their own mind.[e] Prior to an examination of the New Testament world-picture, in the order of scientific thought, must therefore come an examination of the mind of Bultmann, as it is expressed in his text, and prior to this examination must come a scrutiny by the reader of his own mind to the extent that it can be made an object of realistic inquiry. Simple scientific reflection will show that the 'modern man' whom Bultmann claims to be convinced that the New Testament world-picture is obsolete can only be in the first place the reader him-

---

8.1:2a By the sources of belief I mean the places in which the authentic teaching of the Church is expressed: the canonical Scriptures, the formulae of Ecumenical Councils, the *ex cathedra* pronouncements of the Popes, and then the graduated teachings of the ordinary magisterium. Cf. H. Denzinger, *Enchiridion Symbolorum*; the doctrinal references footnoting the articles in the Marietti edition of Aquinas' *Summa Theologiae* (*supra*, 2.1:4a), and various other collections and treatises.

8.2:1a *Supra*, 8.0:2.
8.2:3a *Supra*, 2.4:3.
8.2:3b *Supra*, 2.4:8; 2.6:12.
8.2:3c *Supra*, 2.4:1; 2.4:2; 2.4:12.
8.2:3d *Supra*, 2.6:20.
8.2:3e *Supra*, 2.2:2b.

self, in the second place Bultmann, and only in the third place an allegedly valid abstraction called 'modern man.'

4. In the light of this scientific awareness it should be clear that there can be no *direct* observation of the presuppositions underlying the New Testament text. There can be no real vision of a picture projected within the mind of a New Testament writer. What actually takes place in such an imagining is an uncritical comparison of elements in the text of the New Testament with a world-picture formed in one's own mind and then a projection in one's own mind of what one has inferred to have been the world-picture of the New Testament writer. But because the comparison is uncritical, the result of the comparison is not scientific. It is a fact of scientific thinking that one cannot validly examine the presuppositions of another's thought except by a study of the visible results of the other's thought in terms of the consciously recognized presuppositions of one's own thought.[a] Bultmann expresses his argument without adverting to this fact and its implications. He assumes that his reader is convinced of the obsoleteness of the New Testament world-picture prior to and apart from any critical awareness of the *place and function* of the world-picture in his own consciousness. Bultmann's argument cannot, therefore, be accepted as valid scientifically.[b]

5. The image which Bultmann projects as the seemingly obvious modern picture of the world (*Weltbild*) is a flat, two-dimensional object of cognition. He says that, when we include in this picture an objectification of ourselves as elements of the world, we acquire a world-view (*Weltanschauung*). Bultmann, on the basis of his existentialist philosophy, ultimately rejects the validity of this seemingly obvious picture, on the ground that the genuine self of the observer cannot be objectified, but he misses the real cause of its invalidity. The real reason is that on the level of science there is no flat, two-dimensional picture of the world.[a] Flat pictures are creations of art; they cannot be embodiments of science. The 'view' contained in a 'world-view' is constituted by the medium through which the observer looks at the world, and that medium is the third dimension of the world-picture he is observing.[b] To the extent that the observer is not conscious of his medium he cannot think scientifically about the world.[c] Bultmann's definition of a world-view is, therefore, prescientific.[d]

8.2:4a *Supra*, 2.6:11; 2.2:2b.
8.2:4b *Supra*, 2.6:16.
8.2:5a *Infra*, I.19.
8.2:5b *Supra*, 2.6:6.
8.2:5c *Supra*, 2.6:12.
8.2:5d Bultmann presupposes, not only a certain rigid *uniformity* of the 'modern' point of view, but also a radical *conformity* of the modern mentality with his own view of the world, and this to such an extent that contemporary thinkers who disagree with him on this point are summarily disqualified on the ground that "these are not modern men." (Bultmann, quoted in J. Macquarrie, *The Scope of Demythologizing* [London: SCM, 1960], 236.) If modern man were constrained by very force of his modernity to reject as mythical the New Testament cosmology, his act of rejection would be either prescientific or scientific. As prescientific, it would be a simple and direct act of cognition which could be assumed to be true only until corrected by a more refined scientific argument or by a valid scientific intuition, and it also runs the risk of being so influenced by blind passion or crude bias as to become unscientific. If the act by which modern man is said to reject the New Testament cosmology were truly scientific, the medium of his modernity would have to be formulated in explicit terms. Bultmann has attempted to formulate the medium of his thought in terms of the question of existence, and he has reduced the entire problem to this question, so that he himself cannot be immediately disqualified on this point. But he does not raise the question of existence here. He appeals rather to the sophisticated awareness of all modern men, without reflecting upon the medium of this modernity. By drawing attention to a claimed presupposition of the New Testament thinking while ignoring the operative presupposition of modern thinking, Bultmann is inviting the reader to perform a prescientific act.

Presuppositions do not announce their presence; they are recognized by the mind of an inquirer. Since each science approaches according to its own special viewpoint the object of its inquiry, it is clear that the kind of presuppositions it uncovers will be parallel in some way

6. Bultmann's conception of the New Testament world-picture is expressed in the context of his own literary style. A scientific interpretation of what Bultmann says about the New Testament world-picture requires a prior assessment of the literary genre in which Bultmann's thinking is cast.[a] The judgment that the New Testament narrative is based upon a mythical picture of the world means that the literary genre of the New Testament narrative is recognized to be that of mythology. But this judgment is itself situated within the literary genre of Bultmann's exposition. In other places Bultmann touches indirectly upon the problem of his own genre of thought.

But the fact that he never takes up directly the fundamental problem of his own literary genre shows that his exposition of the literary genre of the New Testament is prescientific.[b]

7. Bultmann explicitly recognizes that there can be no interpretation without the framing of specific questions and the use of a specific terminology. He claims that the right question with regard to the Bible is the question of human existence framed in the terms provided by existential analysis, because everyone is impelled to inquire existentially about his own existence, and this is what ultimately determines our approach to all historical documents.[a]

to the basic presupposition of its own viewpoint. A science will examine a presupposition uncovered in a text under study by a process of comparison and contrast with the presupposition of its own viewpoint. In the case at hand, Bultmann has taken the 'modern mentality' to be a kind of viewpoint which sees the invalidity of the cosmological presupposition of the New Testament. If by the 'modern mentality' he means an empty peering into the past, his criterion is scientifically worthless. When he speaks of an obsolete world-picture presupposed by the New Testament, with what valid world-picture of today is he contrasting it? Within the special viewpoints of modern physics, astronomy, etc., the problem assumes an entirely different form, for modernity as such is not an element of any of these sciences. It is an element within history, and its whole meaning lies in its time-relation to the past. The absence of an articulated awareness of the presupposition of modernity renders vacuous an inquiry into the presupposition of antiquity.

8.2:6a See *infra*, Appendix I/B.

8.2:6b Bultmann refers to the world-picture behind the New Testament preaching as though the text of the New Testament were the prime and immediate object of consideration for the reader of his program. But the immediate object of consideration is not the text of the New Testament; it is rather the text of Bultmann. Only on the prescientific level can a reader assume that the text of Bultmann is merely guiding him in a direct examination of the New Testament. On the scientific level it becomes apparent that the text of Bultmann is a literary genre of its own, whose characteristics must be identified and weighed before

its significance in relation to the New Testament can be ascertained. Bultmann has, in fact, made the validity of his program of demythologizing a presupposition of the modern understanding of the New Testament.

The text of Bultmann represents a thought-medium introduced between his reader and the New Testament. The reader can avoid using that medium unscientifically only by first determining its nature and its validity, because the text of Bultmann is constructed according to a particular viewpoint having its own characteristics and presuppositions. The scientific approach begins with a study of this viewpoint in terms of itself and in terms of four of its presuppositions, namely: a) an underlying notion of history and historicity; b) an underlying conception of existence; c) an underlying idea of mythology; d) a general conclusion concerning the results of New Testament research. The demythologizing of Bultmann is determined by these four presuppositions, unified into a single viewpoint by his unique conception of faith. The first presupposition concerns the relationship of faith to history on the cosmological level. The second concerns the relationship of faith to history on the ontological level. The third concerns the relationship of faith to its expression in human language. The fourth concerns the relationship of faith to the scientific exegesis of the New Testament. It is obvious that such a study must be pursued in terms of a medium of thought. The method of the present study is the critical analysis of this Bultmannian thought-medium in terms of a higher criticism whose principles are consciously formulated as historical science.

8.2:7a *Supra*, 1.3:2; 1.3:3.

8. This reasoning is fallacious. Bultmann's use of the word 'ultimately' is deceptive, because it is not unambiguously defined. What ultimately determines our approach to an historical document is the genre of our thinking about that document. On the level of the *finis operis* of the genre of thought we have chosen, what ultimately determines our approach to the document is the key concept or attitude of that genre.[a] And on the level of the *finis operantis* of our whole concrete self, what ultimately determines our approach to the same document is the command of our will either as enlightened by wisdom or as darkened by attachments opposed to wisdom.[b] Bultmann avers that the right approach to the Bible is identified with the decision to be human in the sense of being a person who accepts the responsibility for his own being as brought to light by existentialist analysis, for otherwise not a word of Scripture has the existential relevance which can make it meaningful.[c] This description of the 'right approach' is erroneous on both levels of finality. It is mistaken on the level of the *finis operis* for the reason that theological thinking is a reality-genre of thought whose approach is ultimately determined by the concept of reality.[d] And it is mistaken on the level of the *finis operantis* for the reason that existentialist analysis as intended by Bultmann does not bring to light the genuine being of man nor does it provide a viewpoint according to which the Scriptures become meaningful.[e]

9. Bultmann distinguishes between reality as the objective representation of the world in which man finds himself and the reality of the historically existing man.[a] In affirming through this distinction that there are two correct ways of understanding the word 'reality' (*Wirklichkeit*), he is searching for the realization that 'reality' is a synthesis of two kinds of objects, one kind being subjective in relation to the other. But, because he has not achieved the synthesis, he does not identify the second kind of object correctly. To the first kind of object he gives the name '*Realität*,' which he uses most frequently in the expression '*empirische Realität*,' and which he thus distinguishes from the other kind of *Wirklichkeit*.[b] This distinction is a step towards philosophical science, but, to the extent that Bultmann uses the other kind of *Wirklichkeit* without having identified its common element with *empirische Realität* in the same genus of reality, his thinking falls short of science into the sophisms of pseudo-science. The other species is not reality until its common element with *empirische Realität* has been identified, so that Bultmann's claim, in the absence of this identification, that existential reality is the prime reality is a merely paradoxical claim which admits in the very affirmation that this is not *really* so.[c] When Bultmann declares that all talk of reality (*Wirklichkeit*) which ignores one's own existence is self-deceit, the 'self-deceit' has to be taken in a very existentialist sense. The reality of empirical sci-

---

8.2:8a   *Supra*, 6.2:5.
8.2:8b   *Supra*, 2.2:7.
8.2:8c   *JCM*, 56-57.
8.2:8d   *Supra*, 2.6:12; *infra*, I.18.
8.2:8e   *Infra*, I.10.
8.2:9a   *Supra*, 1.3:10.
8.2:9b   *Supra*, 1.3:3; 1.3:8; 1.3:10; *infra*, I.17.
8.2:9c   When Bultmann divides '*Wirklichkeit*' into '*Realität*' and another (unnamed) kind of reality, he is obviously dealing with two species of a genus. His logical failure lies in not isolating the specific characteristic from the element which both species have in common. Reality is admittedly more than a universal of the Aristotelian type; it is the real universe. Reality is therefore also a concrete concept, a species of objectivity, which is the most general concept of human consciousness (*supra*, 2.3:18a), containing all objects that are real (*supra*, 2.6:6). Bultmann tries to make the self-awareness of the knowing subject the other kind of reality. The 'existence' of the knowing self is indeed self-evident in human consciousness, in all its uniqueness and singularity, but the knowing self is not a concept and can never transcend the limits of pure experience as such. It has no meaning; it presents nothing for insight; it makes possible no new understanding (*supra*, 2.3:14; 2.6:14). The other 'species' of reality is the intellectual dimension of reality in its objectivity as a visible medium located between the knowing subject and sensible objects (*supra*, 2.6:15). Bultmann's notion of the 'historically existing man' distracts attention from the vision of intellectual objectivity and concentrates it anti-intellectually upon the pure experience of one's self.

ence is real enough apart from the whole discourse of existentialism. Confusion regarding the essence of reality as a genus is shown in Bultmann's statement that we call 'real' (*wirklich*) in ordinary parlance what we can understand in relation to the unified complex of this world, but that such reality prescinds from the reality of man as seen from inside.[d] Since vision is only of objects and man is the subject of his vision, there can be no sight of man from inside.[e] Therefore, both genuine species of reality pertain to the realm of objectivity. The first species of reality is the data of sensory experience. The other species is the meaning of the data as grasped in the intuitive certainty of intellectual experience, and the understanding of it is in the essential constitution of human intelligence.[f] The ultimate criterion of reality is what is given in experience. We do not admit data as real because they fit our scheme of the world. According to valid science, we necessarily admit all data of experience and derive therefrom our notion of the world.[g]

10. While the verification of objects of thought in the data of sense experience is the task of empirical science as empirical, the recognition and comprehension of the second species of reality is the proper task of all science as science. Inasmuch as the content of the concept of reality as a genus is the simple reference to the data of experience, one cannot speak of its second species as an intellectual dimension, for the concept serves merely as the boundary of objects that are known to be real. It is the real objects within the boundary that contain the intelligibility of the intellectual dimension of the human mind.[a] We call the complex of those objects the sphere of reality. Reality is a multidimensional continuum, and fitted into it there are intellectual objects transcending the data of sense. The recognition of these intellectual objects *as real* is a valid intellectual experience made possible by a single generic concept of reality. Each insight into intellectual reality is itself 'real' in the subjective order purely and simply as a unit of human experience; its meaning is real in the objective order as the extramental source of a reproduction in the understanding of the person who has the experience.[b]

11. From the viewpoint of historical science Bultmann's question concerning the New Testament world-picture has to be restated in terms of the concept of reality. Everyone knows that the Word of God is real in the sense that it consists of words written in books and visible to bodily sight. We know that the images commonly suggested to our imagination by those words are real images as components of our consciousness. The prime question is whether the images suggested in the form of historical narrative correspond to events which took place in objective reality. More precisely, the question is whether the description of the events can be seen and understood in terms of our own articulated awareness of reality as our medium of thought. A second question, arising after our sighting of the New Testament narrative within our concept of reality, is whether the genre of that narrative is the same as the genre of our scientific thinking about the narrative, or whether our scientific thinking transcends the *finis operis* of the inspired writing. At this point the particular question of the New Testament world-view comes into the field of vision. Was the New Testament written from the viewpoint of an articulated concept of reality? Do the phantasms suggested by the New Testament text contain meaning that is intended to be seen in the light of scientific concepts?

12. On the level of historical science, the imagery of the New Testament narrative cannot be dismissed without a trial. Bultmann's three-storied universe is just a caricature of what the text portrays. Whether or not heaven and hell, angels and miracles could pertain to the real world is a question that bears reflection. What should be the point of departure of this reflection? If our thinking is to be in the genre of science, we must not begin by testing the plausibility of the discourse in terms of the pleasure it gives

---

8.2:9d *Infra*, I.17.
8.2:9e *Infra*, I.19; 8.2:15.
8.2:9f *Supra*, 2.6:6.

8.2:9g *Infra*, I.18.
8.2:10a *Supra*, 2.3:20.
8.2:10b *Supra*, 2.6:7; 2.6:10.

us, for that is the genre of fiction. We must not begin by asking ourselves whether or not heaven and hell, angels and miracles could possibly occur in the world as it is known to the natural scientist. The only place from which we may begin is from the data, and the only way in which we may proceed is according to the data. We may not dictate what reality can be; we may only recognize what reality is.[a] Bultman says that the possibility of heaven and hell, angels and miracles, is excluded by the scientifically formed mind. This assumption is false. The scientific mind does not exclude any possibility. It simply draws conclusions from the data. The historical scientist cannot declare that an angel could not have appeared to the Blessed Virgin Mary. He can only examine what the data show. If the data show that an angel appeared, he must reason on the basis of the data. Historical reason does not stem from the predictability characteristic of classical laws; it can and must take into account the meaning of the unique,[b] and it is precisely with such meaning that the present question is concerned.

13. Bultmann's exclusion of interventions by God in the mind and heart of man is equally unscientific. What Bultmann means ultimately by 'the mind and heart of man' is the uniqueness of the knowing subject as such. Obviously the conceptual notion of what is totally and absolutely unique and proper in the knowing subject as such excludes any intrinsic composition with outside beings or forces, but this notion does not correspond to anything in objective reality. Bultmann assumes the total independence of this 'self'; he does not establish it. The mind of man does not consist in any sense simply of the knowing self as such. The mind is always composed of the knowing self and the object of his knowledge.[a] The question for science is whether and in what way God can and does intervene in the objective sphere of the mind of man. The heart of man in the sense of the experience of desire does not consist in a subject isolated from the object of desire. The actual existence of the 'heart' is always so composed. Hence, the question for science is whether and in what way God can and does intervene in the volition of man, not from the aspect of what is totally unique and proper to that man, but from the aspect of the object of the volition and the objective capacity to will. Modern scientific man is thus aware that of his very nature he is a divided being: divided between 'himself' in his pure subjectivity and the objectivity which makes self-experience possible.[b]

14. It is Bultmann's opinion that the New Testament, being mythological, is not concerned with the expression of empirical reality, but intends rather to express the reality of existential living.[a] The error here stems from a mistaken notion of mythology. The concept of reality arises in opposition to fancy, illusion, and deception. The reality-genres of human thought are totally segregated from the fiction-genres, of which mythology is one. Of its very nature as a genre of fiction, mythology does not directly intend to express reality at all, although indirectly (*per accidens*) it may express reality too.[b] What Bultmann calls the "reality of existential living" is not reality; it is pure experience not having passed through the test of the concept of reality. Because Bultmann has not grasped the concept of reality as a genus,[c] he cannot assess the reality-content of the New Testament. The first problem is not one of finding the reality beneath the mythology; it is one of determining whether the New Testament is reality or mythology.

15. Bultmann affirms that myth knows a reality (*Wirklichkeit*) different from the reality (*Wirklichkeit*) of scientific regard: myth expresses the awareness that the world in which we live is filled with mysteries. Myth, he says, knows that the world and human life have their ground and limits in a power lying on the other side, but it speaks of this reality in the unsuitable terms of spatial distance.[a] Bultmann does not here succeed in expressing the other species of reality. The awareness that life is

---

8.2:12a   *Supra*, 5.1:4.
8.2:12b   *Supra*, 5.2:18.
8.2:13a   *Supra*, 2.2:2.
8.2:13b   *Supra*, 2.2.
8.2:14a   Bultmann, *op. cit. infra*, 8.2:21a,

page 63 (ET, page 61 = *KaM* II, 185).
8.2:14b   *Supra*, 2.3:20; 2.4:2; 6.3:2; *infra*, 8.4:5; IV.6.
8.2:14c   *Supra*, 8.2:9b.
8.2:15a   *Supra*, 1.3:8.

filled with mysteries is a problem, not an answer. Again, if the myth-maker happens to know that the world and man have their ground and limits in a power lying beyond the world and man, the fact cannot be gathered from the myth, because the genre of myth is not *per se* an expression of reality. But the difficulty here is even greater. What Bultmann means by the 'other side' is identified with the self-awareness of the knowing subject as such, which does not pertain to the realm of reality at all.[b] There is no escape from this dilemma of subjectivity. Bultmann holds that the imagery of the New Testament is a projection into the empirical world of the awareness that our ground and limits lie beyond the world; he says that the Resurrection and other episodes are so projected, but they are not purely imaginary, for they are directed towards an external object, which operates as a reality (*Realität*) within man.[c] But this is not the issue. An external object can operate as a reality within man in the form of a phantasm of the imagination aroused by spoken or written words, but the presence of a phantasm in the imagination does not establish the existence of a corresponding external thing: the imagination can present things which do not exist externally.[d] Again, an external object can operate as a reality within man as an object of appetitive satisfaction, but the experience of appetitive satisfaction does not of itself tell us what the object is or whether the object as it is in external reality corresponds to what it seems to be as an object of appetitive satisfaction. A man's love for his friend is not purely subjective, but the feeling of love does not of itself tell him that his friend exists in external reality. The friend may never have existed; he could be a character in a novel of fiction. When Bultmann defends his characterization of the New Testament narrative as a projection of the imagination, he fails to see that the question is not whether such a projection is purely subjective. The genres of fiction stand on the side of objectivity.[e] The real question is whether such projections are in the realm of objective *reality*. For Bultmann the Resurrection and all other miraculous events narrated in the New Testament are not real events on the other side of the subject-object relationship. According to his conception they are objective as products of the mythologizing process of the imagination, but they are real as identified with the self-awareness of the knowing subject as such. For Bultmann the phantasms suggested by the New Testament narrative are real phantasms, but the things seen in the phantasms are not included in the realm of reality. His basic error lies in identifying the other species of reality with one's own self-awareness as the subject of experience, and thus excluding the formal concept of reality as such.[f]

16. If the genres of fictional thinking are about objects that exist only in the imagination, then the theology of Bultmann is a genre of fiction, for the theology of Bultmann is about phantasms in the imagination which do not support meaning in objective reality.[a] The precise subject of Bultmannian theological reflection is the knowing-self; it does not exist in objective reality. The viewpoint of Bultmann's reflection is that of existential self-understanding; this does not exist in objective reality. The central phantasm of Bultmannian reflection is the vague image of the 'self' who understands 'himself'; to this image there corresponds nothing in objective reality. The direction of Bultmann's train of thought is towards the awareness that the 'self' in the image is not *objective*; the level of Bultmann's train of thought is prescientific because it does not transpire in the awareness that the 'self' in point is not *real*. Bultmann endows the unreal objects of his thought with the adjectives 'genuine,' 'authentic,' and 'real.' On the level of fiction these adjectives affirm a plausibility which adds pleasure to the train of thought; on the level of science they are a misstatement of the truth.

17. Science-fiction has its attractiveness from the plausibility of the unreal. The idea that our 'real' life is to be gained on the other side of the

---

8.2:15b *Infra*, I.14; I.38ff.
8.2:15c Bultmann, as quoted by Helmut Thielicke in *KuM* I, 171 (*KaM* I, 152).
8.2:15d *Supra*, 3.1:10.

8.2:15e *Supra*, 2.3:20.
8.2:15f *Supra*, 8.2:9b.
8.2:16a Cf. *infra*, I.68ff.
8.2:16b Cf. *supra*, 2.2:6; 2.3:14.

looking glass has in itself only the attractiveness of a child's fantasy. It does not appeal to the adult mind. But the adding of plausibility by a train of thought can tickle the fancy of an adult. It is crude to imagine the propulsion of one's body through the looking glass into a world on the other side. The proposal becomes more plausible when one begins to consider that living on the other side may not mean to bring over to the other side the things of this side, but to exist there simultaneously — one could easily say 'paradoxically.' And this line of thinking, as one continues to add from ingenuity more and more elements of plausibility, could become overpowering if it provides an escape from undesirable realities of this world. Now, Bultmann's theological reflection is of this kind. It begins from a proposal that is fantastic in its very notion: that the knowing self can turn his gaze from objectivity and understand himself. The course of Bultmannian reflection is to find plausibility for this idea, and its attractiveness lies in the escape it provides from the awareness of objective reality — beginning with the reality of God. The intellect of man is a kind of mirror in which the meaning of the really existent is reflected. The central image of Bultmannian fiction is not exactly the idea of crossing through the mirror into a world on the other side; his proposal is rather that of turning away from the mirror and disappearing into oneself. The image is different, but the genre of fiction is the same.

18. Bultmann's description of his existential self shows that it is fanciful. To describe his 'self,' to speak of it at all, he must somehow see it as an object. Yet he declares that 'my genuine self' is not at all visible or ascertainable.[a] Bultmann imagines himself transcending the entire picture of objectivity as he speaks of himself, but this is a purely imaginary transition.[b] He imagines an existential encounter with the love of another which includes an understanding of the genuine self,[c] but this is inconceivable, for understanding is the medium between the knowing self and the object of his knowledge.[d] Bultmann imagines that this genuine understanding of one's genuine self illumines the whole world in a 'new light,'[e] but this is nonsense, for illumination is only of visible things and the self of Bultmann is invisible. When Bultmann declares that this new understanding turns the whole world into another world, he betrays the fanciful character of his thought. He does not mean that it *really* becomes another world; he does not even mean that his existential encounter bestows any *real* understanding of oneself. He says, on the contrary, that the new insight into my past and my future has no abiding validity, no truth in itself, but is valid only for the instant in which it takes place.[f] It is not real insight, because insight is the act of intellectual sight,[g] and Bultmann's genuine self is not visible. It is not real understanding, because understanding is the abiding factor of an insight, and Bultmann's insight has no abiding factor, as he himself declares.

19. Furthermore, the past and the future are notions derived from a dichotomy of the objective representation of time.[a] There can be no past or future in the uniqueness of pure subjectivity. 'My past' and 'my future' are nonsense terms on the subjective side of the subject-object relationship or in any world which is said to transcend this relationship. The self of pure subjectivity cannot be a medium in which the past or the future is understood. When Bultmann says that his insight into his own self enables him to see his past and his future in a new sense,[b] he is misreading the essential structure of human understanding, for a 'sense' is nothing else but one half of the dichotomy of a concept, used as a *medium* between the knowing self and the object of his knowledge.[c] Bultmann implicitly admits the fanciful character of his thinking when he says that my past and future become my past and future in a new sense (*in neuen Sinne*), for 'my past' and 'my future' are identified by Bultmann with the act of existential insight in all its uniqueness and non-objectivity

---

8.2:18a   *Supra*, 1.3:4.
8.2:18b   *Infra*, I.17ff.
8.2:18c   *Supra*, 1.3:5.
8.2:18d   *Supra*, 2.6:9.
8.2:18e   *Supra*, 1.3:5.

8.2:18f   *Ibid*.
8.2:18g   *Infra*, I.14.
8.2:19a   *Supra*, 4.1:3; 4.3:3.
8.2:19b   *Supra*, 1.3:5.
8.2:19c   *Supra*, 2.3:10.

and in its non-abiding actuality, which is just what past and future can never become, since the present is their opposite.[d]

20. The notion of historicity underlying Bultmann's conception of the act of faith is no less elusive. What he calls the correct conception of the historicity of man requires that a man, in order to achieve his genuinely historical self, transcend his subjectivity so as to be outside of himself as well as in himself, in good as well as in evil.[a] It requires a renunciation of *Historie* and of *Geschichte* and of all encounter with the phenomena of past history, including an encounter with the Christ after the flesh, in order to encounter the Christ of the kerygma as it confronts me in my historic situation.[b] The eschatological encounter, as proclaimed by Bultmann, is clearly fanciful, for it violates the first principles of reason. A thing is itself, and therefore the knowing subject is himself in all his uniqueness and unity.[c] Yet Bultmann's 'genuinely historical' self is at the same time in himself and outside of himself; nor is he in himself in one sense and outside himself in another, for this genuinely historical self transcends all relationship with any object, and senses are objects of thought. Opposites cannot be predicated of the same thing at the same time and in the same way by scientific man, and yet Bultmann's genuinely historical self is both in good and in evil at the same time and in the same way. Good and evil become synonymous, for by suppressing ('transcending') objectivity there is no longer the means of distinguishing between them.[d]

21. Bultmann expounds a mythological act of faith.[a] According to his exposition there is no revelation in itself; revelation is revelation only *in actu* and only *pro me*; it is recognized only in personal decision. Neither God nor revelation is an objective entity. His declaration that justification is a gift to the man who has fallen on his knees before God is in context purely mythological. 'Before God' is an irrational expression, since for Bultmann God is not an object, and we know that a preposition cannot really function without an object; 'fallen down on his knees' is a meaningless metaphor, since for Bultmann faith cannot prescribe definable acts distinct from the sheer decision to believe, and there is no down or up in a sphere which transcends all relation to objects; the 'gift' is totally identified with what the man must achieve of himself; God is identified with the self-understanding of the self, who consequently cannot in any reasonable way be said to be down on his knees before God. For Bultmann revelation is not an object of intelligence; it is revelation only *in actu*, and the *actus* is my own decision to believe; revelation is not the object of this decision, but is totally contained *in* the decision as aimed squarely at myself (*pro me*); that is what makes the decision *personal*.[b] But the decision to believe *in* a 'revelation' whose only act is the act of disappearance is mythological, because revelation ('uncovering') as a concept is in the opposite genus of appearance, and the decision to believe that it is recognized as being its opposite can only be make-believe.

22. Bultmann does not deny that his interpretation of Christian faith is mythological. He admits that belief in a kerygma whose credibility can in no way be established beyond the sheer decision to believe is an offense to human thought as such, and he coun-

---

8.2:19[d] *Supra*, 4.2:4; 4.3:4.
8.2:20[a] Bultmann, in *KuM* I, 130 (*KaM* I, 113).
8.2:20[b] Bultmann, in *KuM* I, 133-134 (*KaM* I, 117). Cf. *supra*, 1.2:3; 1.2:7; 1.3:10; 1.3:12.
8.2:20[c] Bultmann seems to admit this in his reply to Thielicke: "While modern man may be wrong in identifying his ego (*Ich*) with his subjectivity (*Subjektivität*), he is undoubtedly right in regarding it in its subjective aspect (*als Subjektivität*) as a unity, and in refusing to allow any room for alien powers to interfere in his subjective life (*in sein inneres — subjekti-*

ves — *Leben*)." (*KuM* I, 136; *KaM* I, 120.)
8.2:20[d] *HE*, 150-154. Cf. *KuM* I, 35-39 (*KaM* I, 26-32).
8.2:21[a] R. Bultmann, "Zur Frage der Entmythologisierung: Antwort an Karl Jaspers," in K. Jaspers - R. Bultmann, *Die Frage der Entmythologisierung* (Munich, 1954) [reprinted in *KuM* III, 9-46], 68-70; ET, "The Case for Demythologization," in K. Jaspers - R. Bultmann, *Myth and Christianity* (New York: Noonday Press, 1958), 67-69; or "The Case for Demythologizing," in *KaM* II [181-194], 190-191.
8.2:21[b] *Ibid.*, 70-73 (ET, 69-71; *KaM* II, 191-194).

ters with the observation that the demythologized kerygma at least is not mythological in the original sense of mythology.ᵃ Therefore the demythologized kerygma can be accused of being mythological in some sense — precisely in the sense of Bultmannian fantasy. Bultmann plainly admits that modern man (and he is a modern man) *will* dismiss his kerygma as absurd, "but that is another matter." ᵇ Why is it another matter? Obviously because the decision to enter the dream-world of his genre of thought is a decision to accept the absurd and to find plausibility for it. Bultmann exchanges belief in supernaturally revealed mystery for an irrational mystery of his own creation. He says that the offense to human intelligence of the 'genuine' kerygma is not that the kerygma challenges the intelligence to grasp meaning which exceeds its natural power, but rather that the offense (*skandalon*) pertains to the natural self-understanding of man.ᶜ This means that Bultmann's 'faith' coincides with absurdity in the sense that, in the very act of seeing the absurdity of what his kerygma proposes, one embraces ('paradoxically') the self-understanding called forth by the encounter with absurdity. But this is the genre of fiction and the consciousness of day-dreaming insofar as the decision to indulge in the dreaming takes place in coincidence with the realization that the content of the dream is fictious. This fact explains Bultmann's statement to Jaspers that anyone is free to regard a revealed faith as absurd.ᵈ If Bultmann did not think that his notion of revealed faith is absurd, he would not be prone to make this concession, since his whole style has the form of reasoning discourse. He does not admit, for instance, that anyone is free to regard his modern world-view as absurd. To the contrary, he avers that modern man is enslaved to it. But entry into the world of fiction is only through the free decision to indulge in it, and that explains why Bultmann can say to Jaspers that those who reject his revealed faith as absurd have no business talking about it. Reason and fiction are two different universes of discourse. You have to choose one or the other.

23. How, then, are we to understand Bultmann's claim that, because existentialist interpretation operates on facts, it is *not* a product of fancy? ᵃ We must understand it in keeping with the genre of his thought. His existentialist interpretation of history 'grasps the historical meaning of an event' in 'dialectical relationship' with the objectifying view of the past that accompanies it; its 'understanding' depends upon a 'careful ascertaining of the facts' bearing on the case. But careful research is a prerequisite of good science-fiction. The research in Bultmann's genre has a twofold function: it increases the attractiveness of the genre by refining its plausibility, and it demythologizes the genre of conventional theology. But this function of research does not pertain to the universe of discourse of existentialist interpretation as far as the *finis operis* is concerned; it pertains only to the extrinsic motivation for turning away from the real world into the existentialist universe of discourse. To the extent that the Bultmannian interpreter does not lose his awareness that he is dreaming, he stands simultaneously in both worlds and he is able to make fact and fancy function in dialectical relationship, but he cannot destroy the fact that his existentialist interpretation is fancy.

24. This means that, when Bultmann says that an act of God cannot be posited apart from its existential reference, he is placing the act of God entirely within the realm of fancy.ᵃ For Bultmann, faith makes sense only when it is directed towards a God with a real existence outside the believer, but Bultmann does not insist that his faith makes sense. To the contrary, he admits that it is absurd. He admits also that the encounter of faith with the act of God cannot defend itself against the charge of illusion.ᵇ It is not a question here of a charge of illusion by some other unbelieving person; rather it is the very believer who cannot surmount the thought that it is an illusion. This state of mind stems from the nature of faith as Bultmann conceives it: if faith tried

---

8.2:22ᵃ *KuM* I, 48 (*KaM* I, 43).
8.2:22ᵇ Bultmann, in *KuM* I, 134 (*KaM* I, 118).
8.2:22ᶜ Bultmann, in *KuM* I, 137-138 (*KaM* I, 122-123).

8.2:22ᵈ Bultmann, *op. cit. supra* (8.2:21ᵃ), 68-69 (ET, 67-68 = *KaM* II, 190).
8.2:23ᵃ *Supra*, 1.3:12.
8.2:24ᵃ *Supra*, 1.3:13; *infra*, I.1.
8.2:24ᵇ *JCM*, 40, 70-72.

to refute the charge, it would be misunderstanding its own meaning. This is a puzzling explanation until we have realized that Bultmannian faith is a genre of fiction, whose nature as a genre of fiction is fantasy, and whose meaning as a genre is missed if one tries to make it out to be real. It is only as a fiction-genre that the true strength of faith lies in its indemonstrability with regard to its object. Bultmannian faith refuses to place God on the level of the world in the sense that it refuses to withdraw the object of its thought from the world of fancy and place it in the world of fact.

25. While Bultmann admits that there can be only one true proposition regarding a single phenomenon (in the sense that its contrary cannot be true), he explicitly denies that faith contains any true propositions.[a] Bultmann's faith, then, does not pertain to the realm of truth. Once we have realized this we are in a position to understand why he can say that faith cannot demonstrate itself by logic, seeing that logic binds only in the realm of truth. Arguments for the credibility of the object of faith are indeed an exercise in futility in the case of a faith whose ground and object are both identified with the free decision to make-believe. It is precisely the decision to make-believe that Bultmannian faith can call its own 'nevertheless.'[b] Such faith does not need the 'false security' of ascertainable knowledge within its own universe of discourse, even though it functions in dialectical relationship with a total dependence upon ascertainable knowledge as far as the real world is concerned. Its security, as an escape from reality, truly lies in the abandonment of all security and in the readiness to plunge into the inner darkness of an imaginary world in which the light of intelligence is turned off.[c] As fiction, it enjoys the thought that its security is no real security at all. Bultmann plainly tells us that its mode is to consider oneself *as if* one were not what one really is.[d]

26. Fictional faith is not real faith. Bultmannian faith is not Christian faith. The virtue of faith is situated in the area of real human understanding. It is not the entry-way into a dream-world. The act of Christian faith, not as it is intrinsically in itself, but as it is related to human understanding, is most basically an act of affirmation of the reality of its proper object. Reality is a concept of human understanding. To use this concept as a means of excluding the object of faith from the continuum of reality is tantamount to a denial of faith.[a] Anyone is free (but not reasonably so) to deny that the object of Christian faith is real. By Bultmann's own declaration, anyone who does deny the reality of the object of faith has no business talking about it. Yet Bultmann has made his reputation by talking precisely about this. To the extent that Bultmann has made the unreality of the object of Christian faith the subject of his discourse, it becomes our business and our duty to refute his arguments. The need of carrying out this task is based only negatively and in preliminary fashion upon the menace to faith posed by his mythology. The fuller reason for taking up the task is the need to develop the whole area of theological understanding in the realm of science and reality which is missed by concentration upon the unreality of Bultmannian theology.[b]

8.3. *The Question of Historical Science*

1. Bultmann's argument regarding the New Testament view of history may be expressed as follows:

On the level of modern historical science it is known that history must follow a smooth, unbroken course. But the course of history as depicted in the New Testament is broken by the intervention of supernatural forces. Therefore, the view underlying the New Testament description of history is unscientific.

In corroboration of the minor premise Bultmann points out that in the New Testament the course of history is depicted as being set in motion and guided by supernatural forces, and that history is regarded as being divided into the aeon of the present world enslaved to Satan, sin, and death, and the coming aeon of eternal blessed-

---

8.2:25[a] *Supra*, 1.3:13. Cf. *HE*, 63.
8.2:25[b] Cf. *JCM*, 41, 64-66.
8.2:25[c] Cf. *KuM* I, 29-31 (*KaM* I, 19-22); *JCM* 40-41, 84-85.

8.2:25[d] Cf. *KuM* I, 38-39 (*KaM* I, 30-32). Cf. *JCM*, 83-85.
8.2:26[a] *Supra*, 7.7:4.
8.2:26[b] *Supra*, 7.8:2.

ness to be inaugurated by a cosmic catastrophe.[a]

2. Among the questions provoked by Bultmann's argument are the following: On the level of historical science, what is history? Does historical science know that the course of history must be unbroken? Does the scientific concept of history exclude the intervention of supernatural forces? The first requirement for discussion on the level of historical science is an adequate definition of history. To talk about history without sufficiently defining the term does not rise above the level of the prescientific.[a] Bultmann has made various attempts to define history, among which are the following.

3. History may be defined as "a sequence of critical actions which bring a new present into existence making that which was present irretrievably past."[a] This definition, taken from Erich Frank, correctly indicates that history is a sequence of actions bringing a new present into existence. The word 'critical' suggests a relationship of significance to the knowing subject in terms of what he sees to be essential or important, which implies that history is in the genus of knowledge. This definition, therefore, approximates the precise definition of history. However, it fails to show that history is as much a state of understanding as it is a sequence of actions,[b] and it does not take into account where and how the new present exists,[c] while the expression 'irretrievably past' is confused inasmuch as history is of its very essence a retrieving of the past, so that what is irretrievably past cannot be history.[d] In his use of the definition, Bultmann does not show an awareness that the element 'critical actions' applies to history in both its objective and its subjective aspects;[e] he tends to recognize only the subjective aspect.

4. "Historical narrative proper arises when a people experiences the historical processes by which it is shaped into a nation or a state."[a] Bultmann holds that anything prior to this is myth. The definition shows a lack of awareness of history as a genus. Myth is a species of history in the genus of historical fiction. Historical fiction is a species of history in the genus of historical narrative.[b] Historical narrative is a species of history in the genus of the past as such.[c] History, therefore, in its full definition, is "the past as such."[d] The past as such may be actual or potential.[e] The actual past as such is divided into the knowledge of the past and the expression of the past. The expression of the past may be realistic or fictitious. Bultmann's definition of historical narrative fails to show that such narrative may be realistic or fictitious; because it misses the common element in both species of narration, it does not identify the correct difference between them. By 'historical narrative proper' Bultmann seems to mean historical narrative as a reality-genre, because he distinguishes it from myth. But he does not identify the distinguishing characteristic of 'reality.' He places it rather in the consciousness of becoming a nation or a state. Now, this definition is erroneous, for it does not include the many forms of factual historical narrative whose medium of understanding is not the set of concepts pertaining to nationality or statehood, e.g., autobiography, genealogy, the account of a trial. Furthermore, historical knowledge exists primarily in individual persons and only by derivation in a people as a collectivity. The experience of historical processes is history only when these processes are recognized to be historical. Therefore, it is more correct to say that historical narrative arises from historical knowledge, and historical knowledge is the awareness of the past as such. Scientific historical narrative arises from knowledge of the real past as such, and it requires in addition that the account be a truthful expression of that knowledge. What must intervene is an *intellectual* experience of the genetic relationship between the past and the present.[f]

5. History is "event in time" (*zeitliches Geschehen*).[a] This definition does

---

8.3:1a *Supra*, 8.0:3.
8.3:2a *Supra*, 3.2:1; 3.2:3.
8.3:3a *HE*, 2.
8.3:3b *Supra*, 3.2:10.
8.3:3c *Supra*, 4.2:2.
8.3:3d *Supra*, 4.3:4.
8.3:3e *Supra*, 5.2:13; *infra*, III.10-17.
8.3:4a *HE*, 14.

8.3:4b *Supra*, 6.3:2.
8.3:4c *Supra*, 6.2:3; *infra*, IV.5.
8.3:4d *Supra*, 3.1:12; 3.1:13.
8.3:4e *Supra*, 6.3:1.
8.3:4f *Infra*, III.18.
8.3:5a R. Bultmann, *Jesus* (Tübingen, 1951) (first published in 1926 and hereinafter referred to as *Jesus*), 11; ET, *Jesus*

not bring out that an event is a unit of history, and that these units are related to each other by genetic causality.[b] What the definition means to say is that history is composed of events in time. But the expression 'event in time' is also pleonastic, for an event can be only in time, and it is so closely identified with time that it is actually a unit of a species of time. Bultmann is here searching for the way in which the present enters into history, but he does not find it, for he concludes by this definition that (the object of) history cannot be present. Bultmann does not recognize that the concept of the past is formally present in the mind, even though the content of that concept refers to what is not materially present.[c] Bultmann's definition excludes the formality of historical *knowledge* constituting actual history from his definition of history (which itself applies only to potential history) and thus shows that his thinking about history is prescientific.[d]

6. "(The core of history is) man."[a] This definition substitutes man as the subject of history for the specific character of history as a concept; it exchanges the essence for the 'core.' But what is required for a definition is the essence. The question is not 'Who is the subject of history?' but 'What is history?' To say that history is man is to suppress all intelligent discussion about history, because history is in itself something other than man, and to substitute the word 'man' wherever the word 'history' occurs would not retain the meaning of historical discussion. By defining history existentially, Bultmann identifies man with his history and finds that man is his history. To do this he has to distinguish between history and the core of history, which is irrational, because the real core of history is its essence, and there cannot be a core of the essence.[b] Man is at the center of history for the one and only reason that he is the knowing subject of historical knowledge. Man is equally at the center of nature in the sense that he is the knowing subject of natural science. But man is not nature and man is not history, for man is not identified with the object of his knowledge. Bultmann sees [c] that the human being is a personal subject, an 'I', which decides and has its own vitality, but he fails to see that this psychological self is the subject of knowledge as well as of decision, that it is mysterious in the sense that it can never be brought directly into view, that it is merely the exposed tip of the substance of the man, directly unknowable to him, as it grounds and sustains his conscious awareness. Bultmann does not distinguish between the substantial 'I' and the intellectual 'I'. He confuses the intellectual 'I', which does not grow or degenerate or change in any way, with the substantial 'I', which does grow and degenerate, but not in

---

and the Word (New York: Scribner's, 1958) (hereinafter referred to as *JW*), 8.

8.3:5[b] *Supra*, 5.1:4.
8.3:5[c] *Supra*, 4.1:4; 4.2:4; 4.2:16; 4.3:4.
8.3:5[d] *Supra*, 6.3:1. Bultmann's destinction between *Historie* and *Geschichte* (*supra*, 1.2:6), and his use of this distinction for the conception of 'time as now' (1.2:7), show his awareness of the problem of history as a mode of understanding, but his reduction of historical understanding (*Geschichte*) to the trans-objectivity of the existential encounter, his identifying of historicity (*Geschichtlichkeit*) with being able ever to decide anew in the 'reality' of the concrete here and now (1.2:6c) is the wrong solution, because it fails to find the objective meaning of historical understanding. The vital act of insight to which Bultmann limits his conception of historical understanding (*Geschichte*) has its objective fundament proximately in the concrete historical concepts present in the medium of human understanding and remotely in the genetic meaning adhering to the extramental reality which is being regarded.

8.3:6[a] *HE*, 139.

8.3:6[b] A step is missing in Bultmann's reasoning about the 'core of history.' In excluding a concept of history as expressing a universal essence, he moves straight into the subjective side of the subject-object dichotomy, overlooking the concrete concept of history which is located within objectivity, but closer to the knowing subject than are open classes in the Aristotelian sense. Bultmann's 'core of history' is an undifferentiated use of the subject-object distinction; it is not a scientific expression. What is needed to solve the problem is the 'concrete' concept of the 'essence' of history, which is the *form* as seen by the intellect to be the necessary and sufficient *meaning* corresponding to the identity of history as a thing in itself.

8.3:6[c] *HE*, 145.

a manner that is perceptible to consciousness. He fails to see that as far as historical science is concerned the growth of the 'I' is not in itself but in the medium of its knowledge, which is the human understanding. Bultmann does not clarify his thinking about man as a personal subject when he says that the *Weltanschauungen* are permanent possibilities of human self-understanding.[d] The only possibility of human understanding is the understanding of objects, since understanding is the medium between the knowing self and the objects of his knowledge. There is no possibility of self-understanding in the sense intended here by Bultmann.

7. Bultmann endorses Collingwood's theory of history and repeats its central ideas. While he does disagree with Collingwood on some points, his style of summarizing the general theory leaves the impression that he is in agreement. Bultmann quotes Collingwood's definition of history as "re-enactment of the thoughts of the past in the historian's mind." He again quotes Collingwood to say that history is self-knowledge of the mind and the historical process exists only insofar as the minds which are parts of it know themselves to be parts of it. He quotes Collingwood once more in saying that history does not presuppose mind: "It is the life of the mind itself, which is not mind, except so far as it both lives in historical process and knows itself as so living."[a] He states with Collingwood that thought is not merely the act of thinking, but is the act of man in his entire existence, that is, the act of decision.[b] But this violates the principle that a thing is itself.[c] To say that thought is not merely the act of thinking, but includes the act of man in his entire existence — especially the act of willing — is a contradiction in terms. Burying the act of thinking in the act of willing and in the whole vital activity of man is the destruction of science.[d] The knowing subject of history does not in this act of knowing know himself as living in historical process; only to the extent that he succeeds in turning elements of his existence into objects of knowledge can he fit himself into the historical process; the process is not historical until it recedes from the present, and he as a knowing subject is always present.[e]

8. Collingwood errs in saying that history does not presuppose mind, for the mind is the synthesis of the knowing subject and the object of his knowledge, and history is an object of knowledge.[a] Furthermore, the mind cannot know itself to be part of the historical process, for the historical process is in the objective part of the mind and the act of insight is in the subjective part, identified with the knowing subject as such.[b] Finally, history cannot be an absolute re-enactment of the thoughts of the past, for the absolute medium in which the thoughts of the past occurred was the minds of persons in the past, while the absolute medium of historical knowledge is the mind of the historian.[c] A thought necessarily includes both the act of insight and the object of the insight. The historian may be able to reproduce the object of a past insight, but he can never reproduce the past insight itself.[d] The difference is that the mind of the historian can never coincide with a mind of the past, not only because the assortment of elements in his general medium of understanding will be different, but also and specifically because as an historian he is looking at those objects from his own place in history, which modifies the viewpoint of a previous place in history.[e] Collingwood's definition should therefore be corrected to say that historical knowledge is the reproduction of the past as an object of understanding. Historical knowledge may be of the objects of past thinking, but it is not restricted to this, since it may be of anything past. I do not deny that each mind has its own history, and that the acquisition of history is a process in the mind of the historian. The proximate formal object of history is in the mind of the historian; the remote formal object is in external things. Collingwood does not err in stressing the importance of the formal object

---

8.3:6d *HE*, 147.
8.3:6e *Supra*, 2.2:6a; 2.3:14, 2.6:8-10.
8.3:7a *HE*, 130-134.
8.3:7b *HE*, 135.
8.3:7c *Supra*, 2.3:18.
8.3:7d *Infra*, I.8.

8.3:7e *Supra*, 3.2:5; 4.2:4; 4.2:9; 4.3:8.
8.3:8a *Supra*, 2.2:2; 3.1:13.
8.3:8b *Supra*, 2.6:9.
8.3:8c *Supra*, 2.2:2b.
8.3:8d *Supra*, 4.2:2; 4.2:21; *infra*, II.60.
8.3:8e *Supra*, 4.2:4.

of historical knowledge; his error lies in not seeing that the proximate formal object is a medium between the knowing subject and the remote formal object of his knowledge; he thus identifies the proximate formal object with the knowing subject and misses the essential constitution of the mind.[f]

9. The insufficiency of the five foregoing definitions of history attempted by Bultmann shows that he has not grasped the essence of history, and this explains his inability to see that the essence of history *is* grasped by viewing it.[a] History can be encountered as an object of intellectual vision. Since historical knowledge is knowledge of the past, and the past is an object of thought, Bultmann errs in affirming that in the realm of historical knowledge an ultimate distinction between the knower and the object of his knowledge cannot be maintained.[b] Bultmann's reason for thinking that the historian cannot stand opposite history in a subject-object relationship is that the act of viewing the historical process is itself an historical process and the image of the historical event always bears the imprint of the individuality of the viewer.[c] But this reasoning is not supported by historical science. The act of viewing an historical development involves a double object of vision: the remote object of a development reconstructed out of the past and the intermediate object of the present conceptual complex through which the remote object is being viewed. The remote object is usually the center of attention in the production or reviewing of an historical narrative, and the knowing subject stands in a subject-object relationship to it. The intermediate object is only vaguely perceived by the unskilled, and they may not reflect on its presence at all, but as knowing subjects they stand in a subject-object relationship to it as well. It is the task of the scientist of history to focus his attention upon the historical medium and turn it into an effective instrument.[d] As the historical medium is perfected, it loses the imperfections of excessive subjectivity; it becomes perfectly objective. At the stage of developed historical science, there is no more of the imprint of personal subjectivity in historical knowledge than there is in the knowledge of natural science.[e]

10. Bultmann's theory of pre-understanding stems from a lack of reflection on the medium of science. To say that each interpretation is guided by a certain 'putting of the question'[a] does not explain where the questions come from. Interpretations arise from the interposition of ideas and principles between the knowing subject and the text or object he is studying. When the remote object is seen in a valid concept placed over it, the grasping of certain meanings of that remote object becomes possible. Questions are the striving of the knowing subject to move to a new level of understanding regarding the object of his study. A new insight produces a development of the conceptual mass of the understanding. Bultmann's doctrine of pre-understanding[b] is a crude expression of the fact that the meaning of things cannot be grasped unless one already has a certain affinity for that meaning and asks the right questions about the things. It takes understanding to achieve understanding, and the intellect moves from one degree of understanding to another. In this sense every man is involved in his own history, but he is not trapped in his own subjectivity. Because the knowing subject is radically distinct from the medium of his understanding, Bultmann and Collingwood err in affirming that historical science is objective precisely in its subjectivity.[c] A thing cannot be its opposite. It is a violation of the first principles of science to say that anything is objective precisely in its subjectivity.

11. Since science involves a studied awareness of the medium of its thought, it follows that each science must base its approach on its own method and presuppositions. This does not mean, as Bultmann assumes,[a] that objectifying historical science tries to fit the facts into the possibilities it anticipates. Rather, it accepts all the data and understands them in the light of its own valid concepts. Historical science does not understand the data

---

8.3:8[f] *Supra,* 4.2:11-17; *infra,* III.10ff.
8.3:9[a] Cf. *Jesus,* 7 (*JW,* 3-4).
8.3:9[b] *HE,* 119-120.
8.3:9[c] *Supra,* 1.3:10.
8.3:9[d] *Supra,* 3.2:3; 4.2:11.

8.3:9[e] *Infra,* II.17.
8.3:10[a] *HE,* 113.
8.3:10[b] *Supra,* 1.2:4.
8.3:10[c] *HE,* 133.
8.3:11[a] *Jesus,* 8-10 (*JW,* 5-6).

by seeing them as particular instances of general laws; that would be misplaced natural science. Historical science colligates the data under concrete concepts which are themselves derived from facts. But wisdom is an important species of science. It functions to exclude bias and moral ignorance as obstacles to the development of true historical science. While Bultmann in the name of historical science opposes the pronouncing of value judgments in historical study,[b] he is not speaking scientifically. Where wisdom is not functioning, supreme ignorance will step in to take its place. Bultmann has pronounced many value judgments of his own in the course of his historical research.[c] If he had been more fully aware of the function of wisdom in the method of science,[d] the results of his historical research would have been different.

12. The conclusion that the New Testament narrative is prescientific and mythological is a value judgment of colossal importance for Bultmann and his followers. Bultmann bases this judgment upon the conviction that the man who is liberated from mythical thinking cannot countenance the intervention of outside forces in himself or in the world. For Bultmann historical science cannot admit interventions of God or of angels in history.[a] But Bultmann's reason does not pertain to historical science. The assumption that miracles are impossible[b] is not historical science; it is misunderstood natural science. To exclude data on the ground that they could not have happened is not a valid method in any science.[c] If miracles happen, science cannot rule them out. The reason why natural science cannot formally recognize miracles is that the formality of miracles exceeds the conceptual bounds of the medium of natural science. If on a certain island in the Pacific Ocean there were trees whose fruit was peanut butter *in the can*, genuine natural scientists would have to consider this an extraordinary phenomenon for which they would be honestly open to the valid explanation of historians. Miracles, as supernatural events, are not objects of natural science. But miracles, as events, are objects of historical science, and they are a proper object of historical theology.[d] Bultmann sees the error of excluding data as unhistorical on the ground that they do not fit under the general laws that the historian has formulated.[e] But in making his crucial judgment about the New Testament narrative he misuses this insight. He assumes from natural science that miracles are impossible, that no intervention by God or angels is possible, and on the basis of this assumption he declares the New Testament narrative mythological. His fatal error lies in not realizing that what is not an object of one branch of science may be an object of another.[f] Modern science in itself, apart

---

8.3:11b *Jesus*, 10 (*JW*, 7).
8.3:11c R. Bultmann, *Die Geschichte der synoptischen Tradition* (Göttingen, 1970^8) (hereinafter referred to as *GST*); ET, *History of the Synoptic Tradition*, trans. by J. Marsh (Oxford: Blackwell, 1963) (hereinafter referred to as *HST*). Cf. *GST*, 51 (*HST*, 49); *GST*, 58-60 (*HST*, 56-57); *GST*, 92-93 (*HST*, 88); *GST*, 106-107 (*HST*, 102); *GST*, 222 (*HST*, 205).
8.3:11d *Supra*, 2.6:13; 2.5:5.
8.3:12a *Supra*, 1.3:8.
8.3:12b *Supra*, 1.3:8a.
8.3:12c Since science is the comprehension of what is given in sensory and intellectual experience (2.4:2), the exclusion of data is antiscientific, even though the excuse made is that scientific thinking finds certain data inadmissible.
8.3:12d *Infra*, 8.4:16; 8.5:9.
8.3:12e *Jesus*, 10 (*JW*, 6).
8.3:12f Bultmann expressly states that, if science attributes natural phenomena to non-natural causes, it has ceased to be science (R. Bultmann, "Zu J. Schniewinds Thesen, das Problem der Entmythologisierung," in *KuM* I [122-138], 123; ET, "A Reply to the Theses of J. Schniewind," in *KaM* I [102-123], 103). More correctly, natural science of itself cannot attribute 'natural' phenomena to 'non-natural' causes within its own sphere of thought. But there is only one Universe of reality, and it is synonymous with Nature validly understood. God and supernatural events pertain to the one world of reality. The 'natural scientist' cannot find characteristic meaning in supernatural events because these events are not explainable from laws formulated according to the ordinary pattern of response of material things. But the historical scientist can find meaning in these events, as happenings in Nature (the world of tangible phenomena) and as elements of historical cosmology. In failing to see the full proportions of Nature, as extending beyond the inherent tendencies of created

from an abusive specialization of the scientist in the sciences of nature, does not exclude the miraculous. Natural science as natural science cannot recognize meaning in a supernatural intervention, but the natural scientist can and must recognize the valid conclusions of another science whose medium of understanding is suitable for picking up meaning in the supernatural. That is what the present discussion is all about.

13. Just as there is a natural and a supernatural level of classical (universalizing) science, so there is a natural and a supernatural level of historical science. Natural historical science can recognize the occurrence of data whose meaning is supernatural; it cannot grasp the meaning of such data except in the most inchoative manner. The science of historical theology is equipped to delve into the meaning of supernatural events. The term 'historical science' applies equally to both levels of thought. Bultmann's assumption that history must follow a smooth, unbroken course does not pertain to historical science on either level. This assumption is based on the mistaken idea that the pattern of history has meaning because it conforms to scientific anticipations and unfolds according to ascertainable natural causes. But the meaning of history does not arise from the conformity of a narrative to what the informed person would expect to have happened. Its meaning derives from knowing what actually did happen.[a] Even on the natural level free wills are at work to ruin any degree of accuracy in the prediction of events. Human freedom determines lives and influences the natural environment too. Every time a free human act influences a natural event, it in some way 'breaks' the smooth course of that event, as far as the blind forces of nature are concerned, but that is what history is all about. The course of history is not really broken, for history is not made or broken in terms of efficient causes and their predictability. What constitutes the course of history is the relationship of cause and effect, where the cause in point is the final cause, and the effect is its antecedent.[b] The historical theologian is first concerned to determine *what happened*. His next concern is to find the historical meaning of what happened. If a supernatural intervention has genetic meaning, the event is history and deserves to be recognized as such.

14. Bultmann's argument of the two aeons has no solid basis in historical science. It is history's nature to be divided into eras and epochs. History is divided of its very nature, since its basic unit is the event, and an event is a genetic relationship of two states of a thing to one another. Bultmann and Collingwood are correct when they state that history ends with the present. Sheer prediction of the historical future is fiction, not science. But Bultmann, in his rejection of the coming aeon, misses the basis of its reality. The future of mankind is present to God. It is out of God's present, in retrospect as it were, that real prophecies of the future can be granted to men. Real prophecy is inspired by God, and that is what makes it a species of history. Historical science allows for a division of history into two aeons on this basis.

8.4. *The Question of the Genre of Mythology*

1. Bultmann's argument regarding the obsoleteness of the New Testament world-picture may be expressed as follows:

An obsolete world-picture cannot be accepted by modern man. But the New Testament world-picture is obsolete. Therefore the New Testament world-picture cannot be accepted by modern man.

In support of the major premise Bultmann avers that modern man is convinced that the mythical picture of the world is obsolete, and he cannot force himself to accept an obsolete view. Bultmann claims also that our picture of the world is already determined for us by our place in history. In support of the minor premise Bultmann contends that the kerygma of the New Testament is wrapped in the language of ancient mythology, and mythology is the world-picture of a prescientific era.[a]

---

natures, as formulated in statistical laws, Bultmann overlooks the entire realm of historical science.

8.3:13[a] *Supra*, 5.1:4.
8.3:13[b] *Supra*, 4.2:12.
8.4:1[a] *Supra*, 8.0:4.

2. Among the questions for historical science aroused by Bultmann's argument are the following: What is mythology?[a] What is the scientific basis of the conviction that the mythological view of the world is obsolete? Why is it that our picture of the world is determined by our place in history?

3. Bultmann defines myth as "an account of an event or happening in which supernatural, superhuman forces or persons are operative." A myth is a story of the gods.[a] He defines mythology as the representation of the divine as human, of the non-worldly as this-worldly;[b] in mythological expression the non-worldly and divine appears as worldly and human, what is of the other side appears on this side, and the othersideness (*Jenseitigkeit*) of God is thought of as spatial distance.[c] Myth, he says, knows another reality than that which science has in view; it knows that the world and human life have their ground and limit in a force lying on the other side of everything tangible to human thought and activity, but it speaks of this other reality in the unsuitable terms of spatial distance;[d] myth speaks in terms taken from the visible world about forces which man feels at the periphery of his known world and at the edge of his capacity to do and to suffer; myth is an expression of man's belief that the origin and purpose of the world are to be sought beyond the realm of the known and the tangible; it expresses the belief that in his dependence upon forces operating beyond the borders of the known world man can be delivered from forces within the visible world; but the real purpose of myth is to express man's understanding of himself and to speak of a transcendent power which controls the world and man.[e] Myth objectifies the other side into this side (*objektiviert das Jenseits zum Diesseits*); it objectifies God and the other side into worldly phenomena of this side; it gives immanent, this-worldly objectivity to transcendent, unworldly reality.[f] Bultmann contends that this anthropomorphizing of transcendent power is unacceptable, but within this inept imagery he recognizes the intent to express a certain understanding of human existence. What myth truly affirms does not pertain to the realm of objectivity or to the concreteness of verifiable affirmation; it is ahistorical, not having transpired as a fact, not demonstrable as a fact.[g] Bultmann's conclusion is that mythology is in part primitive science, whose intention is to explain strange or frightening phenomena by attributing them to supernatural causes, that is, to gods or to demons, but it is also a vehicle of the awareness that human life and the world in which man lives are full of riddles and mysteries, whose intent therefore is to express a certain understanding of human existence.[h]

4. Bultmann's definition of myth is fundamentally ambiguous: myth, he says, is an unacceptable objectification of the other side into this side, a representation of the non-worldly as this worldly, *and* it is an expression of man's understanding of himself. Because a thing is itself and nothing else, in its intrinsic essence myth must be one or the other or neither. Until myth is defined unambiguously, there can be no talk about myth on a scientific level. The ambiguity of Bultmann's definition of mythology necessarily makes his demythologizing a confusing and contradictory operation. It is aimed at eliminating the mythology of the New Testament and it is aimed in the very same act at retaining the mythology of the New Testament. *Webster* says that a myth is "a story that ostensibly relates historical events and serves to explain a practice, belief, institution, or natural phenomenon."[a] It is not to our purpose at this point to

---

8.4:2[a] The response given here is limited to an examination of mythology in its relation to science as a genus of thought and expression, staying within the ambit of the principles expounded in the preceding articles.
8.4:3[a] *Supra*, 1.3:7.
8.4:3[b] *Supra*, 1.2:3.
8.4:3[c] *KuM* I, 22, note 2 (*KaM* I, 10, note 2).
8.4:3[d] *Supra*, 1.3:8. Cf. Bultmann, *op. cit.* (8.2:21a), 63 (ET, 61 = *KaM* II, 185).

8.4:3[e] *KuM* I, 22 (*KaM* I, 10).
8.4:3[f] *Supra*, 1.2:1b; *JCM*, 19.
8.4:3[g] *Supra*, 1.2:3.
8.4:3[h] *Supra*, 1.3:8.
8.4:4[a] A myth is: 1) "a story that is usually of unknown origin and at least partially traditional that ostensibly relates historical events usually of such character as to serve to explain some practice, belief, institution, or natural phenomenon, and that is especially associated with religious rites and beliefs"; 2) "a story invented as

scrutinize the definition; it will do at this stage of discussion, and Bultmann prefers not to quibble over the definition.[b] But the fundamental ambiguity of Bultmann's definition is not a quibble; in its presence there can be no reasonable discussion.

5. As a narrative, a myth is a genre of literature and of thought. Because a myth is a narrative of seemingly factual events which are not factual, it is a fiction-genre of history.[a] The *finis operis* of myth, which is the intrinsic purpose of what myth is in itself, is the narration of fictitious superhuman events. The *finis operantis* of the creator of myth may be to 'explain' a practice, belief, or natural phenomenon, but not in the realm of real explanation, because mythological thinking is a studious concentration upon the unreal. Myth unfolds part of the world-view of a people, but only the part that is enclosed in the fictional realm of consciousness and only of those people who have chosen to think fictitiously. Myth is not an explanation; it is an escape from explanation. Explanation is reasoning in terms of reality; myth is a distraction of the mind from the world of reality. The hearers of myth can be deceived into thinking that the events it narrates are real, but to the extent that they have chosen to think realistically about its content they will consciously seek to discover whether it is real or not. The scientific process is of its very nature a demythologizing process.

6. Since myth and science are in two radically opposite universes of discourse, it is not true to say that myth is in part primitive science. Because of the deception accompanying the transmission of myth and ignorance regarding its origin, primitive science was mixed with mythology. But the mythology was not the science. Nor is mythology the world-picture of a prescientific era, except to the extent that people were deceived by its false credentials. In every era each individual and each community has had the freedom of choice to think fictitiously or to think realistically about the world and its causes. Those who chose to think realistically in the eras preceding the rise of modern times did not willingly accept the world-picture of mythology; its fictional climate of opinion was never very central in their view of the world.

7. Bultmann's definition of myth as an account of an event in which supernatural forces or persons are operative omits myth's essential distinguishing characteristic. Myth is a *fictious* account of supernatural events. An account of real supernatural happenings is not myth; it is historical science. It can easily be supposed and to some extent verified that makers of myth and hearers of myth do not clearly distinguish between the real and the fictitious modes of thought. That is a question for historical research. But what is of initial and paramount importance for us on the level of historical science is that *we* make a clear distinction between the two and that we maintain it by ever-renewed acts of decision in the now of our historical thinking. And as we maintain this crucial distinction which keeps us in the realm of science, the first question we must ask of Bultmann's account is whether it is a narrative of fact or of fiction. The cardinal question in this discussion is the question of reality. The fact that Bultmann does not locate myth in the realm of fiction is already an indication that he has not distinguished his own thought from the world of fiction. He gives us other clues as well. He says, for instance, that myth gives immanent, this-worldly objectivity to transcendent, unworldly reality, thus raising the question of reality, but by distinguishing an "unworldly reality" from the realm of (worldly) *objectivity* he shows that his question is not up to the level of science, for the concept of reality is the necessary intermediate *object* of all science.[a] Again he says that myth "knows another reality" than that which science has in view, but that takes his own discourse out of the realm of reality altogether, for the 'reality' of myth is by the very nature of myth a fictitious and imaginary 'reality,' which is not reality at all, so that when Bultmann calls myth's focus of reference 'reality' he associates his own thinking with the thought-form of myth. On the level of science this

---

a veiled explanation of a truth"; 3) "a person or thing existing only in imagination or whose actuality is not verifiable." (Sub verbo, *Webster's Third New International Dictionary*, unabridged ed., 1965.)
8.4:4b *Supra*, 1.3:7.
8.4:5a *Supra*, 6.3:2.
8.4:7a *Supra*, 2.4:7.

evidence shows that Bultmann's thinking is a fiction-genre, for if it were a reality-genre he would have to dissociate it consciously from the genre of fancy and unfold it within the concept of reality, that is, within the comprehensive concept of all science. Nor could Bultmann say in the realm of reality-thinking that myth *knows* another reality, for myth does not know in the realm of real knowledge but only in the realm of fanciful knowledge. Science knows that there is no other reality, for it knows that a thing is itself and therefore that there can be no 'other reality' not included within the concept of reality which is the universal object of science. The myth-teller can tease the fancy of his hearers by affirming that what he is narrating is "really true and is more real than reality"; that is his poetic license. By indulging in this kind of poetic license, Bultmann reveals to the critical reader that the genre of his account is fictional.[b]

8. The same conclusion follows from Bultmann's use of the word 'mythological.' He postpones any direct confrontation with the genre of his own thinking, assuring himself that even Kamlah finds it philosophically justifiable to use the mythological language of an act of God, which is the theme of Bultmann's exposition.[a] In examining Bultmann's discourse at this point we find an equivocation at work. On the one hand, every statement about God or the gods which is not in terms of existentialist interpretation he declares to be 'mythological.' On the other hand, every statement about God or an act of God in terms of existentialist interpretation is also 'mythological,' but in a sense that is philosophically acceptable. This means that Bultmann is unconsciously using 'mythology' as a genus divided into the two principal species of primitive mythology, which is unacceptable to modern man, and philosophical mythology, which is acceptable to modern man. He thus admits implicitly that his thinking is in the species of mythology and the genus of fiction.[b]

9. Is then the mythological language of an act of God acceptable to the man of today? To the extent that a man has chosen to think scientifically, that is, realistically, it is not acceptable, for the realistic thinker recognizes it to be an illicit escape from reality. Fiction has its function in modern thought, but there is no legitimate place for fiction about an act of God, because every man is radically obliged to face the reality of God. It is the ground of acceptability that we question. Bultmann tries to distinguish between primitive obsolete mythology and modern streamlined mythology, but neither kind is now acceptable to the man immersed in reality — or ever has been. To the reality-thinker mythology is simply a lie, and lies have never been acceptable. The man who has made the decision to think in terms of reality is therefore convinced that all mythology — including the mythological existentialist language of the act of God — is a false and harmful expression of human creativity undeserving of the credence of anyone.

10. It is clear also from Bultmann's definition of myth that his own thinking is mythological. He says that myth "objectifies the other side into this side." The immediate question which this statement raises for historical science is "the other side of what?" In some way it is the other side of the world, for in Bultmann's definition myth is the representation of the non-worldly as this worldly, and he clarifies this as being the other side of everything tangible to human thought and activity. But it is not what is beyond visibility in the order of the visible; it is what transcends the order of visibility and objectivity. For Bultman the 'other side' does not contain anything that can even be conceived as being seen, for it is on the other side of objectivity.[a] It is on the recognition of this fact that the question of a wonderland arises. Bultmann defends the plausibility of his non-world with the consideration that non-world is not a locus in a spatial dimension. But this will not take non-world out of the realm of fantasy. The objectivity of the subject-object relationship is in the intellectual dimension; it is the proximate or remote object of

---

8.4:7[b] *Infra*, Appendix I/B.
8.4:8[a] *KuM* I, 40 (*KaM* I, 34).
8.4:8[b] Bultmann, in *KuM* I, 134 (*KaM* I, 118).
8.4:9[a] *Supra*, 7.7:4; *infra*, I.23ff.

8.4:10[a] Bultmann, in *KuM* I, 130 (*KaM* I, 113), avers that only in the factual here (and now) of the concrete life-encounter does the invisible non-world of existence in Christ become clear (*deutlich*).

the act of intelligence, whose relation to physical space is only analogous.[b] Bultmann's transition to non-world requires that the entire subject-object relationship be 'transcended,' and this is where sheer fantasy takes possession. The only thing that transcends objectivity is the subjectivity of the knowing subject as such, and this subjectivity is nothing but the focal point at this end of the act of knowing. It has no differentiation or characteristics. It is totally unified and unique. When Bultmann spins a discourse around this unintelligible opposite of the object of knowledge, speaking of self-understanding, self-knowledge, the invisibility of the act of God, the identity of knowledge about God with knowledge about self, and the rest, he is thus unfolding an absurd mythology, based upon an intellectual contradiction and the false image of a world (a non-world) extended (non-extended) on the other side of the knowing subject from the objectivity he faces. It is this turning into himself of the knowing subject that is totally and irreparably fantastic, for a point of absolute unity has nothing inside of itself to which it may turn. Bultmann's thinking about the 'other side' is simply a confusion. Essentially, his 'other side' has to be on *this side*, for it 'transcends' objectivity, and the 'other side' of objectivity is my own self as a knowing subject. But after positing the imaginary folding in of all objectivity upon the knowing subject, so that the dimension of objectivity is excluded, Bultmann brings in the idea of a force or an act (of God) existing on the 'other side,' thus introducing a vague image of opposition to the knowing self within the very subjectivity of the knowing self. He calls this opposition the existential encounter.[c]

11. To scientific reflection it is known that 'transcending' the subject-object relationship means turning off the light of intelligence.[a] There cannot be even sense knowledge without objectivity.[b] There can be an encounter of man on an extra-intellectual level, but there can be no encounter that is not with an object. The notion that God is not something and that there can be an encounter with an act that is not the act of something is pure fantasy. Bultmann tries to escape this fact by placing the encounter in the realm of decision, as an act of the will. But this does not save him. An act of the will that is not enlightened by intellectual perceptions is subhuman; it is beneath the dignity of man. The folding back of intellectual perception upon the sheer appetite of a man lowers his life to the level of an irrational animal.[c] Bultmann is not calling for the suppression of reason, but he does claim a rightful place for his philosophical mythology as the complement or counterpart of science,[d] and this may explain his whole psychology of decision. Aesthetic pleasure is basically a feeling; it is situated in the appetitive faculties of the soul or of consciousness. When Bultmann appeals from the primordial dichotomy of the intellect to an act of love in the encounter of faith, he is making aesthetic pleasure the prime motive of that decision. This involves a supreme judgment of value, since it moves the consideration of God from the world of fact to the world of fiction and then draws conclusions from the non-existence of God in the world of reality. The person endowed with wisdom will not indulge in pleasure of this kind.[e]

12. It is only in the sense of pre-intellectual aesthetic pleasure that myth can be said to 'know' that human life has its ground and limit in a force lying on the other side of everything tangible and objective. The ground of

---

8.4:10[b] *Supra*, 2.4:8; 2.6:6; 2.6:15.
8.4:10[c] Cf. *JCM*, 40-41, 59, 63, 66-68, 80, 85; *HE*, 119-121, 133.
8.4:11[a] *Supra*, 2.1:1.
8.4:11[b] *Supra*, 2.3:19; 2.3:20.
8.4:11[c] *Supra*, 2.2:3.
8.4:11[d] *Supra*, 1.3:9.
8.4:11[e] The grain of truth in the reduction of the act of faith to an act of love lies in the rôle of the will in the act of faith (*supra*, 7.6:1). But Bultmann suppresses the essential intellectuality of the act of faith (7.6:2). I have described above the act of faith as an affirmation of the reality of the object of faith (7.7:4). This affirmation, like any affirmation, requires a concomitant act of the will (2.2:7), and, as an affirmation of the reality of supernatural meaning from within the focus of natural consciousness, there is an additional need for an act of the will (7.7:6). But it is an error to suppress the objective meaning behind the words of divine revelation (7.7:3). The meaning is there, and it is observable on the level of supernatural insight.

human life is beneath man, not above him. The forces which man feels at the edge of his known world are behind the knowing subject as he stands at the window of intellectual perception; they are the blind forces of his body and emotions.[a] The worship of these forces is an offense to God. Because God is infinite personal being, He must be on the side of form and intelligibility; He can be loved only by a will enlightened by intelligence. Man can look upwards toward God only by the use of his intellect, and looking implies objectivity; it is folly for a man not to look for God or to look for Him in any other way.

13. Bultmann conjectures that the dichotomy between science and mythology is not ultimate. He says that there is a *tertium quid*: a language in which existence is expressed ingenuously and a corresponding science that speaks of existence without objectifying it.[a] The *tertium* would thus be the genre of the *existentiell* and the *existential*. For the level of the *existentiell*, he says that expressions like 'I love you' and 'Please excuse me' are neither scientific nor mythological. But this observation is based on an inadequate definition of science. Any form of language that expresses reality pertains to science. The expression 'I love you' is scientific if I really do love you; otherwise it is fictitious.[b] The fact that the declaration may be on the level of common sense rather than of specialized science does not mean that it is not scientific.[c] The example is misleading, because the act of love is the proper province of the will, and therefore it stands least in the order of science. The same is true of 'Please excuse me,' which is addressed to the will rather than the intellect. One can construct a language of love, closely adhering to the proper act of the will, but the evaluation of that language still pertains to the intellect. In the expression 'I love you,' it makes a great deal of difference whether the 'you' happens to be the medicine that will save a person's life or the poison that will take a person's life; or again whether the 'you' is a person who is going to ennoble one's life or debase it because of that love. For the level of the *existential*, Bultmann distinguishes between the science of objectifying thought and the science of existence, maintaining that the science of existence seeks an understanding that is free of all worldly objectification, whether scientific or mythological.[d] But this observation is based on an inadequate definition of mythology. Since mythology has its character, not from objectification but from fiction, a fictional science of existence is in the realm of fiction, whether it objectifies or not. The proper name for it is science-fiction. The de-objectivized language of existentialist analysis is in spite of itself a kind of objectification, but, because it flees objectivity, it results in a pseudo-objectification enclosed within subjectivity as such. We call this kind of science pseudo-science. To say that the kerygma is not addressed to theoretical reason[e] means to have missed the point of theological science, which, as science, must approach the object of faith in terms of the formal concept of reality.

14. Historical theology helps us to realize that God exists 'on the other side' of the known world, and the *sense* in which this is so. It brings out the way in which the invisible God can render Himself visible. An illustration may elucidate the viewpoint of historical science with regard to God. Natural man finds himself situated on the planet Earth at a certain point with regard to the physical universe. When he studies the sky with his natural eyesight under proper conditions he is able to see a small amount of the external universe. To a very limited extent he is able to move about in that universe and place himself where the seeing conditions are best, thus enlarging somewhat the number of objects that he can see in the sky. He may be able to travel to nearby celestial bodies like the moon, but he has no physical possibility of journeying to distant celestial bodies like the stars. Yet there are three other ways in which his vision of those distant bodies can be increased. The first way is for a distant object to make itself more visible from its own place. This happens in the case of a supernova. The second way is for a distant body to come closer to the viewer. This happens when a comet comes from far away and swings

---

8.4:12a *Supra*, 2.2:5.
8.4:13a *Supra*, 1.3:9.
8.4:13b *Supra*, 2.4:2.

8.4:13c *Supra*, 2.5:2.
8.4:13d *Supra*, 1.3:9.
8.4:13e *JCM*, 36.

around the sun. The third way is for the viewer to interpose a telescopic instrument between the celestial objects and his eye, and this can increase his range of vision immensely. In an analogous way natural man finds himself in a certain place with regard to God. Man's place is at the focal point of his subjectivity as a knowing self, and God is situated beyond the range of his natural intellectual sight. When he looks up to God with his natural insight he can see something of God, but the view is hazy and unclear. Just as the Great Nebula in Andromeda is just a hazy patch to the naked eye, but reveals itself in the telescope to be a great galaxy of stars, so God to natural man is just a hazy notion. By moving about in his natural world man can improve his vision of God somewhat, and this is the natural philosophy of God. But there are three ways in which man's vision of God can be greatly enhanced. The first way is for God to make Himself more visible from his own distant place opposite man. This happens when God works a miracle. The second way is for God to come closer to man. This happened when the Incarnation took place. The third way is for God to interpose a visual medium between Himself and the eye of the viewer. This happens when the light of faith is turned on in human consciousness.[a]

15. God, as God, is everywhere and nowhere at the same time. He is everywhere, because by his power he sustains everything and by his knowledge he gives everything the reason for its existence. He is nowhere, because as an infinite being Who is pure actuality He cannot be located in any place in physical space. But God is somewhere with relation to man. It is more precise to say that God is not related to man, but man is related to God. However, from man's viewpoint in his subjectivity God has a place opposite him. God is 'beyond' the world that is accessible to the natural intellectual vision of man. It is not a question of physical distance. To imagine God as an immense cosmic man hidden behind the sphere of the sky and holding the universe in his hand is a crude anthropomorphism. When theology speaks of God as being beyond the visible world, it means especially beyond the range of human physical and intellectual vision.[a] The knowing subject looks 'down' at material objects, but he looks 'up' to the meaning of objects in the formality of his understanding. God is so far above the world of natural man and the meaning that natural man can derive from his world that God's nature is a hazy and undefinable patch of significance. And just as those without a special interest tend not to pay much attention to the Great Nebula in Andromeda, so does natural man tend not to pay much attention to God, as he thinks about the world.

16. The foregoing illustration may help to clarify the way in which theology 'objectifies' God and the works of God. Bultmann claims that faith has no relation to objectifying thought of any kind, and he maintains that when faith tries to project God and his acts into the realm of worldly objectivity it thereby becomes mythological.[a] But Bultmann does not make the important distinction between creative objectification and recognitional objectification. Mythology fictionalizes by projecting supernatural events in its historical account; it creates these events out of the fantasy of the artistic imagination. In this sense the events of mythology come forth from the subjectivity of man and do not originate in the real world. The historian of reality must also project a description of what he narrates. The account which he expresses is both an object and a pattern of related objects. But this account is an artistic representation of the truth which he knows in his mind. As an expression of art it is creative, but as a representation of reality it is not. The objectification of historical science takes place first in the mind of the historian, and it consists in the development and differentiation of his understanding, which is the formal object of his knowledge. The real historical account is the verbal expression by the use of professional art of the meaning which he sees objectified in his mind.[b] Bultmann's rejection of the New Testament narrative on the ground that it projects God into the field of known reality [c] does not take into account the function of recognitional objectification.[d] He does not realize that the his-

---

8.4:14a *Supra*, 7.5:2.
8.4:15a *Infra*, 8.6:17; 8.6:22.
8.4:16a *Supra*, 1.2:1, 1.3:1.

8.4:16b *Supra*, 6.2:4.
8.4:16c *JCM*, 37-38; *supra*, 1.2:3; 1.3:4.
8.4:16d *Supra*, 2.6:15b; 2.6:18.

torical objectification of the acts of God need not by that very fact be a projection of fiction.ᵉ His mind is not open to the real possibility that the New Testament narrative is the literary expression of real knowledge objectified in the understanding of the sacred writers and based upon data which science must recognize as true. Bultmann admits that a supervening new set of facts can modify our outlook on the world.ᶠ It is precisely the new set of facts narrated in the New Testament that is the subject of his discussion. When he excludes *on principle* that the new set of facts described in the New Testament could even hypothetically be considered as genuine, he is not showing understanding of what this admission means.

8.5. *The Question of the Scientific Mentality*

1. Bultmann's argument regarding the way in which modern man's outlook on the world is shaped may be expressed as follows:

> No world-picture that is in conflict with the scientific world-picture is admissible by modern man. But the world-picture of the New Testament is in conflict with the scientific world-picture. Therefore the world-picture of the New Testament is not admissible by modern man.

In support of the major premise Bultmann affirms that thinking today is irrevocably shaped by science, while it would be unreasonable (and therefore unscientific) to accept a worldview that conflicts with the world-view of science. He further states that as far as reason is concerned the acceptance of the New Testament world-view would only cause a man to be inwardly divided and insincere, following one view in his faith and the other in his life in the world, while as far as faith is concerned the insistence upon its acceptance is a reduction of faith to works.ᵃ

2. Among the questions for historical theology provoked by this argument are the following: What is the shape of scientific thinking? Is the world-view of the New Testament in conflict with the world-view of science?

3. With regard to the shape of scientific thinking it is to be noted that science is knowledge of the real as such.ᵃ Because reality is a concept whose content is reference to the data of experience, it is characteristic of scientific thinking that it adhere to the data of experience.ᵇ The method of science is reflection upon its medium of thought, and the goal of science is to render this medium into clearly differentiated objectivity.ᶜ A mind that is shaped by scientific thinking is therefore a mind in which the intellectual medium of thought is reduced to clearly defined concepts. While scientific thinking begins from the data of sense experience, and never contravenes this data, it proceeds to the incorporation also of the data of intellectual experience, for reality includes the meaning apprehended by valid intellectual insight.ᵈ What scientific thinking excludes from its universe of discourse is the acceptance of ideas and statements for reasons that are not based upon valid intellectual insight, such as pride, self-love, sensuality, and the appeal of aesthetic pleasure. Scientific thinking sifts its real knowledge out of the total experience of consciousness and shapes it organically around basic concepts. Such concepts need not necessarily be the open classes which characterize science of the classical kind. They may also be the concrete concepts of historical science or even those of common sense.ᵉ But it is only the man whose understanding has developed by means of a series of valid intellectual experiences of this kind who can say that his thinking has been shaped by science.

4. Bultmann's argument about the modern outlook does not take into sufficient consideration the actual shape of scientific thinking. The fact that Bultmann does not define science either in terms of reality, which pertains to its essence, or in terms of awareness of its own medium of thought, which pertains essentially to its method, shows that his own mind is not shaped by scientific thinking. He affirms a conflict between two

---

8.4:16ᵉ *Supra*, 3.2:10; *infra*, 8.6:19.
8.4:16ᶠ *Supra*, 8.0:4.
8.5:1ᵃ *Supra*, 8.0:5.
8.5:3ᵃ *Supra*, 2.4:3.

8.5:3ᵇ *Supra*, 2.4:2; 2.4:8.
8.5:3ᶜ *Supra*, 2.6:12.
8.5:3ᵈ *Supra*, 2.6:6.
8.5:3ᵉ *Supra*, 5.2:14; 5.2:15.

views of the world without first having determined what a view is and how it functions in the mind. How, then, can he know that there need be a conflict?

5. One characteristic of scientific thinking is that it takes into consideration the whole of known reality. This consistency requires that a specialist in one branch of science respect and adhere to the valid conclusions of every other branch of science. If the science of historical theology is a science, as in fact it shows itself to be,[a] it is incumbent upon scientists in every other branch of science to respect and adhere to its valid conclusions. Because the scientific medium of thought is differentiated into a great multiplicity of more restricted media, it is not unreasonable to say that the conclusions of different areas of science are located in different regions of objectivity in the mind of the person who possesses them. Diversification does not divide the knowing subject or make him insincere, for he remains one and the same; it is only the objects of his knowledge that become diversified.

6. When Bultmann says that our thinking today is irrevocably shaped by science, he is thinking exclusively of empirical science, and he is not allowing for the possibility that our thinking can also be developed in other areas of objectivity without conflicting with the conclusions of empirical science. All that is necessary for this expansion to take place is to become aware of the concepts which light up the new areas of understanding and to use them on the data provided within the area of reality.[a]

7. The enlightening of a new area of the mind is not a *sacrificium intellectus*. It is rather Bultmann who by his method asks his hearers to sacrifice their understanding by turning off the light of intelligence. The mind is the synthesis of the two parts of the primordial dichotomy of intelligence;[a] the suppression of this dichotomy — the 'transcending' of the subject-object relationship — is the supreme sacrifice of human understanding. It also turns the Bultmannian existentialist into a strangely divided knowing subject, who is no longer in possession of the uniqueness of his identity, for he is both the subject and the object of his knowledge, he is mysteriously identified with his past and his future in the instantaneousness of the present moment, he is and is not his authentic self, he is constantly becoming what he already is, never accomplishing what he is doing, and condemned to continue the same fruitless activity for the rest of his life. Such a being is also insincere, because he cannot bring this mythological preoccupation over into the real world in which he lives, and so he goes through life with two conflicting views of reality — the reality which he has to face in the real world, and the 'genuine' reality of his existentialist dream-world. This is what he gains by giving up the world-view of the New Testament.

8. There is no room for mythology in the world of science — not even the mythology of existentialism. But there is room for the New Testament in the world of science. The science of theology is reality-thinking about the Bible and about the realities which it contains. Using one's natural intelligence to assist in the comprehension of supernatural truth is not a reducing of faith to works in the sense rejected by St. Paul; it is rather the extension of the meaning of revelation to the full consciousness of man, which is a sanctifying process.[a] The total separation of faith and human understanding is an unscientific principle that can only harm the mental life of its followers. To sacrifice one's understanding in this way is totally unreasonable.

9. Bultmann states that science can overlook truths contained in ancient myths. This is true, although the genre of myth does not formally include the conveyance of truth. But Bultmann's observation has much meaning for himself. He is a child of a theological movement that has made many enthusiastic 'discoveries' over the past century and more. He accords to many of its historical conclusions the status of science; they are woven into the fabric of his view of the world. What has been left by the wayside in the course of these discoveries is practically the whole content of the New Testament, as far as serious historical truth is concerned. Bultmann has built his theology upon his belief in these con-

---

8.5:5a *Supra,* 7.8:9; 7.8:10.
8.5:6a *Supra,* 5.3:1.

8.5:7a *Supra,* 2.2:2.
8.5:8a *Supra,* 7.7:19.

clusions. They leave him convinced that the New Testament 'mythology' can never be seriously accepted any more. The considerations about the concept of history and the method of historical science which have been unfolded in the course of the present study should serve to raise serious doubts about the scientific validity of the historical movement to which Bultmann belongs. Prior to the needed thoroughgoing scientific review of the thinking which produced those conclusions, the critical reader of the program of demythologizing will at least suspect that those conclusions are based on enthusiasm and first fervor and not entirely on valid intellectual experience. The honest thinker will even go so far as to give the Bible another chance, and that is what the science of historical theology is about.

10. Bultmann is deceived by what he calls "our situation in history." To the extent that he means by that the situation of *Dasein* as revealed by existentialist analysis, he is simply speaking beside his own point. But he means also the intellectual tradition which we have inherited. Now Bultmann has inherited the tradition of the German Theological School, with emphasis on the liberal tendency within that school. What he fails to see is that there is a standpoint outside of that tradition. Even those who are nurtured in that tradition have the possibility in freedom to develop their minds in terms of concepts of which that school was not aware. It has been the purpose of the present study to indicate some of these concepts. Every knowing subject, as a knowing subject, has a standpoint outside of history.[a] The more that a person realizes this fact, the more scientific he becomes in his view of history.[b] Bultmann claims that there is no distinguishable objectivity in history. Claims like this serve the unfortunate purpose of distracting us from our task of building historical science. If thinking people of today come to realize that our situation in history is precisely to give to history the status of a full-fledged science — in all its objectivity — and go ahead and produce the science, they will raise their own situation in history to an entirely new level of existence.

8.6. *The Question of the Place Where God Dwells*

1. Bultmann's argument regarding the place of God may be expressed as follows:

The Ascension of Christ presupposes the existence of heaven. But there is no heaven. Therefore there was no Ascension of Christ.

In support of his minor premise Bultmann affirms that there is no longer any heaven in the traditional sense of the word, because heaven and hell are components of the mythical three-storied universe, and no one capable of mature thought can any longer suppose that God dwells in a local heaven.[a]

2. Among the questions for historical science provoked by Bultmann's argument are the following: What is mature thought? Does theology in the traditional sense suppose that God dwells in a local heaven? Does the Ascension of Christ presuppose the existence of a local heaven?

3. Let us immediately clarify the first question by defining 'independent thought' as scientific thought in the true conceptual sense of science defined above in Article 2. Serious reasons for doubting the independence of Bultmann's thought on the level of science have been expressed in the preceding sections of the present article. As long as the independence of Bultmann's thought has not been vindicated, the critical thinker cannot take seriously what he says about independence of thought. However, the independent thinker can himself think independently about the questions which Bultmann's argument raises.

4. Mature and independent thought about the place where God dwells means for the purposes of the present discussion thinking about the question of the place of God in terms of the concept of reality recognized within our own individual consciousness and in keeping with the meaningful dimension of intellectual experience.[a] It means the achievement of real (and not fanciful) insight into the question of God's location. Since God is perfect form and infinite act, with no admixture of matter or potentiality of any kind, we know that He cannot be re-

---

8.5:10a *Supra,* 4.2:17.
8.5:10b *Supra,* 2.4:7; 3.2:5.

8.6:1a *Supra,* 8.0:6.
8.6:4a *Supra,* 2.4:2.

stricted to any place in physical space and He cannot be identified with the physical universe or any part of it, including the substance and identity of man. It pertains to the demythologizing function of real science to have this realization and to understand its implications. But the positive function of science is something different. It is to find the place of God in terms of the intellectual concepts which are its medium of understanding and to know that it is doing so.[b]

5. Science knows that understanding is an object of intellectual vision; it is the formal object of human insight.[a] Reality is a species dividing real understanding from non-real understanding: like the genus of understanding in which it stands, its reference is both to an external object and to an internal subject; it seeks real meaning in its external object and it confers real insight upon its internal subject.[b] All talk about God and all thought about God, to the extent that it is on the level of science, is in the order of objectivity: in some way as the reality of a remote extramental object and in some way as a proximate cognitive object. Science knows both from natural reason and from the knowledge of faith that God exists. As science it must search for the place of God in the proximate area of objectivity within human consciousness.[c]

6. Whether or not there corresponds in ultimate external reality the same structure of being as we see in human consciousness is an epistemological side-issue that Bultmann has not raised and that is not central to our discussion. The old question as to whether there *really* exists that chair that we see to be a chair from within our consciousness as we view it with our natural vision is a difficulty that stems largely from a failure to analyze the concept of reality before proceeding to use it. The central point about God is that we can know Him only as an object of our intellectual vision. Since no extramental thing as an extramental thing is an object in itself, it follows that God is not an object in Himself. Furthermore, since only infinite being can comprehend infinite being, it follows that God as He is in Himself cannot become an object of human insight. But this fact can be misleading. Since quality and degree are identified in the infinity of God, in the most fundamental sense God is either known infinitely or He is not known at all, *as He is in Himself*.[a] Traditional theology admits this fact, yet it finds a middle ground. It is possible for God to reveal something about Himself as He is in Himself that is above the natural comprehension of a created intellect and yet does not require infinite intelligence to comprehend it. This is the revelation of God, and, *as revealed*, God *is an object* of man's knowledge.[b] Theological science searches for this object within its own specific realm of objectivity.

7. When we speak about our knowledge of God under the light of unaided natural reason, we are already presupposing its contrast with reason assisted by grace. This means that we are examining *e luce fidei* what our natural reason can know even though it is not unfolding *sub luce fidei*. In contrast with what we know of God from the light of faith, we find that the knowledge of God from pure reason is essentially negative and empty. We find that we know God by inference from the effects to the cause, but the effects are natural, and they do not tell us anything positive about what the cause is like. We know of the existence of God, but we do not know God as an intelligible object. We can form no concept of God, and therefore we have not the means of understanding God. We can learn something about ourselves in our relation to God: we can school our pride and our self-assertion; we can worship God; we can know that God is great and we are tiny. But we cannot know what the greatness of God means.

8. By witnessing miracles appearing in the world, natural man can learn something more about God. Bultmann denies this. He excludes what he calls a "miraculous happening in nature" (*mirakulöses Naturereignis*) for two reasons: it breaks the closed pattern of cause and effect known to science and it is meaningless for philosophical (existentialist) man, since man can see God at work only in the reality of his

---

8.6:4[b] *Supra*, 2.6:12.
8.6:5[a] *Supra*, 2.6:10.
8.6:5[b] *Supra*, 2.6:9.

8.6:5[c] *Supra*, 2.4:7; 2.4:8; 2.6:15; 2.6:18.
8.6:6[a] *Supra*, 7.7:7.
8.6:6[b] *Supra*, 7.7:12.

personal life.[a] But this reasoning results from a confused idea of nature.

9. One cannot speak scientifically of a nature-miracle without first defining the term. It is to be noted that the word 'nature' has more than one denotation in common parlance. Among the definitions of 'nature' given by *Webster* are the following: (a) "essential character or constitution; distinguishing quality or qualities; essence"; (b) "innate or inherent character, disposition, or temperament"; (c) "the system of all phenomena in space and time; the physical universe." Here we have three common meanings of the word, which, when not clearly distinguished, can produce confused thinking about nature-miracles. When Bultman speaks of a miraculous happening in nature, in which of the three senses does he intend the term? One thinks immediately of the third sense, that he is speaking of an intervention by God in the physical universe as the system of all phenomena in space and time. But what is this system, and how is it identified? It is understood that a nature-miracle in this sense would consist of phenomena in space and time; therefore it could not be distinguished as a 'miracle,' for the idea of a miracle is something supernatural, and as a phenomenon it would simply be nature. The equivocation of the nature-miracle becomes clearer when we consider the first common meaning of 'nature.' The nature of a thing is simply its essence; it is what essentially makes the thing to be what it is. By this use of the word, each object in the physical universe has its nature, and God has his nature. Even a miracle has its own nature as a miracle. Reflection on this fact shows that the idea of a miracle is based upon the distinction between the natural and the supernatural, where by the supernatural is meant what is above that essential constitution of a thing which makes it operate in the way that it characteristically does. It would be supernatural for a tree to feel pain, or for a horse to give a lecture. When Bultmann speaks of a miraculous happening in nature, he is also imagining something occurring that is above the characteristic operation of the things involved, and thus he is using all three definitions of the word 'nature' at once:

he is assuming what is the characteristic operation of the things or forces involved; he is assuming that the things or forces involved have been identified in their essence, for it is *their* characteristic operation that is concerned; and he is vaguely identifying these things or forces with the entire composite of phenomena in space and time. It is obvious that, if by 'nature' is meant all of the phenomena in space and time, and if by 'miracle' is meant a phenomenon in space and time that is above nature, there cannot be a nature-miracle. But there is a hidden equivocation in the argument.

10. Once the ambiguity of the word 'nature' has thus been clarified, it should also be clear that we are not speaking very exactly if we say that "empirical science studies the phenomena of nature." Empirical science studies phenomena; it bases its conclusions upon the data of sense and the use of sound reason. But the phrase 'of nature' in 'the phenomena of nature' creates an essence where there is no essence and sets up a false attitude regarding things that transcend the data of sense. It also produces an unscientific approach to the data of sense. Science does not decide what the data of sense are or can be. It observes the data and constructs its conclusions from them. A miracle can be observable to the senses; as such it pertains to the data of sense. Empirical science does not have within its medium of understanding the capacity to find meaning in miracles, for a miracle by its very notion is an exception to what empirical science has found characteristic of the forces involved, and empirical science is interested only in what is characteristic. But empirical science cannot infer from its own medium of thought that the phenomena associated with miracles cannot occur or that these phenomena could not have meaning from the viewpoint of another science. When Bultmann says that miracles are meaningless, he is not keeping this possibility in mind.

11. The implication of the reflection in the preceding paragraph is that the conception of 'nature' as the entire collection or series of phenomena perceived or perceptible to the senses is unscientific and misleading. It is worse to use it as "the system of all phenomena in space and time," where the 'system' is strictly limited to the or-

---

8.6:8[a] *KuM* I, 20 (*KaM* I, 8); *supra*, 1.3:4-8.

ganic meaning that empirical science can find in it. Such usage is unscientific, because it closes the mind of its user to the perception of meaning that is not contained within the sphere of empirical science. It is misleading, because there is no such thing as nature in this sense. Bultmann admits that the world-picture of empirical science is a construct produced from unifying ideas in the minds of scientists. To say that all of the meaning of all of the phenomena of sense is contained in such constructs is contrary to the open-mindedness of science; to suppress data for the reason that they are not meaningful within the constructs of empirical science is contrary to the honesty of science.

12. Empirical science does not postulate a closed pattern of cause and effect in the entirety of sensible phenomena; it simply restricts its attention to those phenomena which manifest the kind of causality that is meaningful to its own medium of thought. A biologist, for example, knows that the plants which biology characteristically studies and describes do not produce lemon juice in glass jars. But the biologist does not say that there could not be a jar of lemon juice, on the ground that such a phenomenon would violate the closed pattern of cause and effect of his science. He rather infers that this phenomenon was caused by the intervention of a free human force, but not really by an intervention, for it is a part of the world of possible phenomena. The face of the earth is covered with phenomena that cannot be explained within the sphere of empirical science, as every empirical scientist knows. The intervention of free causes makes it impossible for empirical science to include all the data within a single system that could be called 'nature,' and empirical scientists are accustomed to exclude the interventions of man from the idea of 'nature' that they have. It is but a small step of understanding to allow for phenomena attributable to free causes above man, including the interventions of God in the form of miracles.

13. Bultmann's second reason for the exclusion of miracles is that man can see God at work only in the reality of his personal life. This reason is based upon a mistaken notion of human nature and the failure to recognize the objectivity of God, as has been shown above.

14. The witnessing of a miracle in the world of sense-phenomena can enable natural man to know something further about God, because miracles are meaningful events within the context of history. A miracle says something about man's relation to God and about the relationship of particular men to God in their historical situation. But natural man can see only the surface meaning of miracles; he cannot grasp their deeper significance. God does not need to resort to miracles in order to govern the universe, but He uses miracles to make his presence felt by man. God, as infinite intelligence, can create a universe that will run without the need of occasional miracles. But in addition to the blind forces of subhuman nature there is also the fact of sin and the existence of evil forces even on the level of man and apart from the influence of evil angels. God has a special love for each individual human being, and He has two special ways of exercising his care in the midst of the operation of free evil forces. The usual way is by his Providence, and this is a subtle intervention that is not recognizable as an exception to the ordinary course of phenomena in the world. The extraordinary way is by working miracles.

15. In order that the deep significance of miracles in the order of phenomena may be grasped, man needs the help of a further miracle: a miracle of grace. The chief purpose of miracles in phenomena is to carry meaning which can be recognized only in the light of faith instilled in a man's consciousness. This light is supernatural to man in a sense that needs to be clarified. When we say that the supernatural is what is above that essential constitution of a thing which makes it operate in the way that it characteristically does, we are speaking of the absolutely supernatural, and we know that a thing cannot act in a manner that exceeds its essential constitution without ceasing to be that thing and becoming something else. A horse cannot give a lecture while continuing to be a horse. But it is not impossible for a thing to be enabled by grace to do something that exceeds the actual capability of its nature but does not exceed its radical potency.

The light of faith exceeds the actual capability of man, but it is turned on within the natural intelligence of man and is contained within the order and operation of intellectual insight. The man retains his identity as a man, but he begins to share in the life of God.

16. The man who experiences the operation of the virtue of faith is witness to a miracle taking place within his very consciousness. The light of faith is the formal object (*lumen quo*) of insight into heavenly meaning. It is an obscure awareness of the presence of God and of the nature of God. God dwells in the direction of this awareness; He is somehow contained within the formality of that light. But God has provided for faith also a material object in which the meaning of God is carried. This object is, above all, the human nature of Jesus Christ, in his visible life and activity. In the second place, the material object of faith is situated in the word of Sacred Scripture, whose divine meaning is related to the central figure of Jesus Christ.[a] By the light of faith the boundless *meaning* of the Incarnation and of the Resurrection comes to light. These and other miracles provide the intelligibility which faith recognizes and the science of theology comprehends.[b] It is with an awareness of this special double objectivity of faith that the historical theologian approaches the word of Scripture. He knows that God dwells in Jesus Christ and that in the recognition of the meaning of Jesus Christ God dwells also in himself. He strives to know God in his own consciousness by finding the historical meaning of Jesus Christ both in the Scriptures and in the teaching of the Church.[c]

17. This is where God is as far as human knowledge is concerned. God is the Creator and sustaining force 'behind' and 'above' all of the phenomena of sense. God is therefore also the ultimate force behind the phenomena of miracles. God is a non-material being located at the summit of form, intelligibility, and all perfection.[a] God is situated in the direction of the meaning of miracles. This meaning can be penetrated only in the light of a supernatural power of vision instilled in human consciousness.[b] God is visible at the summit of the understanding made possible by the light of faith; his location is thus in the direction of supernatural understanding, not as being identified with supernatural understanding but as being the concrete extramental source of the meaning visible within supernatural understanding. The clear vision of God takes place in the life after death; in this life it is our vocation and our duty to move obscurely towards the state of clear vision. Heaven is the clear vision of God and the happiness that accompanies it. Heaven is up, for it is only by ascending the intellectual gradient of meaning that a knowing subject can reach it.[c] The meaning which is heaven is real meaning; it stands at the top of the continuum of reality.[d] The beatific vision of God is a datum of real experience; it is the undeniable sight of the ultimate Reality.

18. Because God is the ultimate Reality and is totally distinct from all lesser realities, it follows that God is not a product of human fantasy or a being existing only or primarily in human consciousness, for knowledge is a form of consciousness, but the reference of the knowledge of God is to God in external reality. God is pure spirit, with no material element admixed, and therefore cannot directly be a material object of knowledge. Whether and how the beatific vision is of God as the object of glorified understanding is a question that so exceeds our experience in this life that we are at a loss to discuss it. Only those who have the beatific vision can answer this question.[a] But historical theology has a way of approaching the question of the knowledge of God in this life. God has made Himself indirectly (and obscurely) visible in a pattern of supernaturally consituted material objects of faith. These material objects are phenomena perceptible to the senses, but their positive specific meaning is evident only in the light of faith. To natural man they have only the negative meaning of a puzzle

---

8.6:16a *Supra*, 7.7:1; 7.7:9.
8.6:16b *Supra*, 7.7:3.
8.6:16c *Supra*, 7.8:10.
8.6:17a Cf. 1 Tim 6:16: "Who alone has immortality and dwells in light inaccessible (φῶς ἀπρόσιτον)...."

8.6:17b *Supra*, 7.5:2.
8.6:17c *Supra*, 2.6:14; 2.6:15.
8.6:17d *Supra*, 2.6:16; 2.6:20.
8.6:18a The answer to the question of how God can be an object requires intuitive vision of the Blessed Trinity. We

or a 'wonder.' It is with the positive meaning of these phenomena that the science of historical theology is chiefly concerned.

19. The first and principal material object of the science of historical theology is Jesus Christ in his life on earth both before and after his Resurrection from the dead. This life is a miraculous life, and it is precisely in its miraculous character that historical theology finds its specific intelligibility. It is the purpose of historical theology to discover and expound the meaning of the life of Jesus Christ. The kind of meaning that historical theology discovers is historical meaning; it knows Jesus Christ as past. This means that it knows Jesus Christ as a reconstruction in memory of how his life appeared as it actually took place. It is not the direct witnessing of the life of Jesus Christ as it is actually taking place, for that occurrence is forever past. But the historical theologian does not simply reconstruct the life of Christ from scattered vestiges that he may happen to find around him. He finds the life of Christ already reconstructed in the scripture of the New Testament. The text of Sacred Scripture is a literary medium constituting a more immediate material object of his consideration as an historical scientist.[a] In the New Testament he finds the miracle of the life of Jesus Christ narrated, and he penetrates the meaning of this miracle. The meaning of this miracle is the meaning of God for man. It is a meaning that is set in the reality of the life of Christ, and it is from within the formal medium of reality and reality-concepts that the meaning of the life of Christ becomes clear. As real meaning it has reference to the data of sense, which are remotely the historically credible data perceptible in the life of Christ as it actually took place and as narrated in the New Testament, and proximately the presently visible phantasms induced by the literal meaning of the words of the New Testament itself. What the historical theologian soon realizes is that the New Testament (and the whole Bible) is an *intellectual miracle*, and he has the opportunity to search for its meaning while actually experiencing the miracle that it is. The more he becomes aware of the miraculous nature of the New Testament, the more definitely can he also test his knowledge of God in the reality that he sees before him. The meaning of God is contained in the miracle of Jesus Christ, and the miracle of Jesus Christ is contained in the intellectual miracle of the New Testament, which is an abiding reality. The dimension of understanding of God which develops in the mind of the historian as he delves into the meaning of God contained in the New Testament is therefore a dimension of real understanding, firmly rooted in the data of sense and in continuity with the world of reality that we associate with the data of sense.

20. While the material object of historical theology is Jesus Christ as past, the formal object in which the historical knowledge is contained must be present. It has been pointed out that the formal object, as the general formal object of all theology, is the obscure concept of God.[a] Jesus Christ is both God and man. The formal object of historical theology is the God Who is Jesus Christ, while its material object is the Man in Whom divinity dwells corporally.[b] Because God is in Jesus Christ yesterday, today, and forever, it follows that the abiding presence of Jesus Christ is an object of theology; it is the material object of mystical theology. As such it does not pertain directly to the present question, but what does pertain is the history of the departure of Jesus Christ from visible presence in the world of sense phenomena, and that is the story of the Ascension.

21. By the Ascension of Jesus Christ into heaven is meant the real event which took place forty days after his real Resurrection from the dead and which is historically narrated in the New Testament. It was a real event, because it took place within the world of reality, as distinguished from the

---

know from revelation that in some sense God the Son, as the divine Word, is the 'objectivity' within God, that God the Father is the origin (and 'subject') of this 'objectivity,' and that God the Holy Spirit is the 'dynamism' of the life of God.

8.6:19a The text of Sacred Scripture is an expression of material historical science (2.6:28) and, as such, contains the 'proposed' formal object (guaranteed by divine authority) of historical theological understanding.

8.6:20a *Supra*, 7.7:6.

8.6:20b Cf. *supra*, 7.7:1; 7.7:9; 7.8:9.

world of fantasy, and we know that it was real because it is rooted in the data of sense. Knowledge of the Ascension, to the extent that it is a meaningful intellectual experience in the order of real intellectual comprehension and increases our organic understanding of the whole of reality, is by that very fact a part of our knowledge of reality, but God has been so good as to solidify our knowledge of the reality of this event by concretizing it in the data of sense. How this event is so concretized is a question for historical theology.

22. It is obvious from the notion of heaven made available by the light of faith that an ascent into heaven could not be seen with the eyes of physical vision, since heaven is where God dwells, and God dwells at the summit of being along the intellectual line of ascent.[a] In the divinity of his Person, Jesus Christ always was in heaven. Therefore, in his substantial Person and in his life as God He did not ascend. But his human nature is united hypostatically to the divine Person by a special relationship that makes it the Body of his divine Person without mixture with the identity of his divine nature. By the Incarnation the Second Person of the Blessed Trinity assumed the Body of Jesus Christ to Himself by a process that was perfected in two stages: in the first stage the life of the human nature was not totally pervaded by the association with the life of God that is possible in heaven, while in the second stage it is. The Ascension of Jesus Christ was the transition from the first to the second stage. This transition may be regarded whether with respect to his human conscious life or with respect to his Body. In the one case the ascent into heaven meant the total immersion of his human consciousness in the beatific vision in the sense that no obscurity or suffering of any kind was any longer possible to it. This transition took place in the objective part of his cognition and his appetition. There was no change in Jesus Christ as a knowing subject, for the knowing subject as such is unique and unchangeable. We do not know what the self-awareness of a man whose Person is divine is like; we can never know. There is no valid analogy from what our self-awareness is like to what the self-awareness of Jesus Christ was like. The only bond of similarity is in the objectivity which we share. The reason for the delay in the total assumption of the human nature of Jesus Christ to the life of the Trinity and the beatific vision was to enable Him to suffer for our redemption and to teach us by his example. The second stage in the assumption of the consciousness of Christ is called an 'ascent,' because its objective medium of knowledge and love moved upward to total transfiguration by the meaning and goodness of God.

23. In the other case the Ascension of Christ means the total glorification of his Body. By the Resurrection his Body was already partially glorified; this glory was consummated in the ascent into heaven. Is heaven, then, a place in physical space? This question is the stone of stumbling for thinkers like Bultmann. If heaven is in physical space, so that the glorified Body of Jesus is *somewhere*, then where could heaven be? According to the New Testament narrative the Apostles saw Him flying upward into the sky until a cloud took Him out of their sight. But is this not pure mythology? If He moved upward at the indicated rate (about fifty kilometers an hour, let us say), He would still not be out of the solar system. And we have scrutinized the whole area rather thoroughly.

24. A visual answer to this question is not possible. To expect an answer in terms of bodily vision is to mistake the entire notion and meaning of heaven. To expect an answer in terms of intellectual vision is to require the expression of an intellectual experience that we cannot have in this life. It is like expecting a dog to tell us why he cannot work a problem of calculus. The whole framing of the question is wrong. We do know that, when asked by the Apostles where He was going (in the sense of physical space), Jesus refused to give them an answer, simply stating that where He was going they could not come. They could not come physically until death; they could not come intellectually in the sense of grasping the place in their minds. Because the place is in heaven, the rec-

---

8.6:22[a] I repeat for the sake of clarity that this line of ascent is primarily extramental; it is the formal structure of the reality which contains and sustains the order of meaning. Meaning, in turn, is the outer perimeter of intellectual vision.

ognition of the place requires that one be in heaven. And that is why the very question is mistaken. We can know that bodies can ascend into heaven, but we cannot know where they are when they are in heaven.

25. But how do we know that heaven is a place where even bodies can go? We know it from the historical narrative of the New Testament and from the witness of the Apostles therein contained. The Apostles did not witness the actual ascent and entry of the Body of Christ into the place where the bodies of glorified human beings exist after their resurrection and ascension; it could not be seen with bodily eyes in this world. But they did witness with their bodily vision the miracle of the ascent of the Body of Jesus Christ into the physical air until a cloud took it out of their sight, and this miracle is the testimony of God that the Body of Jesus was transferred into heaven in a way that could not be physically witnessed. The visual miracle was not the ascension; it was the divine testimony of the fact of the ascension. This miracle was the most graphic way of illustrating the ascension that could not be seen, and the kindness of Jesus in performing it for us is a final indication of his boundless generosity in our behalf. Meditation upon the miracle gives a sufficient phantasm for intellectual appreciation of the fact of the invisible event, and we know that the phantasm is based in reality.

26. There can be no scientific speculation on the mode of the actual ascension of Jesus into heaven, because science has reference to the data of experience, and we have no experience of how this ascent took place. We can speculate on the possibility of bodily ascension (in addition to our historical knowledge of the fact), but any explicitation of the mode is necessarily a product of the creative imagination. We know that a bodily substance can collapse into itself to the point where it becomes invisible; we can imagine the transformation of a body into energy; we can imagine the transfer of such a body at incalculable speed to any point in the universe; we can imagine the disappearance of a body from physical space and its reappearance somewhere else in space; we can imagine the existence of a body in an unextended state (even though we cannot imagine such a body); we can imagine the existence of a supernatural dimension of a body according to which it would be extended into another dimension of space that we cannot now know or see; and we can imagine the existence of a physical place visible even to the bodily vision that is so far away that it cannot be reached by our bodily vision. All of this is unscientific speculation. As far as the physical place of heaven is concerned, we do not know, and we do not need to know.

8.7. *The Question of Angelic Spirits*

1. Bultmann's argument regarding the rôle of angelic spirits may be expressed as follows:

No world-view projecting a belief in angelic spirits is acceptable to modern man. But the New Testament world-view projects a belief in angelic spirits. Therefore the New Testament world-view is unacceptable to modern man.

In defense of the minor premise Bultmann notes that the world at the basis of the New Testament narrative is a world of spirits and miracles in which sickness and disease are attributed to the causality of evil spirits. In support of the major premise he observes that the laws of nature have now been discovered and that even those physiological and psychological phenomena whose causes remain mysterious become scientifically intelligible only to the extent that we are gradually assigning natural causes to them.[a]

2. Among the questions for historical science provoked by Bultmann's argument are the following: Does the New Testament project a belief in spirits? How do phenomena become scientifically intelligible?

3. It is Bultmann's contention that modern man cannot ascribe to any phenomena the causality of angelic spirits, because phenomena are intelligible only in the light of scientific causality. The fundamental scientific error in his reasoning lies in his limiting scientific causality to the causality which is intelligible to empirical science. It is by no means true that even empirical science always knows the nature of the forces to which it attributes the causality of certain phenomena, but the names it gives to forces

---

8.7:1[a] *Supra*, 8.0:7.

that remain mysterious do necessarily reflect notions that remain within the borders of its own viewpoint, or thought-medium. Empirical science cannot identify and correctly describe the causes operative in an historical event for the simple reason that the characteristic causality of empirical science is not historical causality. While it is true that empirical science from within its own field cannot ascribe any phenomenon to angelic causality, it is equally true that empirical science cannot ascribe any phenomenon to free human causality. The same argument, therefore, which is intended to sweep the world of reality clean of angels also sweeps away free human causality and man himself as far as his essential freedom is concerned.

4. The deceptiveness of Bultmann's argument stems from the failure to recognize historical causality. To ascribe a malady to the influence of an evil spirit is certainly the recognition of a kind of efficient causality at work, but not in the sense of empirical science. When a woman and a man marry, the exercise of their freedom is the efficient cause of the marriage, but this is in no way visible to the observation of empirical science. Historical science approaches problems from the viewpoint of final causality, and it is in the light of final causality that it recognizes efficient causes at work on the level of its observation.[a] When Bultmann says that phenomena remain mysterious except to the extent that science can assign causes to them, he is therefore oversimplifying the case. A wedding is a totally mysterious event as far as atomic physics is concerned, but it is not totally mysterious to the atomic physicist who is able to rise above the purview of his own specialty and see historical causality at work.[b]

5. Science faces reality head-on. When Bultmann says that psychological phenomena remain mysterious until their cause has been found, meaning the efficient cause as determinable by empirical science, he is not speaking scientifically, because he is not facing reality squarely. He is surrounding the whole world of human activity with an aura of mystery that is not there. By this reasoning, a game of chess, a season-ticket of admission to a football stadium, the candles on a birthday cake, are totally mysterious things whose causality has not been scientifically assigned. Bultmann takes this kind of human causality for granted, without realizing that it is the *purpose* that is the causality, and without relating the assumption logically to his argument about the causality of angels.[a]

6. When Bultmann says that modern man cannot believe in spirits because "the forces and laws of nature have been discovered,"[a] he is unscientifically restricting the statement to the blind forces of material nature and the laws thereof. He is not letting his vision expand to include the free forces of intelligent beings, such as angels and men. His existentialist notion of human freedom locates genuine freedom in the fanciful world of existential self-understanding and does not take into account the obvious presence of freedom in the real world. He cannot find a place for freedom in objectivity. Now, while it is true that the source of freedom is the subjectivity of intelligent beings, the exercise of freedom produces effects in the world of objectivity, and that is a fact for which the existentialist theory of history cannot account. Bultmann's theory of history does not allow that the exercise of freedom can be recognized as *the* cause by which an historical event becomes scientifically intelligible, and this false presupposition rules out any real causality by angels or by human beings as human. Historical science does not seek the laws that are supposed to explain an historical event; the historical scientist knows that laws cannot explain any historical event as an historical event.[b]

7. Empirical science of both the classical and the statistical types is heavily dependent upon mathematics as the medium of its reasoning.[a] What mathematical evidence is there for Bultmann's affirmation that modern man can no longer believe in spirits, whether good or evil? He does not offer any statistical proof. He does not refer the reader to any scientific studies made. He does not indicate that there is any more scientific justification for his declaration than the mere fact that this is what he thinks about the matter. His own view is made the standard of what all modern men can think.

8.7:4a *Supra*, 5.1:2.
8.7:4b *Supra*, 5.3:5-6.
8.7:5a *Supra*, 5.1:4.

8.7:6a *Supra*, 8.0:7.
8.7:6b *Supra*, 5.2:20.
8.7:7a *Supra*, 2.6:27.

To draw a conclusion of this kind without having reviewed any inductive evidence is from the viewpoint of empirical science an unscientific operation. It is not therefore worthy of scientific credence. Actually there are so many people around in the modern world who do believe what the New Testament says about spirits, good and evil, that Bultmann's blanket declaration can be statistically refuted at will. But that puts Bultmann in contradiction of the method by which empirical science discovers intelligibility. Bultmann says that it is impossible to make use of the radio or of modern medical discoveries and at the same time believe in the New Testament world of spirits and miracles. Is this statement based on scientific fact? Clearly it is not, for the historical fact is that the man who discovered the laws of genetics (Gregor Mendel) believed what the New Testament says about spirits and miracles; so did the man who discovered bacteria (Louis Pasteur); and so did the man who discovered the radio (Guglielmo Marconi). Bultmann is not, therefore, adhering to the historical facts as they have really taken place, when he declares that it is impossible to be functionally aware of modern scientific instruments and at the same time "to believe in the New Testament world of spirits and miracles." [b] Bultmann's conclusion is scholarly, but it does not represent the facts; rather, it is fiction fabricated out of a preconceived idea, which, as both modern science and common sense agree, cannot be changed into solid fact by intervening powers of literary erudition.

8. John Macquarrie relates that he once asked Bultmann about the thousands of people who go to Lourdes every year. Bultmann's reply was that "these are not modern men." They are not modern men because they do not conform to what Bultmann thinks modern man must be.[a] But many of the people who go to Lourdes are eminent in empirical science and in technology. Some of the people who have gone to Lourdes (not all of them in a spirit of belief) have been eminent doctors of medicine who have written scientific descriptions of the miracles they witnessed. Bultmann has not gone to Lourdes. He shows no evidence that he has ever studied the testimony of these doctors of medicine or reviewed the data at hand. Yet he feels that he as a modern man can exclude the possibility of miracles and declare the scientists who know about miracles not to be modern men. From this it is clear that Bultmann's 'modern man' is not modern scientific man, but modern biased man, and modern unscientific man. His argument about the existence of angelic spirits is therefore open to the following rebuttal:

Nothing that is impossible can happen. But it happens that some modern men accept the New Testament belief in angelic spirits. Therefore it is not impossible for modern man to accept the New Testament belief in angelic spirits.

9. The science of historical theology sees the problem of the activity of angelic spirits as described in the New Testament to be an historical one. It seeks the historical intelligibility of this activity. It does not embark on the false method of *assuming* that the episodes involving angelic spirits are creative objectifications of human fantasy, but it rather remains *open to the possibility* that the episodes are a recognitional objectification of the facts as they are understood in the light of their historical intelligibility. The historical meaning of these facts is essentially tied to the historical situation in which they took place. A detailed examination of particular episodes would be a digression in the present discussion; it is the object of the historical exegesis of the New Testament. But it is important for the present discussion at least to advert to the historical character of their meaning. Bultmann draws attention to the ascribing of illness to the influence of evil spirits, and there are several episodes of this kind. But it is nowhere stated in the New Testament, either explicitly or implicitly, that evil spirits are the *ordinary* cause of illnesses, and it says nothing at all about the causes on the level of empirical science. Bultmann's case for pseudoscience in the New Testament is not factual. It is not stated in the New Testament that the stars are demonic beings which enslave mankind to their service, or that the stars and planets are pushed around by angels. This is pseudosci-

---

8.7:7[b] *KuM* I, 18 (*KaM* I, 5).

8.7:8[a] *Supra*, 8.2:5[d].

ence that Bultmann is projecting into the New Testament in order to prove his case. There is in the New Testament a conscious distinction made between ordinary illnesses, due to strictly physical causation, and those with an added supernatural cause. Bultmann overlooks the clear indication in the New Testament narrative that the diabolical phenomena are extramentally related to the operation of sin and that the miracles are related to the life and mission of Jesus Christ. The presence of Jesus Christ on earth is a special historical reason why the phenomena described could occur repeatedly then and yet be extremely rare in other historical situations.

10. It is the task of historical theology to examine the phenomena precisely in their relation to this unique historical situation. As a science, historical theology must also take into account the intelligible medium in which the operation of diabolical forces is claimed to have been recognized. In the creation of the New Testament as a recognitional objectification there were operating insights that are not available to ordinary men. Bultmann provides for no difference between the way in which Jesus as the God-Man saw the factors operating in a given situation and the way in which they were seen by the people around Him. Yet this difference is the very point of the New Testament. We know that even in ordinary human experience there is a vast range of difference in the depth with which various persons will see the same situation. We might consider, for example, the difference between the views of a given surgical operation as witnessed by a relative of the patient, a student nurse, an attending colleague of the surgeon performing the operation, and the most eminent specialist in the world for that kind of operation. This is but a tiny example of the range of viewpoints implied in the New Testament narrative. One has to have a sufficient familiarity with certain modes of activity in order to be apt to recognize them. Would Bultmann be able to recognize an angelic intervention if he encountered one? Does he have any idea at all of the place in human consciousness that an angel would be likely to appear, if he did appear? Bultmann's argument assumes that everything in science is as obvious as black and white, but science knows that it is not. Exquisite arguments for the possibility of angels and angelic influence in the world have been expressed by writers like Thomas Aquinas.[a] Bultmann seems acquainted with the defects in medieval cosmological theory as regards the movement of planets by angels, but he does not seem to have read what these writers have said about the activity of angels in history.[b] He rejects their arguments without giving them a hearing. The theory of history was largely neglected in medieval times; it has been left to our era to develop it. But the medieval writers had no little insight into the historical situation described in the New Testament. They unfolded their historical explanations regarding the angelic interventions in the human episodes around the twofold awareness of the rôle of sin and the vocation to eternal happiness. Bultmann's rejection of angelic activity does not take account of this awareness.

11. Neither the New Testament nor traditional theology presents good and bad angels as stage-characters in a world of fiction. They are presented as forces in the real world. The scientist of history, as he examines the data of the past, must always keep account of the relationship of this data to his own present, that is, to his own historical situation.[a] The Bible tells us that Lucifer, the prince of devils, was cast eternally into hell for actuating by his free decision a philosophy based upon his own self. Lucifer identified divinity with his own self; he refused to worship the God who exists in objective reality. The Bible tells us also that Lucifer suggested his own philosophy of self-understanding to our first parents, and that was what brought Original Sin into the world. Bultmann rejects Original Sin as untenable mythology. He rejects the whole idea of sin, except within the confines of his notion of existence. But the notion of sin is the only means we have of avoiding the sin of pride that was Lucifer's, as far as that sin is related to our situation in the real world of scientific reality and to the God Who exists in objective reality. How does Bultmann recommend that we avoid Lucifer's sin, in the light of existentialist analysis? Bultmann's theology of existence

---

8.7:10a *S.Th.*, I, qq. 59-64.
8.7:10b *S.Th.*, I, qq. 106-114.
8.7:11a *Supra*, 4.2:17.

is ingenious from the viewpoint of its mythology. But from the viewpoint of historical science, it is lethal.

8.8. The Question of the End of the World

1. Bultmann's argument regarding the end of the world may be expressed as follows:

> The idea that history will end in a cosmic catastrophe is mythological. But the New Testament idea is that history will end in a cosmic catastrophe. Therefore the New Testament idea is mythological.

In support of the major premise Bultmann affirms that every sensible person (*jeder Zurechnungsfähige*) knows that history will continue to run its course. He allows for the belief that the world as we know it may come to an end, but this end is expected to take the form of a natural catastrophe. In support of the minor premise Bultmann points out that according to the New Testament expectation a cosmic catastrophe will come very soon, the Judge will come from heaven, the dead will rise and be judged, and men will enter either into eternal life or into eternal damnation.[a]

2. Among the questions for historical science provoked by this argument are the following: How does every sensible person know that history will continue to run its course forever? How do we know how this world will end? What is the difference between a cosmic catastrophe and a natural catastrophe of these proportions?

3. The historical scientist knows that as an historical scientist he has no vision of the future at all, the reason being that history is knowledge of the past, and the future is not past.[a] The limited predictability that the knowledge of the past gives is only a probability, and it is always liable to be nullified by what actually takes place.[b] The historical scientist is a specialist in particular situations; he can gather the factors entering into a past situation; but he can never be sure of all the factors that will enter into a future situation. History is filled with surprises; new forces constantly arise to change the balance that might have made some prediction possible. But the past and the future pertain to the historical sphere of discourse (since the future will at some time be past) and not to the sphere of empirical science as such. The empirical scientist cannot predict what the future will bring. He can predict the effect that a given material force will produce according to the laws he has defined; he can predict how a given blind instinct will operate; but he cannot predict what forces will be operative in a future situation.

4. Bultmann's prediction of a 'natural catastrophe' is a confusion of history with empirical science. The point of the question about the end of the world is not what material forces will be operative to do the job. That is entirely beside the point. The real point is that, when it happens, empirical scientists will find themselves out of a job. No natural scientist is going to take the measurements of that catastrophe. And when it comes to the question of the destruction of the cosmos, there is no difference between a natural catastrophe and a cosmic catastrophe. Bultmann's argument is therefore based upon a distinction that has no meaning.

5. It follows from the nature of historical science as well as from the nature of empirical science that no one knows from his own knowledge of the world how the world will end. The historical scientist knows that he does not know, and every sensible person is an historical scientist, at least on the level of common sense. We do not know that history will continue to run its course; we may consider ourselves fortunate if we know how history has run its course up to the present.

6. Bultmann assumes that the New Testament prediction about the end of the world is a guess based upon the mythical knowledge of the world available at the time. There is no rational basis for this assumption. The New Testament reports the prediction of Christ about the end of the world, and Jesus had divine knowledge concerning this subject. Bultmann does not believe that Jesus is God, he does not believe that God exists in external objective reality, and he therefore does not believe that the New Testament could be reporting divine knowledge about the end of the world. But Bult-

---

8.8:1a *Supra*, 8.0:8.
8.8:3a *Supra*, 3.2:9.
8.8:3b *Supra*, 6.1:3.

mann's nonbelief is based upon confused reasoning, and it is no more valid than it can demonstrate itself to be.

7. The prediction of the end of the world reported in the New Testament is possible for the reason that what is future for man is already past as far as the divine knowledge of history is concerned. God knows what the future will be, and He is able to reveal it to man. Eschatology in the proper sense of the word is knowledge of the future, based upon what God has revealed about it. Prophecy in the full sense of the word is a genus of historical expression that is unique in its own right. It is history in the sense that it is knowledge of the past from a future viewpoint that is not within the direct reach of man; it is unique in this that its absolute present is not our absolute present. Eschatology has its own medium of understanding and its own hermeneutical rules by which that meaning is recognized. To judge the truth of New Testament prophecy without first determining the literary genus in which it is composed is doing the New Testament an injustice.

8. The New Testament does not predict that history will come to an end. What it does predict is that on the level of mankind as a whole the present phase of history will end and a new one will begin. How the transition will take place is obscure, but some knowledge can be gained by a correct exegesis of the inspired text. What Bultmann says about the New Testament prophecy is not worthy of credence, because he has missed the literary genre in which the prophecy is cast. Nor does he analyze the text in the light of wisdom, which is the first requisite for the interpretation of Sacred Scripture. We know that Bultmann does not bring wisdom to his work, because the fear of God is the beginning of wisdom,[a] and Bultmann is fearless of the judgment of God, as he tells us himself.[b]

9. The New Testament predictions about the future are not only about the future of the cosmos and of mankind as a whole. They also regard the future of each individual man. Every man moves from this world to the next at the moment of his death, and it is then that his eternal destiny is determined. God warns us in the word of Scripture that we must face particular judgment at that time. Bultmann seems to devote little attention to this fact in his consideration of the end of the world, even though it has significance for his problem of the parousia. Bultmann holds that the mythical eschatology of the New Testament was exploded by the fact that the parousia of Christ never took place as the New Testament expected. When did the New Testament expect the parousia to take place? It does not specify the length of the interval. It does not venture a date. It says rather that "no one knows."[a] What some or many people in the early Christian Church may have thought about the event is only what those people thought; it is not what the inspired word of the New Testament says, and that is the subject of our discussion. Bultmann says that the New Testament expected the parousia to come "soon." Does the New Testament say that it will come soon? And how long is soon? On the one hand, a thousand years are as a single day in the courts of the Lord. That makes less than two days since the Ascension. On the other hand, the Lord has come in judgment upon all the preceding generations of people, and for them He came 'soon.' The members of the first Christian community were beginning to face death from the very moment of the Ascension. In the order of values, was this not a far more vital and imminent consideration than speculation about the end of all mankind? Bultmann should have given more thought to this aspect of the parousia.

10. Bultmann's destruction in his own mind of what the New Testament says about the future, with his substitution of the existentialist dream, perfectly eliminates the virtue of Christian hope.[a] When the idea of heaven loses its meaning,[b] human life loses its meaning, and the future becomes no future to look forward to. 'God's future' is the real knowledge of our future that God has revealed to us. If we mythologize God's future by transforming it into the mythology of existentialism, we destroy our own future in the process.

8.8: 8a Ecclesiasticus, 1:16.
8.8: 8b *Supra*, 8.0:6-8; *KuM* I, 38-39 (*KaM* I, 30-32).

8.8: 9a Mark 13:32.
8.8: 10a *Infra*, I.35.
8.8: 10b Cf. *JCM*, 20.

APPENDIX I

CONCERNING RUDOLF BULTMANN'S CONCEPTION OF NEW TESTAMENT THEOLOGY

A. *The Insight Underlying Bultmann's Theology*

The Understanding of Faith

EXPOSITION

1. Rudolf Bultmann holds the conviction that faith can be nothing else but response to the kerygma, and that the kerygma is nothing else than God's word addressing man as a questioning and promising word, a condemning and forgiving word. He maintains that the word of God does not offer itself to critical thought, it speaks only to our concrete existence. The kerygma can never be spoken except in a human language and as formed by human thought. Therefore the kerygma never appears without already having been given some theological interpretation. This fact makes it clear that the statements of the kerygma are not universal truths but are personal address in a concrete situation, and that is why they appear only in a form shaped by the way in which an individual understands his own existence or by his interpretation of that understanding. The statements of the kerygma are understandable only as a word addressed to the individual in his situation — as a question asked to him and a demand made of him. "The kerygma is understandable as kerygma only when the self-understanding awakened by it is recognized to be a possibility of human self-understanding and thereby becomes the call to decision." [a] This is so because the theological investigator obviously (*offenbar*) cannot presuppose his own faith as a medium of understanding (*Erkenntnismittel*) and make use of it as a presupposition for methodical work. What he can do is keep himself aware of the questionability of all human self-understanding in the knowledge that existential (*existentiell*) understanding of one's self is real (*wirklich*) only in the act of existence and not in the isolated reflection of thought (*in der isolierten denkenden Reflexion*). [b]

CRITIQUE

2. The insight that faith can be nothing else but response to the word of God as a questioning and promising word, as Bultmann describes it, is a 'negative insight.' It is a claim to know what faith cannot be. Negative insights are often illusions; they yield before the occurrence of the positive insights they have ruled out. The blind might agree among themselves that the sight of color is impossible, but it would be evident to anyone with the power of physical sight that their conclusion was false. Similarly, Bultmann's claim concerning what faith can never be is evidently false to those who see what he claims cannot be seen. Bultmann's claim is obviously false to those who have seen that the word of God *does* offer itself to rational thought and that the object of faith is understandable in the form of objective truth. Those who have this insight will necessarily regard Bultmann's claim about faith as a testimony of his blindness.

3. The mind of the theological sci-

I.1a "Anders ausgedrückt: das Kerygma ist als Kerygma nur verständlich, wenn das durch es geweckte Selbstverständnis als eine Möglichkeit menschlichen Selbstverständnisses verstanden wird und damit zum Ruf zur Entscheidung wird." *Theologie*, 589.

I.1b R. Bultmann, *Theologie des Neuen Testaments* (6th rev. ed., Tübingen, 1968) (hereinafter referred to as *Theologie*), 588-589; ET, *Theology of the New Testament* (New York: Scribner's, Vol. I, 1951; Vol. II, 1955) (hereinafter referred to as *Theology* I-II), II, 240-241.

entist is his chief instrument for the attainment of truth — both the knowledge of the truth and the understanding of the truth. If he is a theologian in the true sense defined above (7.7:12), he has within his mind both the light of natural reason and the light of supernatural faith. Subjectively speaking, the 'light of faith' is the instilled capacity to 'see' the object of faith, and it is also the act of seeing this object with reference to the insight of the knowing subject as such. Objectively speaking, the 'light of faith' is the supernatural intelligibility of the object of faith as a source of illumination detached from the knowing subject as such and thus constituting the illuminated object. The objective light of faith is also the meaning of the object of faith as this meaning is perceived by the knowing subject (2.4:8). The middle ground between the knowing subject and the object of his faith is occupied by the supernatural understanding, which absorbs both the insight and the meaning into itself as an abiding state, fusing the intellectual dichotomy of this supernatural light into the unity of the intellect in its own concrete existence.

4. Once understanding has been recognized to be the existential ground on which the act of insight occurs, it becomes possible to resolve the question raised by Bultmann as to when the understanding of one's self is real. It has been shown above (2.3:14) that the knowing self can never directly understand itself, because by very definition it cannot be an *object* of knowledge and therefore there is no middle ground for understanding to occupy. There is no real understanding of the knowing self as such. But since the understanding possessed by the knowing self is existentially one and the same with the insight of the knowing self, there is, in the act of understanding truth, an indirect understanding of the self who understands the truth. Such understanding is 'real,' although one must keep in mind that reality has its modes.

5. Bultmann affirms that "obviously the theological investigator cannot presuppose his own faith as a means of understanding." In this he errs, for he confuses the deceptive knowledge of negative insight with the sound and genuine knowledge of positive insight. What is meant by the word 'obviously'? Only that is obvious which is given in sense intuition or which informs the mind by intellectual intuition. Negative insight does neither; it is merely the absence of a hypothetical insight. Furthermore, what seems to be obvious from sensory intuition may actually be a false conclusion concerning that intuition. It may seem obvious to the uninstructed that the sun and the moon are almost equal in size, yet this is only an illusion. Such illusions remain 'obvious' until they are eliminated by the intellectual insight derived from rational thought about what sensory experience makes present. Bultmann's affirmation labors under a similar kind of illusion. It may seem obvious to the uncritical theological investigator that his faith cannot be used as a presupposition for methodical work, but to the extent that he critically thinks about this seeming fact he will find that it can. Faith is a means of understanding; it is precisely that kind of means which we call an 'intellectual medium,' and this fact indicates that faith pertains somehow to human understanding (7.7:15). Scientific work, to the extent that it is scientific, requires an explicit awareness of its own medium of thought (2.4:7). And theological science, to the extent that it is scientific, requires an explicit awareness of the formal object of faith as its specific medium of thought (7.7:6). It is therefore clear that the 'methodical work' predicated by Bultmann of his 'theological investigator' is not scientific work in the true and proper sense of the word 'science'.

6. Because Bultmann does not see that faith is a medium of understanding, he draws a false conclusion regarding what it means to understand the Christian message (called by him the 'kerygma'). He says that the Christian message is understandable in its specific character only as a recognition that it is a possibility of human self-understanding and is therefore able to be interpreted as the call to decision. The first regards the nature of human understanding and the second regards the relationship of a cognitive message to human volition. To 'understand the kerygma as kerygma' means to achieve an insight of abiding quality into the distinctive meaning of the kerygma. Now to the extent that human intelligence is allowed to function at all it becomes clear that the 'self' of the kerygma is not my own self. I know this from the evident fact that I am

talking *about* the kerygma, and therefore I am necessarily distinguishing my own self from the self of the kerygma. It is consequently irrational to say that understanding the kerygma is identical with understanding myself. This is not to imply that understanding the kerygma does not include with it (*per accidens*) in some way an understanding of myself. No one should deny that. What it does imply is that for an intellectual and scientific grasp of the subject of Bultmann's discourse when he speaks of 'understanding the kerygma,' it is absolutely necessary to recognize the difference between the self of the kerygma and my own self and then to penetrate the meaning of the self of the kerygma. It is unscientific to confuse these two intellectual elements.

7. It is to be admitted that my own self is the most 'unobjectifiable' of all objects of my own understanding. In this sense my own self — the knowing subject of all my understanding — is absolutely incapable of being understood by me. But the existence of my self as the knowing subject (prescinding from its grade of existence) can at least be posited, and once it is posited it can, as the symbol of my identity, serve as an object of cognition (2.3:18a). Only as an object of cognition can the self be spoken of at all. Bultmann says that the kerygma is recognizable as kerygma only when (to the extent that) it awakens the understanding of my self. Here my 'self' appears as an objective genitive, since the possessive genitive would in this statement be meaningless, and therefore the 'self' is being used as an object. When the kerygma awakens the understanding, as Bultmann puts it, it is awakening an object. What is this object? Where the knowing self is not functioning, the understanding is not awake. Bultmann's situation is the self understanding itself, and this necessarily means that the self is somehow the object of the understanding. But the self understanding itself is not the same as the self understanding the kerygma, if the kerygma is anything at all in itself that is therefore distinct from my own self. Now, if the kerygma is not anything in itself distinct from my own self, then it is a mistake to talk about the kerygma at all except for the sake of eliminating it. But the Christian message is something in itself. When we speak of the kerygma, we must logically speak of the self of the kerygma and not of the self of ourself. Bultmann's error is thus elucidated.

8. The second error in Bultmann's affirmation regarding the understanding of the kerygma regards his transformation of the kerygma into a call to decision. The scientific realization that the understanding of the kerygma requires insight into an object other than one's self does not yield to a plunging of the question into a vague encounter with life, for the imagery of such an encounter is preconceptual and prescientific. Nor does this scientific realization yield to a transformation of the kerygma as an object of cognition into "the call to decision," which would be an object of appetition. The question of the specific character of the kerygma is a cognitive question; its proper answer can be given only in the cognitive order (7.6:1). It is unscientific in face of this question to divert the consideration into the appetitive order. What Bultmann misses here is the fundamental distinction between the intellect and the will, which as a distinction, belongs to the cognitive order. To keep the two orders distinct does not mean to deny that the will has its rôle to play. Any act of the intellect involves an act of the will, and the attractiveness of meaning which draws the intellect to embrace it, as an attractiveness, is in its essence an appeal to the will, which is defined as the intellectual appetite. Bultmann's error does not lie in his calling the kerygma a call to decision, for that it must be as an object offering itself to be understood. The error consists in confusing this appeal of the kerygma with its understandability in the cognitive order. The precise fact missed by Bultmann is that the intelligibility of the Christian message, as of any object of knowledge, is intrinsically dependent upon the meaning it contains and only extrinsically or transcendentally upon the appeal that it has for the will.

9. While every advance in understanding is extrinsically a call to decision in the sense described in the preceding paragraph, insight into the kerygma as kerygma involves the will in a special way, since every such insight is a deepening of the act of faith. As explained above (7.7:6), the act of faith is intrinsically an act of the intellect, involving extrinsically an act of the will. It is the prerogative of the will

in the ordinary functioning of the mind to permit or to forbid the intellect to recognize meaning, for the will is free to embrace the alternative to insight in a given situation and to choose non-meaning. If the direction of meaning is called 'objectivity,' the alternative direction of non-meaning may be called 'subjectivity.' Since the center and focal point of subjectivity is one's own self, Bultmann's characterization of the kerygma as a possibility of human self-understanding could be interpreted as a turning away from meaning, because his "call to decision" in this context could be a mere temptation to refuse to recognize the objective meaning of the kerygma in order to indulge in love of self. Bultmann's description of the meaning of the kerygma unfortunately contains no element by which it can be distinguished from a personal decision in the concrete circumstance of one's own situation to sin against truth. The temptation to sin against insight into the truth arises in proportion to the proximity with which the meaning of the truth threatens inordinate love of self. That is why the love which comes from understanding the truth can never be identified with the love which comes from contemplation of one's self. Truth as truth resides only in objectivity. The kerygma as kerygma can be understood only in its objective meaning. The meaning of the kerygma can be seen only by means of the light of faith, which shines softly in the consciousness of those who possess it. The danger of Bultmann's formula is that it recommends a search for meaning in obscure areas of human consciousness, illuminated by no intellectual light, and it excludes precisely that act of the will included in the act of faith by which the human intellect opens in understanding of the meaning of revealed truth. While faith requires that the intellect of man be directed to that area in his consciousness that is illuminated by supernatural light, Bultmann's formula turns the attention away from the locus of faith and preoccupies it with one's own self.

10. For matters pertaining to the understanding of faith, Bultmann rejects the distinction between subject and object of knowledge (8.4:10), but in valid theological science this distinction must be used. We know that the dichotomy between subject and object is the first act of human intelligence as such. It is the primordial distinction of human reason. It is vain to attempt to circumvent this dichotomy by appealing to a realm of consciousness which transcends it, for every such appeal, every such reflection is already based upon a use of the dichotomy. Supposing that there is a realm of consciousness in which strictly supernatural reflection can take place (an hypothesis that Bultmann will not grant), one could imagine that the subject-object awareness may be considerably modified. We might call its characteristic reflection 'mystical theology.' But the fact remains that the discussion raised by Bultmann is taking place within the realm of natural awareness, where the subject-object relationship is always present. Our theological reflection is unfolding within natural consciousness as influenced by the virtue of faith. To natural consciousness the object of faith will appear initially as a set of truths recognizable by the intellect and to be accepted by the will. These revealed truths are objects of our cogntion as knowing subjects. It would be an error in methodology to overlook this whole area of scientific thought on the ground that mystical theology may be different. Right method requires that we develop the area that is the proper subject of our thought. If we call this area 'lower-level theology,' we may rightly say that the first step in the process of lower-level theology involves the application of the primordial dichotomy of reason to the revelation of God in order to see from the viewpoint of natural consciousness (7.7:13) what meanings can be brought to light. Every subsequent step in the process of lower-level theology will necessarily take place within this dichotomy, so that all theological science is a study of revealed truth as an *object* of consideration. Science is the knowledge of the real as such (2.4:3). (Lower-level) theological science is the knowledge of revelation under the aspect of what is real in this revelation, and the real is a distinction based upon the primordial subject-object dichotomy (2.4:1). Bultmann's claim that *existentiell* self-understanding is real in the act of existence but not in the isolated reflection of thought is thus a meaningless assertion, for the notion of the real is of its very nature enclosed within the objectivity of the subject-object relationship. The 'reality' of Bultmann's 'understanding of

self' is therefore nothing more or less than the reality of whatever experience the self is having while in a state of blindness with regard to rational thought. Such experience is preconceptual and subhuman. It has no rational justification. A human being can by his free decision embrace even the 'understanding of self' contained in such *existentiell* experiences as a fit of rage, a narcotic stupor, or total intoxication with one's self.

11. Bultmann claims that the kerygma can never be spoken except as formed by human thought. But 'human thought' is here used in a misleading way. There is human thought which is strictly limited to the natural reflection of man; there is human thought which is influenced by the principles of faith while transpiring within natural awareness; and there is human thought on the strictly mystical level. Bultmann assumes that the specific intelligibility of the revealed word of God stands entirely within the bounds of the natural reflection of man, a position that cannot be assumed, because it is the point at issue. Just as the specific meaning of Jesus Christ, Who is both God and man, stems from a recognition of how He differs from man rather than from how He is identical with man, so the specific meaning of the Scriptures stems from a recognition of how they differ from merely human language rather than from how they are the same as other human language. And the assertion that the meaning of the Scriptures is shaped by the way a man understands his own existence is false for all of the reasons stated above. In fact, the specific meaning of the Scriptures is not in the illusory understanding that a man may have of himself, but in the true understanding of how God understands Himself and his creatures. The true understanding of myself is identical with the understanding that God has of me; it is correlative to the developed understanding of objective revealed truth; it is coincident with love for God as the 'totally other.'

12. Bultmann claims that the statements of the kerygma are not universal truths. I have shown above (5.2:21) that the word 'universal' is ambiguous. What is especially important here is that the specific meaning of divine revelation is an object of conceptual insight with all of the qualities that such insight must have. God is not a universal truth in the sense of an open class of things. But what God reveals to us of his essence, or of his being, is presented to be conceptually grasped to the extent that we are able. This essence of God is somehow 'contained' in the sensible data of his revelation. When faith seeks the specific meaning in this data, it moves in the direction of God as He has revealed Himself. This is the task of theology. Therefore, as far as theological science is concerned, it is not true to say that the word of God speaks only to one's concrete existence. It speaks also and especially to the living awareness of the individual human person, and that means to a subject knowing God as an object of cognition, knowing Him abstractly, because He is totally other than everything that is material and finite, and yet knowing Him concretely in the sense that He is individually existent and not just an open class of things.

Theological Thinking

EXPOSITION

13. Bultmann holds to be most important the "basic insight" (*Grundeinsicht*) that the theological thoughts of the New Testament are the unfolding of faith itself growing out of that new understanding of God, the world, and man which is one's new self-understanding conferred in and by faith. He says that this new existential understanding of myself is inseparable from my understanding of God and the world. It is not of myself as objectified into a phenomenon of the world, for I am I only as inseparably bound up with God and the world and not as distinguishable and objectifiable in a world-phenomenon. Hence, the scientific exposition of the theological thoughts of the New Testament presents, not the object of faith, but faith itself in its own self-interpretation. It is not possible to distinguish clearly and sharply the kerygmatic statements in the New Testament from the theological ones, but the scientific expounder of New Testament theology must keep this distinction constantly in mind and must interpret the theological thoughts as the unfolding of the self-understanding awakened by the kerygma, if he is to avoid conceiving them

as an objectifying kind of thought, cut loose from the act of living.[a]

CRITIQUE

14. As explained above (8.4:10), what Bultmann calls a new understanding of God, the world, and man is not an understanding of God, of the world, or of man, because it has no object at all and it therefore cannot be located in that region of the human mind where understanding lies (2.6:9). The 'myself' to which all true understanding is related is not myself objectified into a phenomenon of the world, but myself as the knowing subject of intellectual awareness. Since the first and fundamental act of intellectual consciousness is the dichotomy of self from all objectivity (2.3:18[a]), there can be no act of insight or understanding in which the knowing self is inseparably bound up with God or with the world. Faith itself is an inanimate object; it is a virtue adhering to an intellectual being. Therefore, faith cannot interpret itself; it can only serve as a means by which an intellectually knowing self may interpret the object of his knowledge. Intellectual thought is a vital activity of intellectual beings, but intellectual thought is impossible for any man apart from the distinction of subject from object. Therefore, objectifying thought is not something cut loose from the act of living; it is one of the finest manifestations of human living. The very word 'insight' (*Einsicht*) implies a seeing into an object. Bultmann's 'basic insight' that the thought of the New Testament is not about objects is not an insight; it is the antithesis of insight.

EXPOSITION

15. Bultmann is of the opinion that there can be no normative Christian dogmatics, for the task of the science called New Testament theology, which is to set forth the theological thoughts of the New Testament writings, permits only ever-repeated attempts at solution, each in its own particular historical situation. The continuity of theology consists not in the retaining of propositions formulated once and for all but rather in the constant vitality with which faith understandingly masters its constantly new historical situation. Theological propositions can never be the *object* of faith; they can only be the unfolding of the understanding which is inherent in faith itself. This is true, he says, even for propositions stated in the New Testament.[a]

CRITIQUE

16. Bultmann's confusion about the *object* of the science of theology has been explained above (I.5), where it is shown that his epistemological reason for excluding an object of Christian faith is without foundation. A further confusion regards the question of theological propositions. All propositions are objects of knowledge. If they embody universal truths they are objects of speculative knowledge. If they embody historical truths they are objects of historical knowledge. In either case they can be objects of the science of theology, as is clear to those who have grasped the concept of Christian theology and the distinction between speculative and historical theology. Speculative theology pertains to timeless truths. The objects of historical theology have an intrinsic relationship to time and to the historical situation. But this historical situation needs to be approached on the level of historical science, not on the level of a prescientific romanticism that misses the important rôle played by Christian dogmatics in the understanding of revelation.

Faith without Object

EXPOSITION

17. Bultmann observes that 'reality' (*Wirklichkeit*), as the term is ordinarily used, reflects a world-picture that has dominated our thinking since the Renaissance.[a] We call something 'real' (*wirklich*) if we can understand it in relation to the unified complex of this world. This world-picture is conceived without reference to our own existence;

I.13[a] *Theologie*, 587-588 (*Theology* II, 239-240).

I.15[a] *Theologie*, 585-586 (*Theology* II, 237-238).

I.17[a] R. Bultmann, "Welchen Sinn hat es, von Gott zu reden?" in *GuV* I, 26-37 (reprinted from *Theologische Blätter*, 4 (1925), 129-135); ET: "What Does It Mean to Speak of God?" in *FaU*, 53-65.

we merely see ourselves as an object among other objects. The world-picture completed by the inclusion of man is referred to as a world-view (*Weltanschauung*). Such world-views are defective. They cannot be repaired by labeling ourselves as 'subject' in opposition to the other objects, because man is still being seen from outside. The question of our own existence is outside of the subject-object distinction. A Christian world-view based upon some idea of the dependence of our existence upon God is equally defective, because God is thus regarded as an object. The work of God is not a universal process or an activity that can be seen by us apart from our own existence; to conceive of it as a process into which we can insert our own existence is an abandonment of the primary concept of God as the reality determining our existence (*die unsere Existenz bestimmenden Wirklichkeit*).[b] The existence of God is not a general truth with its place in a system of cognitions (*in einem System von Erkenntnissen*). God is not something objectively given with which there could be a relation of understanding accessible to us and achievable at will.[c] God or his existence is not something in respect of which we could establish an attitude or undertake anything at all. We can say that God is the Lord of reality (*Wirklichkeit*) only in the sense of knowing one's own existence to be claimed by God, because all talk of reality which ignores one's own existence is self-deceit. God is the 'wholly other,' not because He is somewhere outside the world, but because this world, being godless, is sinful. When the question is raised of how any speaking of God can be possible, the answer must be, it is only possible as talk of ourselves.

CRITIQUE

18. We call 'real' those objects which are given either in sensory or in intellectual intuition, and which therefore fit under the universal concept called 'reality' (2.4:8). As an open class of things, the concept of 'reality' is not limited to sensory objects; it is not limited at all except by its own specific character as a concept. We derive our concept of reality from an intellectual intuition suggested by our associations with things in the sensible world; we tend to test it there; but it would be a mistake to suppose that the concept will apply only to material things. Because 'reality' is the first concept of nascent science, it necessarily unfolds according to the unified complex of extramental existence, to which corresponds the vital complex of scientific understanding. We assume that anything given in sensory or in intellectual intuition is part of the unified complex of the real. We cannot, therefore, scientifically reject anything given in experience simply because it does not immediately fit into the framework of principles that may be erected in our minds. Bultmann tends to identify reality (*empirische Realität*) with material objectivity. This is a methodological error. God is wholly other from 'this world' in the sense that He is not a creature. But in the more fundamental scientific distinction between the real world and the fictitious world, God is a being in the real world. The Christian distinction between this world and heaven is not to be confused with the distinction between the real world and imaginary worlds.

19. The world-picture of science is not like a flat, two-dimensional painting with our own selves included in it (2.6:6). It is rather like a three-dimensional object whose third dimension reaches towards us and culminates in the focal point which is our concept of reality (2.6:15). The scientist is the knowing subject on this side of the concept of reality. Whether or not he sees himself also as an object within the focus of reality is a special question. The general fact is that he is a subject beholding an object which is his notion of the world. No question can be outside of this subject-object relationship, for questions derive from functioning intelligence, and there is no functioning intelligence that is not based upon the primordial dichotomy. Sight is only of objects. Bultmann therefore errs in supposing that there can be sight of one's self 'from inside.'

20. It follows that God can be regarded only as an object. Bultmann errs gravely in affirming that God cannot be regarded as an object which is

---

I.17[b] *GuV* I, 31-32 (*FaU*, 58-59).
I.17[c] *GuV* I, 32 (*FaU*, 60): "Da wäre Gott eine Gegebenheit zu der eine Erkenntnis- relation möglich ist, die nach Belieben vollzogen werden kann." Cf. *GuV* I, 36 (*FaU*, 63).

wholly other from the world of creation, that God is not the ultimate external cause of the existence of the world and of our own existence, that the reality of God is not prior to and independent of our own existence. Bultmann's conclusion about God is based upon a mistaken epistemology. While Bultmann tries to extinguish the light of objectivity and turn the attention to the self at this end of the subject-object dichotomy, he can never succeed, because when the dichotomy is suppressed the knowing subject is no longer humanly conscious. Bultmann is so obsessed with the desire to eliminate the dichotomy that he fails to see that it is always functioning in all his arguments. It is he who has concentrated so totally on the idea of the dichotomy that he makes it the theme of his theology, but he has chosen to attend only to the dark end of the distinction. The self at the dark end is indeed not an object in its distinction from all objectivity, but to the extent that it is made the center of attention it becomes an object — dark, obscure, and meaningless though it is (2.2:6). It is sheer fantasy to suppose that God, Who is perfect form and infinite meaning, the light of all light, could be found at this end of the dichotomy. What Bultmann says about God as being real only in relation to the existence of one's self has no meaning or intelligibility in itself; it is a pure hypothesis set up to counter the objectivity of God, which Bultmann, on the basis of his epistemology, denies. The truth is that God is the supreme object of our intelligence; He is the antithesis of the darkness which is our knowing self as such.

21. But how can God be 'an object'? Bultmann makes the valid observation that an object is somehow at our disposal; it is somehow beneath us as living intelligent beings. The answer is that God is not an object in Himself; He becomes an object in revealing Himself to us. Therefore the object which we call God is not God in his own infinite mode of being; it is God as He appears to us. This difference is very significant, but it does not gainsay the fact that the revealed God is a true image of God as He is in Himself. Furthermore, we know from revelation that God is three Persons, and that the Second Person is the Son of the Father and the Word of God. This fact gives us an indication of how God is in Himself, as well as of the way in which we, as intellectual beings, resemble Him. This is not a proof, but it is an indication that God, even as He is in Himself, as He 'sees Himself from the inside,' knows Himself as the object of his knowledge.

22. Bultmann says that God is the reality determining our existence, but not as the efficient or the final cause. His use of the word 'determining' is vague and undeveloped. Somehow it seems to suggest a vital identity between God's determining our existence and our determining our own existence. This suggestion cannot be elucidated by rational argument or insight, because it refers to an unintelligible notion of the 'self.' What Bultmann seems to mean by 'my own existence' is pure conscious experience in its most individual and incommunicable aspect. This is the lowest and most unstable grade of human existence — lacking all form and quality. The damned in hell possess it fully, and it avails them naught. To identify God with the exclusivity of this kind of chosen existence is vainglory; it is the exclusion of all true value because of an exaggerated love of self. It is the task of Christian living precisely to escape from all such preoccupation with one's self in order to embrace the shining objectivity of God's being as it is revealed to us.

23. The sin according to which we are free to reserve to ourselves the worship that belongs to God does not adhere to the substance of the world or to our own substantial existence; it is located precisely at that point of consciousness that Bultmann calls 'our own existence.' This is the area of the will unenlightened by the objective truth of God's revelation. To turn towards the objectivity of God entails a recognition of the importance of God in the world we face. God is the efficient, final, and exemplary cause of that world. We may find traces of God in nature and the image of God in man. But, above all, we find God revealed in the word of Sacred Scripture, presented to us by the Church. We achieve the fulfillment of our existence by searching for God in the word of his revelation and by finding Him in the objective meaning of his existence, as it is hidden within the word of divine revelation. This is the task of theology.

24. It follows that the revealed God

is an object in respect of which the Christian must establish his basic attitude towards life. The supernaturally instilled concept of God is the basic viewpoint of all theological research (7.7:7). Vague and obscure though it is, in comparison with the beatific vision, it is nevertheless of greater intellectual value and use than any concept of natural reason. The presence of God contained in this concept has its place in our system of cognitions; it is, in fact, the head and source of all theological cognitions. God is present both to the intellect and to the will of the sanctified believer. He is present to the intellect through the light of faith, which is appropriated to the divine Word of God, the knowledge and wisdom of the Father. He is present to the will through the unction of the Holy Spirit. The light of faith as the abiding concept of God is a concrete intellectual object situated between the knowing subject and whatever he knows by faith (7.7:12). It is given by grace to be at our disposal for the proper orientation of our thinking in terms of our vocation to eternal life with God.

25. The supernaturally instilled concept of God is a *real* object of intellectual intuition. It is true that 'reality' is a concept of natural reason and that the presence of God in faith is somehow prior to reality in the order of cognitions, so that it would seem that the awareness of God prescinds from the question of reality. The subject-object distinction is also a dichotomy of natural reason, so that it would seem that the awareness of God transcends this distinction. And yet this is not so. The awareness of faith is seated in human intelligence; without the subject-object dichotomy it would not be so situated. Even if we grant that the awareness of God in faith may be able to function when we are not in a waking state of consciousness, we need not admit that the subject-object dichotomy is not yet operative. Certainly we need not admit the existence of a new supernatural ego functioning beneath the awareness of rational consciousness and thus truly splitting our personality in twain. Nor can we conceive of an objectless and subjectless thought that could be taking place in the supernatural consciousness. Thought is there, the ego is one and the same, and the objects are *seen* in the supernatural medium of the revealed presence of God.[a]

26. The question of the reality of the object of faith is a theological question. It is the first and most fundamental question that lower-level theology asks (7.6:4). It is precisely the gathering of the objects of thought according to faith under the concept of reality that constitutes all methodical theological science. The question of reality can elucidate the explanation of the nature of faith given by Thomas Aquinas. In the exposition of St. Thomas, faith seems to have a double essence. Intrinsically it is an act of the intellect, and yet it is formally constituted by an act of the will. As has been shown above (2.2:7), any act of the intellect is extrinsically dependent upon an act of the will for its motive power. It is clear that the intellect will not open to an insight when the will chooses to obstruct the insight (2.2:6). This fact applies to all understanding. But St. Thomas makes a special case for faith. He sees faith as an act of the intellect consisting formally in obedience of the will to the testimony of God. Here a distinction is necessary. On the level of sheer faith — prior to any intervention of natural understanding apart from the basic subject-object dichotomy — there can at least theoretically be an act of supernatural understanding without the extraordinary obediential command of the will. But faith is situated in human consciousness, and it must operate in conjunction with the light of natural understanding. The first question which natural understanding asks of the object of faith is the question of its reality. Therefore every act of faith performed with natural consciousness includes an answer to this question. The science of theology is rooted in the affirmation that the object of faith is real. St. Thomas, in speaking of what is intrinsic to faith, is speaking of the pure act of supernatural knowledge. When he requires the formal intervention of the will, he is speaking of faith on the theological level of natural consciousness (7.7:15). Faith is called faith, not because it is intrinsically unenlightened, but because it needs the protection of the will, lest its light be overwhelmed by the negative insights of unenlightened natural consciousness.

I.25a *Supra,* 7.5; 7.7; esp. 7.5:2; 7.7:5; 7.7:12; 7.7:16.

Whether God Is Something

EXPOSITION

27. Bultmann claims that it is impossible to speak 'about God' as of an object of thought, because 'speaking about' presupposes a standpoint external to what is being talked about, while there cannot be a standpoint (*Standpunkt*) external to God. Therefore it is not legitimate to speak about God in general statements of universally valid truths. There can be no convincing argument for the reality (*Wirklichkeit*) of God; there is no proof of God's existence. It is no more possible to speak meaningfully about God than to speak meaningfully about love, and any talk of love that is not itself an act of love is not about love, because it stands outside of love. It would be no more atheistic for science to deny the reality of God than to affirm it. To speak about God in scientific propositions, that is, in general truths, is *sin*.[a] God is wholly different from man only as determining man's existence. God is not *something* — whether a metaphysical being or a kind of immaterial world or a creative source. God is not something apart from me for which I must search; there is no escaping from oneself in order to find God. The actual situation of man is that of the sinner who wants to speak of God and cannot.[b] God does not inspire us. We can never possess certain knowledge of God or of our own reality. Wholly fortuitously, wholly as specific event does God's Word enter our world. No guarantee comes with it by virtue of which it is to be believed.[c]

CRITIQUE

28. The intellect and the will are opposite faculties of characteristically human activity. Since love is the act which pertains most properly to the will as distinct from the intellect in man, it is clear that love is the characteristically human act which it is most difficult for the intellect to visualize in the form of an object. The example is marginal. Nevertheless, we can know or talk about love to the extent that we can see it as an object, and Bultmann in fact makes it an object when he talks about it. When Bultmann says that something stands "outside of love," he is somehow visualizing 'love.' It is true that his aim is to empty that visualized object of all content and validity, but his point will not carry. The real reason why we cannot visualize love with any clarity as an object is because the objectivity of the human act of love is contained in the intellectual objects of that act. Having made the distinction between the intellect and the will, it is obvious that we cannot expect to find in the will what is characteristic of the intellect. Yet, even love as an object is an important element of thought, and to say that it is not is to miss the point of an enormous amount of speech and literature.

29. God is love (1 John 4:8). God is light (1 John 1:5). The love that is God exists substantially in the Holy Spirit. The light that is God exists substantially in the Son. To the extent that God is known in the form of love, it is especially difficult for us to express this knowledge by an intelligible object. But to the extent that God is known as light (the light of meaning and understanding), our knowledge is expressed in the form of an intelligible object. It is precisely as light that we are discussing God in the present context, for the question of knowledge pertains properly to the intellect, and not to the will. Therefore, Bultmann's comparison with the act of love is ill-chosen.

30. It is erroneous to assert that there can be no standpoint external to God. The whole discussion raised by Bultmann unfolds from the viewpoint of himself and of ourselves as knowing subjects. As knowing subjects we are necessarily external to God. To deny this is to put ourselves in the place of God; it is tantamount to calling ourselves God. Our viewpoint is external to God for the reason

---

I.27[a] *GuV* I, 26-27 (*FaU*, 53-54).
I.27[b] *GuV* I, 29-30 (*FaU*, 56-58).
I.27[c] *GuV* I, 36-37 (*FaU*, 64-65). "Thus we find ourselves led to the conclusion that our own existence, since it depends on our free act, can never be known to us. Is it illusion? Unreality [*Unwirklichkeit*]? Certainly it is nothing about which we have knowledge, about which we can speak. And yet it is only our existence which can, if it really [*wirklich*] is in our speaking and acting, give reality to this speaking and acting. *We* can only believe it." *GuV* I, 35-36 (*FaU*, 63).

that there is intervening between us and God the medium of our mind. When we speak about God, we do so in terms of the medium of our mind, which is less than God and is therefore outside of God. It is from our mind that general statements come which contain universal truths about God. The mode of existence of these statements is human; they are external to God. By the use of our mind we can arrive at the valid conclusion that God exists and that He is real. In denying this, Bultmann fails to see the functional presence of the human mind as the medium of rational proof. The concepts of natural reason are standpoints external to God from which we can speak validly about God. In particular, the concept of reality is a standpoint from which we can study the existence of God and prove that He is real. For the Christian believer there is the additional standpoint of the obscure concept of God instilled by grace, which enables us to speak about God on the level of revealed science. In failing to see the presence and operation of this multiple standpoint external to God, Bultmann shows that his thought is not on the level of theological science.

31. Bultmann goes so far as to designate as 'sin' the propositions and conclusions of theological science. He imagines that to search for God apart from my awareness of myself is sin. The opposite is the truth. Bultmann affirms that *he* knows nothing *about* God. Those who have the concepts which serve as the medium of knowledge about God know that they know something about God. Only *they* are qualified to speak scientifically about God. Bultmann believes that God is not something Then the encounter with the act of God, which is the theme of Bultmann's theology, is an encounter with the act of nothing. To declare that God cannot inspire us, that we can never do or say anything that rises above the level of the merely human is the pseudoscientific conclusion of a pseudo-theology. Such declarations may claim to be in accord with science, but they are not. We can know that we have certain knowledge of the God Who is real. God's Word enters our world as specific event, not only as the historical event of the Incarnation, but also as such historical events following from it as the composition of the inspired word of God and the mental events by which individual men are given insights into the revealed essence of God and into its meaning by concrete acts of understanding in the actual historical circumstances of their concrete lives. These insights may abide for the duration of their earthly lives and on into eternity.

The Act of Faith

EXPOSITION

32. Bultmann observes [a] that the word 'faith' is used in the New Testament in its specifically Christian sense to mean an acceptance of the Christian kerygma. It is thus saving faith which recognizes and appropriates God's saving work in Christ.[b] The element of confident hope is less prominent in this usage than it is in the Old Testament. Christian faith looks primarily to what God has done, not to what He will do. Acknowledgment of Jesus as Lord is intrinsic to Christian faith, along with acceptance of the miracle of his resurrection as true.[c] Bultmann notes [d] that the New Testament concept of hope is essentially determined by the Old Testament concept of hope, the difference between the two lying in the situation of the one who hopes. Christian hope rests on the divine act of salvation accomplished in Christ; hope is thus an eschatological blessing, for now is the time when we may have confidence. Hope does not regard the realization of a human dream of the future; with a confidence directed away from the world to God it waits patiently for

---

I.32a R. Bultmann, "πιστεύω, πίστις, κ. τ.λ.," in G. Kittel ed., *Theologisches Wörterbuch zum Neuen Testament*, Vol. VI (Stuttgart, 1959), 174-182; 197-230 (hereinafter referred to as *TWNT* VI); ET: same title in G. Kittel ed., *Theological Dictionary of the New Testament*, Vol. VI (Grand Rapids, 1968), 174-182; 197-228 (hereinafter referred to as *TDNT* VI).

I.32b *TWNT* VI, 209 (*TDNT* VI, 208).
I.32c *TWNT* VI, 209 (*TDNT* VI, 209).
I.32d R. Bultmann, "ἐλπίς, ἐλπίζω," in *TWNT* II (Stuttgart, 1935), 515-520; 525-531; ET: in *TDNT* II (Grand Rapids, 1964), 517-523; 529-535).
I.32e *TWNT* II, 527-529 (*TDNT* II, 530-532).

God's gift, which is received, not as a possession, but in the assurance that God will maintain what He has given.e Faith in Christ does not indicate the mere presence of a strange and previously unknown divine Person, for the figure of Jesus Christ cannot be detached from its 'myth,' which is the history enacted in his life, death, and resurrection. This history is salvation-history, in the sense that whoever accepts the kerygma in faith recognizes that this history took place for him. To believe in Jesus Christ entails a personal relation to Christ analogous to the Old Testament relation to God, the difference being that New Testament faith is not directed to the God whose existence is always presupposed. Only in faith itself is the existence of the Lord Jesus Christ recognized and acknowledged; faith is not obedience to a Lord Who is already known.f In many typical instances in the New Testament, Christian faith is *fides qua creditur*. But the New Testament usage develops in such a way that it can also in some places mean *fides quae creditur*.g Faith is the act by which a man, responding to God's eschatological deed in Christ, comes out of the world and makes a radical reorientation to God. Faith is the act in which the new eschatological existence of the Christian is established.h God and Christ are not set before the believer as two different objects of faith in coordinate or subordinate relationship, for God meets us only in Christ. In Christ dwells all the fullness of the Godhead; Christ is God's final act, in which the future is also embraced.i

CRITIQUE

33. What Bultmann presents as the concept of faith or of hope in the New Testament writings is vaguely identified with Bultmann's own concept of the same and vaguely different from it. Bultmann, in the course of his theological meditation, eliminates many faith-formulae in the New Testament in order to arrive at what he considers to be the pure conception, but he does not clearly recognize the operation of this 'pure conception' (his own viewpoint) as he passes the text of the New Testament in review. The absence of clarity regarding Bultmann's own conception deprives his exposition of scientific value. For instance, Bultmann affirms that the New Testament conception of faith requires an acceptance of the miracle of Jesus' resurrection as true. Here there is a disparity between the view of the New Testament writers and that of Bultmann, who excludes the possibility of the resurrection of a corpse. The prescientific reader might uncritically assume that Bultmann's own view is not involved, seeing that Bultmann is simply reporting his observations of the New Testament. But the scientifically critical reader will realize that Bultmann's own view is modifying everything he says about the subject. What Bultmann finds to be essential or unessential to the concept of Christian faith is the result of a comparison with his own conception, a judgment concerning what he himself considers to be essential or unessential, and an anticipation of how what he observes in the New Testament will assist his own theory of theology. Because of his theory of demythologizing and its system of reflex principles, he can comfortably retain the conclusion that the New Testament writers in some sense (precisely in the sense defined in demythologizing) saw the miracle of the resurrection as intrinsic to Christian faith. Bultmann admits no miracle of any sort. His faith prescinds from the need for miracles. His view claims to be less naive and more mature than that of the New Testament writers. But the reader who accepts Bultmann's view on faith is by no means mature. Before the mature reader can accept Bultmann's conclusions about the conception of faith in the New Testament, he has the prior task of examining for himself the validity of Bultmann's own viewpoint.

34. Bultmann finds that the conception of Christian faith presented in the New Testament writings is often *fides qua creditur*, but in some instances is also *fides quae creditur*. This conclusion is derived from his own conception of faith. Christian faith is an intellectual light by which supernatural meaning can be seen and understood (7.7:6). Of its very essence, Christian faith is always and necessarily both *fides qua creditur* and *fides quae creditur*, since the *qua* expresses the

---

I.32f *TWNT* VI, 212 (*TDNT* VI, 211).
I.32g *TWNT* VI, 214 (*TDNT* VI, 213).

I.32h *TWNT* VI, 217 (*TDNT* VI, 216).
I.32i *TWNT* VI, 218 (*TDNT* VI, 217).

means of understanding and the *quae* expresses the object of understanding. The act is one; it can be regarded theologically under either aspect. The definition of faith as a supernatural light will fit any use of the Greek word πίστις in the New Testament, where Christian faith is intended. Bultmann's failure to see that both meanings will always apply (the one directly and the other indirectly) is simply due to his erroneous conception of the nature of faith.

35. The same is true of Bultmann's description of hope in the New Testament. He claims that Christian hope does not regard a dream of the future within the consciousness of man, but is the eschatological blessing consisting in the realization that now is the time when we may have confidence. Bultmann's own view of hope is the presupposition of this conclusion. Elsewhere he expounds upon his conception of hope, expressing his eschatology on the basis of the existentialism of Heidegger. By eliminating the object of hope and folding it back into the now of present consolation, Bultmann destroys the nature of hope. His attempt to direct hope away from the world to God is actually the turning off of the intellectual light of faith and the elimination of the motivation upon which hope is based. Hope as confidence in what we are now is not hope at all. In order to be hope it has to be confidence in what will be in the future. Existentialist hope in the eschatological now is nothing more than the entertaining of a negative symbol in place of the real hope which is no longer operative. Bultmann's conception of hope will not stand up under theological analysis. His conclusions about hope in the New Testament, governed by his own conception, are invalidated by the incorrectness of his own viewpoint.

36. In the correct sense, faith is indeed a coming out of the world and a radical reorientation to God. It is not a folding of intellectual objectivity into a flattened state of suspended animation; it is not a radical reorientation towards what is not 'something' apart from ourselves. By faith the Christian is established in a new existence, not as vitalizing in some inconceivable way the area of the knowing self in its incommunicable exclusivity, but as vitalizing the intellect and will with a new supernatural objectivity.

Christian hope presupposes a realization of what God has done and can do; it is essentially oriented by intellectual insight towards a recognition of what God will do for those who love Him. Faith is obedience to a Lord already known. The recognition of the Lord as Lord is a prior insight presented to the intellect as an object that is intrinsic to the act of faith; the actuation of faith within natural consciousness requires an act of obedience by the will to the Lord already known by this intellectual insight.

37. God and Christ are set before the believer as two different objects of faith, for Christ is both God and man. It is possible for faith to distinguish between the divinity of Christ and the humanity of Christ. The principle subject of contemplation in faith is Jesus Christ, in Whom dwells the fullness of the divinity *bodily* (σωματικῶς; Coloss 2:9), that is, as incarnate in his humanity. Through our powers of natural perception we are able to see the humanity of Christ, presented to us as an object. Within the humanity we are able to see something of the divinity of Christ in the supernatural light of faith. Aspects of this divinity become illuminated by reason of the intelligibility contained in the concept of the divinity, which resides in the mind of the believer as the supernatural intellectual medium (7.7:12). This concept of the divinity of Christ is one object of faith — the formal object in scholastic terminology. The human figure of Jesus Christ is another object of faith — the material object in scholastic terms. Thus the two objects are in subordinate relationship; the meaning of the humanity of Christ is referred to his divinity. Our personal relationship to Christ is a relationship to the divine Person distinguishable within and above the humanity of Christ. To know Christ in faith, the existence of God is presupposed; the presupposition is located in the obscure concept of God instilled in the mind as the intellectual medium of faith.

38. Though the experience contained in what Bultmann calls the existential encounter must in some way pertain to the order of reality, nothing that Bultmann has said would indicate that anything which is conceptually distinct from the encounter itself and which is used to describe the encounter pertains to the real order. All else is simply mythological. Thus

Bultmann admits that to speak of an act of God' is to speak in terms of mythology. And he qualifies the declaration merely by noting that it is a type of mythology which is consonant with his notion of faith and with what modern man is disposed to believe.[a] If we speak of an 'act of God' in terms of the configuration of the real world, we necessarily imply the objective existence of God, Who is acting. According to conventional reason, there could not be an act of God without a God, Who is acting with an act distinct from our own existence. Bultmann, however, in terms of this distinction, clearly states that for him there is no *really* existing God. He says, for instance, that "outside of revelation there is no God"[b] and that "God cannot be known apart from Jesus."[c] Again, he says: "The formula 'Christ is God' is false in every sense in which God is considered as an objectively existing entity. It is correct only when God is considered as 'the event of God's acting.' "[d]

39. But to Bultmann even the event of Christ appears to be a residuum of mythology which may not escape the demythologizing process. This is "a serious problem, and if Christian faith is to recover its self-assurance it must be faced."[a] The Christian faith in jeopardy is primarily that of Bultmann himself, and he notes in another place that, whether or not Jesus of Nazareth ever started the movement we call Christianity, however probable on an historical (*historisch*) level, it is not an essential presupposition of his faith: "Should it prove otherwise, that does not change in any way what is said in the record."[b] Hence, according to Bultmann, God does not exist in objective reality apart from Jesus, and Jesus, inasmuch as He is significant to faith, need not ever have existed in objective reality apart from the kerygma in which He is brought to us. The existential encounter, then, is not with God or with Jesus, but with the kerygma, and the way in which it is encountered is known as 'faith.'[c]

40. For Bultmann, 'faith in Christ' does not mean faith in Him Whom the Jews of the Old Testament thought of as God, nor in the figure of Him Whom in his life, death, and resurrection the Christians of the New Testament worshipped as God. It is belief in a Lord whose existence is first known (both in temporal and in ontological order) with the positing of faith itself.[a] This Lord can encounter the believer only in the kerygma, which itself is not an objective message of any kind, a *revelatum* with enclosed linguistic features, but is simply the fundament of belief.[b] The believer believes because he believes, and he cannot offer for his belief any motive conceptually distinct from the act of faith itself.[c] Faith in the kerygma and in the person mediated by it are inseparable, and faith always remains a bold venture in the sense that it is based on the kerygma.[d]

41. For Bultmann, faith has no intelligible content:[a] Faith is a "leap in the dark" based on the desire to obey God's will.[b] Faith is an opening of man's eyes concerning his own existence so that he may once again under-

---

I.38a Bultmann, in *KuM* I, 124 (*KaM* I, 105).
I.38b R. Bultmann, "γινώσχω, κ.τ.λ.," in *TWNT* I, 712 (*TDNT* I, 711): "Ausserhalb der Offenbarung ist Gott nicht da: Jesus aber ist die Offenbarung für den sündigen κόσμος; wer ihn sieht bzw. erkennt, der sieht den Vater."
I.38c *Ibid.*
I.38d "Die Formel 'Christus ist Gott' ist falsch in jedem Sinn, in dem Gott als eine objectivierbare Grösse verstanden wird, mag sie nun arianisch oder nizäisch, orthodox oder liberal verstanden sein. Sie ist richtig, wenn 'Gott' hier verstanden wird als das Ereignis des Handelns Gottes." *GuV* II, 258 (*FaU*, 287). Cf. *KaM* I, 33-34 and 106; *BEF*, 26, 175, 177-178.
I.39a Bultmann, in *KuM* I, 32 (*KaM* I, 23). Cf. Bultmann, in *TWNT* VI, 212 (*TDNT* VI, 211-212), and *supra*, 1.2:5.
I.39b *JW*, 14.
I.39c Bultmann, in *KuM* I, 133-134 (*KaM* I, 117-118).
I.40a Bultmann, in *TWNT* VI, 212 (*TDNT* VI, 211-212).
I.40b *Ibid.*
I.40c *Ibid.* Cf. Bultmann, in *KuM* I, 132 (*KaM* I, 116); *GuV* I, 35-36 (*FaU*, 63).
I.40d Bultmann, in *TWNT* VI, 212 (*TDNT* VI, 211-212).
I.41a Bultmann, in *KuM* I, 39-40 (*KaM* I, 32-33). Cf. *HE*, 96; "How Does God Speak to Us through the Bible?" in *BEF* [166-170], 169; "Die Geschichtlichkeit des Daseins und der Glaube" (*supra*, 1.2:4b).
I.41b R. Bultmann, "Der Glaube als Wagnis," in *Die christliche Welt*, 42 (1928), 1008-1010; ET: "Faith as Venture," in *BEF*, 55-57. Cf. *BEF*, 168 and 215.

stand himself.c Faith is not belief in any divinely revealed fact or doctrine. It is not belief in a communication of supernatural knowledge which can be grasped, possessed, and applied.d
42. Hence, in Bultmann's conception, faith does not involve self-commitment to any norm which can be applied to one's activity in the world of reality.a In this sense it can have no real influence upon one's life except as an aesthetic diversion for the listener or a source of occupation for the preacher. Redemption and justification are not entities added to what the believer knew and possessed from the beginning, and they can make no real change in his soul.b They do not come from the real world 'out there.'c The believer, in taking the blind step into faith, must reassure himself (although not rationally) that in leaving the world behind he is not losing anything of value.d Bultmann admits his great difficulty in trying to show that his step into faith is *genuine*.e His answer is that we must simply *believe* that it is genuine.f He admits also that there is no continuum between his non-world of faith and the real world.g He asserts that the reason why the Resurrection of Christ could never be established as a real event is because, while one is believing, he may not appeal to what has been described in this study as the empirical-intellectual fact of reality.h Nor does he consider it fair to ask if the Cross has any significance within the real world. The only event of the real world which has significance for faith was the decision of the first disciples to take the step (non-step) from the consciousness of (empirical-intellectual) reality to the experience of faith.i Bultmann exclaims: "It is precisely its immunity from proof which secures the Christian proclamation against the charge of being mythological."j Yes, the question of proof is indeed meaningless for those who believe that the Christian proclamation to which they adhere does not pertain to the order of reality in the proper sense of the word. They face the thought that the content of faith is mythological, not by striving to locate the content of faith within the sphere of reality, but by ever-repeated acts of affection for the non-reality which is for them the proper sphere of faith. It is our contention, to the contrary, that the object and content of Christian faith is set in historical (*historisch*) reality. "(Christianity) has from the first had an objective existence, and has thrown itself upon the great concourse of men. Its home is in the world; and to know what it is, we must seek it in the world, and hear the world's witness of it."k

B. *A Tentative Characterization of the Genre of Bultmann's Theological Writing*

43. Rudolf Bultmann's theological reflection is based upon premises which may be called *postulates* to the extent that they have been formulated, explicitly or implicitly, by Bultmann himself, and *radical presuppositions* to the extent that they transcend the formulated exposition.a The radical presuppositions must be formulated before the conclusions can be brought within the context of understanding. They are situated out of focus behind the postulates, which are many in number and of highly graduated importance. Bult-

I.41c Bultmann, in *KuM* II, 197 (*KaM* I, 198).
I.41d Bultmann, in *KuM* II, 202, 207-208 (*KaM* I, 203-204, 210-211). Cf. *KuM* I, 39-40 (*KaM* I, 32-33).
I.42a Bultmann, in *KuM* I, 125-126 (*KaM* I, 107).
I.42b *GuV* I, 36-37 (*FaU*, 64).
I.42c *KuM* I, 42 and 46-47 (*KaM* I, 36 and 42). Cf. "Der Sinn des christlichen Schöpfungsglaubens," in *Zeitschrift für Missionskunde und Religionswissenschaft* 51 (1936), 1-20; ET: "The Meaning of Christian Faith in Creation," in *BEF*, 206-225.
I.42d *KuM* I, 28-29 (*KaM* I, 18-19).

I.42e Cf. *JCM*, 70-71, 83-84; *KuM* II, 196-197 (*KaM* I, 197); *GuV* I, 35-36 (*FaU*, 62-63).
I.42f *Jesus*, 175 (*JW*, 190-191). Cf. *KuM* I, 29 (*KaM* I, 20).
I.42g *JCM*, 65; *KuM* II, 197 (*KaM* I, 197).
I.42h *KuM* I, 45 (*KaM* I, 40).
I.42i *KuM* I, 46-47 (*KaM* I, 41-42).
I.42j *KuM* I, 48 (*KaM* I, 44).
I.42k J.H. Newman, *The Development of Christian Doctrine* (London, 1960 [first published, 1845]), 3.
I.43a Cf. U. Lattanzi, "I Sinnottici e la Chiesa secondo R. Bultmann," in *Miscellanea Antonio Piolanti I* (Romae, 1963 [*Lateranum*, n.s., 29th year]), 141-169.

mann has never presented an organized and integrated scheme of his postulates; they are expressed almost at random throughout the corpus of his writings. Some of the postulates have been singled out by writers like Heinrich Ott and Anton Vögtle as being the central ideas on which his thought is constructed. When Bultmann speaks of his presuppositions, he is referring specifically to the conclusions of the existentialist analysis of being, and it is with reference to this analysis that the discussion of his postulates has unfolded. But there are other postulates as well. For a complete characterization of Bultmann's literary genre, all of the presuppositions would have to be analyzed, but for the purposes of the present tentative description, it should suffice to treat briefly the existentialist principles which Bultmann recognizes as his postulates and the radical presuppositions which Bultmann uses but does not formulate at all.

44. Five of Bultmann's existentialist postulates may be expressed as follows: authentic existence is the existence of the existential encounter; there is no continuum whatsoever between the non-world of the existential encounter and the world; genuine understanding is the existential grasp (and non-grasp) of what one already knows non-existentially; genuine history is the opening of man to his existential future; genuine eschatology transpires within the ever-repeated 'now' of the existential encounter.[a] These five postulates are among the most important, because they represent controlling principles in five fields of enquiry: metaphysics, cosmology, epistemology, history, and mysticism. As here collected and expressed they represent a framework superimposed upon the letter of Bultmann's writings. But they are postulates, and not mere presuppositions. Each postulate is based upon a conception of antithetical nature, so that five central notions emerge from Bultmann's thought: existential existence, non-world, preunderstanding, genuine history (*Geschichte*), and timelessness. It is only by an analysis of these notions that one can arrive at an understanding of the meaning and the rôle of the five postulates in their relationship to demythologizing. Bultmann explicitly requires the acceptance of his existentialist postulates as a basis for discussion of his theological program, in which they emerge in consistent harmony.[b] Bultmann did not create them *ex nihilo*. They are formulations derived from larger thought-currents to which Bultmann had been exposed in his formative years: historical criticism and the history of religions, rationalist liberalism with its *Weltanschauung*, existentialist philosophy with its paradoxical notion of existence, and 'dialectical theology.' In the program of Bultmann, these currents flow together into an imposingly compact internal unity that appears logical and consistent, however terrible may be its implications.[c]

45. *Existential Existence.* Existence, as defined by the scholastics, is the unintelligible substratum of all that can be known.[a] A thing is knowable inasmuch as it has form. Bultmann, in his flight from formal intelligibility, has based his thought upon a metaphysics of the unintelligible.[b] It has no meaning in itself or anything positive to contribute to the development of thought. Only as a symbol of negative quantity (that is, of pure identity) can it be used, and Bultmann has used it effectively within the context of his train of thought.

46. The noesis of the unintelligible, as presented by Bultmann, is replete with paradoxical 'meaning.' First, it is a metaphysical concept which makes possible the fusion of his writings into a highly unified system and enables him to undercut those theologians and exegetes who manipulate terms without reflecting sufficiently on their metaphysical implications. Secondly, it enables him to formulate problems that are unsolved and even unnoticed by many thinkers around him. Thirdly, it enables him to treat in his own paradoxical manner of realities latent in the prescientific consciousness of modern man and not yet recognized

---

I.44a Cf. Vögtle, *op. cit.*, 833-858.

I.44b R. Bultmann, "Heilsgeschichte und Geschichte: Zu Oscar Cullmann, *Christus und die Zeit*," in *Theologische Literaturzeitung* 73 (1948), 663.

I.44c Cf. Vögtle, *op. cit.*, 832.

I.45a *Supra*, 2.4:6a.

I.45b That form is the principle of cognition in the knower is shown in *S.Th.*, I, q. 14, art. 1; Aquinas, *De Veritate*, q. 10, art. 4; ET: *Truth*, trans. J.V. McGlynn, Vol. II (Chicago, 1953).

scientifically. It would be an unnecessary digression to analyze here the ontology of existence, but the complete interpretation of the genre of Bultmann's thought will require an analysis also of the ontological aspect of its expression.[a]

47. The Heideggerian-Bultmannian notion of existence, according to which genuine human existence (*Dasein*) is made the touchstone of its own authenticity, is false for the reason that the grade and value of all existence, human existence included, is measured on the scale of form.[a] Existence is the unintelligible fundament of everything that can be known; form is the intelligible superstructure by which things are known.[b] Since existence is by very definition the unintelligible fundament of the known, it cannot confirm its own authenticity. By locating human existence in pure experience and in the will, Bultmann side-steps the question of intelligibility. But, if genuine existence is made to pertain to the will and not to the intellect, then it is a deception to call it 'understanding.' And the will is a blind faculty. Bultmann takes refuge in pure experience as the fact which ratifies existence and consummates the union of intellect and will. But here a blank wall is reached, since experience is but the undifferentiated 'raw material' out of which all perceived objects are constructed. What we call 'experience' in the realm of human thought is the lowest form of understanding. How can understanding's lowest form be the judge and criterion of the higher forms? Certainly, it is false to aver that experience is more real than reality itself. It follows that, if experience is made its own infallible witness of authenticity, then the 'listening and hearing in a concrete and individual stupor' of the existential encounter has no more qualitative value than has the bliss of an opium dream. If human existence were its own supreme and infallible evidence of genuineness, the one possibility that could never be admitted is the event of death — the annihilation of 'genuine existence' by outside forces, whose existence can hardly be called not-genuine. If death triumphs over 'genuine existence,' then 'genuine existence' is not the supreme reality.[c]

48. *The Conception of Non-World.* There can be no doubt that Bultmann has taken as one basis of his system a radical and peculiar distinction between world and non-world. Because there is a point of contact between them, namely, the existential encounter, no amount of qualifying can entirely eliminate the last residual image of objectivity which must be accorded to non-world in order to be able to think of the encounter at all. Bultmann has discussed in various places this hazy objectivity of non-world and has managed with considerable adroitness to argue away any given features or attributes which could be suggested for it. He does not mind calling it 'God.' The name is as good as any other. But all of the features which go into the common idea of God as presented by conventional Christianity are for Bultmann 'mythological.' Just as the Jesus of Bultmann does not speak objectively of God, his nature, or his attributes,[a] so also can Bultmann tell his hearers nothing objectively of God, his nature, or his attributes.

49. The 'non-world' of Bultmann is something different from the infinite superiority of God to anything created. It is a denial that the analogy of being can in any way be applied to God, and Bultmann has in fact substituted an *analogia fidei* for the *analogia entis*.[a] Even the human self (*Dasein*) in its most intimate center is said to transcend the subject-object relationship presented by rational mind or

---

I.46a Cf. *Philosophiae scholasticae summa*, Biblioteca de autores cristianos (3 vols., 2d ed., Madrid, 1953-1959), Vol. I (authored by L. Salcedo and C. Fernández), Tract. IV, Nos. 29, 75, 88, 190, 191, 212, 249, 263, 566, 730.
I.47a *Supra*, 2.4:2b.
I.47b *Supra*, 2.3:18a.
I.47c Bultmann has not faced this problem. Cf. "Der Sinn des christlichen Schöpfungsglaubens" (*supra*, I.42c). Nor does the speculation of Heidegger, for instance in Division Two, Chs. I and II, of *Sein und Zeit*, really face the problem of genuine existence.
I.48a *GuV* II, 258 (*BEPT*, 287).
I.49a Bultmann, in *KuM* II, 196-197 (*KaM* I, 196); *JCM*, 68. Schubert Ogden, in his translator's introduction to *BEF*, says that he sees Bultmann's analogy of faith to be one of proportionality: God is to his existence as man is to man's existence. I have found no support for this conclusion in the writings of Bultmann, and it seems improbable in view of Bultmann's clear negation of any analogy between this

spirit (*nous, Geist*) and its correlative world. Hence the act of God encountered by the inmost self in the existential encounter has no need to be clothed with objective and 'mythological' characteristics. Thus does the hazy residual image of an objective 'something' which our mind necessarily projects in order to conceive at all of an encounter disappear into itself until only its 'act' remains.[b] The basic error of Bultmann's analogy of faith lies in the transition from the subjective to the objective order. Either the existential encounter is an encounter or it is not. If it is an encounter, then in order to be, or even to be conceived of, it must contain two 'somethings.' To dissolve one something or the other destroys the encounter. Natural intellectual consciousness cannot tell us how God knows Himself in his inmost being, but we can think of God in intellectual consciousness only as an object, and God as object remains even on the higher level of mystical consciousness and beatific vision.

50. *The Conception of Pre-Understanding.* A concept is double if it applies to two distinct phases of a phenomenon in identical fashion with no conceptual difference. If, for instance, we were to say that manhood and boyhood are identical, because a man is not anything which the boy was not and all of the characteristics of manhood are found germinally in boyhood, then we would have a double concept of manhood. Using the concept, we might say: The term 'boyhood' should be eliminated from our vocabulary, because boyhood is really manhood. Now, in this sense Bultmann presents a double concept of understanding. He says, in effect, that understanding and pre-understanding are identical, because anything which we have come to understand was germinally understood by us even before we came to understand it. Hence, the term pre-understanding should be eliminated from our vocabulary. He prefers to call it 'suppressed knowledge.'[a] And this double conception of understanding is again doubled by Bultmann.

In the first phase, the 'genuine' understanding of a 'non-something' like loyalty, friendship, or love, is identical with one's pre-understanding of the same.[b] In the second phase, the genuine understanding is seen as identical with the understanding that one has of himself.[c]

51. While Bultmann is not a total subjectivist in that he retains the notion of the encounter,[a] his tendency to value only the subjective aspect of a given series of situations makes him a subjectivist in regard to every theological object apart from the existential encounter, and even the encounter is thoroughly de-objectivized. The cosmological image of God, which one might have considered the other 'something' of the encounter, disappears inside of itself until only its 'act' remains. The nature of God is then designated 'will.'[b] The cosmic figure of Christ, more proximate party of the encounter, disappears inside of itself until only the Cross remains. Then the Cross dissolves into its Christological meaning. At the end of the process, the only objectivity remaining in the other party of the encounter is the objectivity of the kerygma. Thus the actuality of the real world is reduced to a mere departure-point for the plunge into subjectivity. The very use of the term 'objectifying thought' by Bultmann leaves the impression that the world of empirical-intellectual reality does not exist 'out there' prior to its being known, but is simply a projection of the mind. In the Bultmannian train of thought, the specific differences of all things are gradually dissolved in the 'developing tank' of self-knowledge until the only thing which remains is the question of human existence. In the Bultmannian proclamation, this question alone may be legitimately asked of the figure of Jesus Christ in the New Testament.

52. There are serious contradictions visible in the Bultmannian train of thought. His 'encounter' is not an encounter with anything behind or beyond the words of the kerygma, since the reality behind the kerygma is

---

world and the non-world of faith.

I.49[b] In addition to the references given above, see Bultmann, in *BEF*, 167.

I.50[a] Bultmann, in *KuM* I, 125-126 (*KaM* I, 107).

I.50[b] *HE*, 113-117; R. Bultmann, "Vom geheimnisvollen und offenbaren Gott," in *Die christliche Welt*, 31 (1917), 572-579; ET: "Concerning the Hidden and the Revealed God," in *BEF* [23-34], 26.

I.50[c] *HE*, 144; *BEF*, 92-93 and 221.

I.51[a] Cf. Macquarrie, *The Scope of Demythologizing*, 108-116.

I.51[b] *HE*, 96.

methodically dissolved. And his dissolving process is antiscientific, since it is based upon disregard for specific differences between concepts in the context of historical development. The historical scientist knows that, while it is true that in one sense our eventual understanding of things is undifferentiated from our pre-understanding of them, it is precisely the task of the scientist to discover the ways in which they differ, leaving to the poet, the musician, and the artist the depicting of identical elements in modal diversity. The historical scientist knows how the man differed from the boy, how the French Republic differed from the French Monarchy, and so forth.

53. Bultmann turns inside-out the notion of genuineness. The direct and non-paradoxical notion of the genuine is what conforms with reality as known to sensory and intellectual experience. The non-genuine is identified by contrast with what is known to be genuine. To the direct notion of genuineness can be added a reflex notion based upon the realization that the mind does not necessarily conform with reality in every act of thought. By recognizing instances of lack of conformity, the mind comes to know its own subjectivity, and thus arises a new understanding of reality, expressed by the term 'objective reality.' This is a negative qualification. It means 'reality which does not contain the lack of conformity indicative of the human mind.' And it does not modify in any way the structure of reality; it merely perceives a meaning of reality by reference to the human mind. Bultmann errs in trying to make human subjectivity a reality superior to reality itself.

54. *The Conception of 'Genuine History' (Geschichte)*. Bultmann sees a vital relationship of history to genuine existence and to genuine understanding. "It is only by taking account of the past that man can think and act and be. In this the historicity of his existence consists." [a] He affirms the inadequacy for faith of conventional history (*Historie*), conceived as "a closed continuum of effects in which individual events are connected by the succession of cause and effect." [b] In the transition from history (*Historie*) to eschatology (*Eschatologie*) he uses the vehicle of the Heideggerian notion of history (*Geschichte*), which is not conceptually distinct from the existentialist idea of the historicity of man: History has to do with meaning, and the human self ultimately means to exist, to be confronted with non-being, to be faced ever and again with decision to be one's self.[c]

55. It is easy to see how Bultmann, in his deep awareness of the difficulties underlying historicity, felt attracted to these conclusions. Born into a theological tradition that was not presenting him with satisfactory answers, he saw himself threatened by the historical process and used 'history' to fight history. Recognizing that history is somehow a being of the mind, he located it in the notion of the human self. He did not see that history could be a science with its own objective nature and identity formally distinct from every other science and mode of knowledge.[a] Bultmann did not see that history, as a science, is formally a being of the mind located in a field of objectivity that may be called its subject, or object, or content, while historicity is the fidelity with which historical reality is reproduced. The historicity of man is nothing else but man as an object of historical science.

56. *The Conception of 'Timelessness' (Zeitlosigkeit)*. As a person whose *Weltanschauung* excludes the idea of life after death, Bultmann has erased the future-oriented character of Christian faith with considerable skill.[a] While admitting that for the first disciples, as for conventional Christianity, the ultimate and definitive future of the Christian outlook was embodied in the cosmic conceptions of judgment, heaven, and hell,[b] he adroitly transposes the issue by bringing in the new eschatology, pioneered in the 1890s by Johannes Weiss.[c] According to the new

---

I.54a *HE*, 2 (quoting Erich Frank, *Philosophical Understanding and Religious Truth* [pub. in 1945]).
I.54b R. Bultmann, "Ist voraussetzungslose Exegese möglich?" in *Theologische Zeitschrift*, 13 (1957) [409-417], 410-412; ET: "Is Exegesis without Presuppositions Possible?" in *BEF* [289-296], 290-292.

I.54c *HE*, 143-144. Cf. *HE*, 5; *BEF*, 102-110.
I.55a *Supra*, 3.2:1; 3.2:9.
I.56a Cf. Vögtle, "Rivelazione e mito," 851-858.
I.56b *KuM* I, 15-16 (*KaM* I, 1-2).
I.56c J. Weiss, *Die Predigt Jesu vom Reiche Gottes* (Göttingen, 1892); ET: *Jesus'*

eschatology, the Kingdom of God in the preaching of Jesus is not seen as immanent in the world and does not develop in unison with the history of the world.[d] Since the fulfillment of Christianity does not stand at the end of what is commonly called the process of history, the cosmic interpretation of eschatology has turned out to be false. According to Bultmann it was exploded by the fact that Christ did not come in the heavens as He had been expected to do.[e]

57. Bultmann's notion of the future is based upon what Heidegger calls the 'genuine possibility of being.' [a] For Bultmann, the future is the singularity of one's own authentic possibility of being; to Heidegger's notion he thus adds the characteristic of individuality. This future, he claims, can be known to the philosophers, but its attainment is a gift from God. Bultmann thus attempts to raise the philosophical speculation of Heidegger to the level of a theology.[b].

58. The question now arises of why true selfhood must be received as a gift, why Christ or the kerygma should have any necessary part in the future. It is in answer to this question that Bultmann advances his notion of timelessness. His 'timelessness' is the 'now' of the New Testament; it does not flow like empirical time. It is not the timelessness of the mystics, "as if all events in time were but parables of eternity"; it is the entrance of man into encounter with non-world. Timelessness becomes an event for the individual by virtue of an encounter which is also (paradoxically) in time. The encounter with non-world beyond the flow of empirical time is the mystery of the New Testament eschatology.[a] Bultmann thus reduces his conclusions to a mystery, and on that ground he calls his system a theology.

59. Since the present is the cognitive principle of time and is not formally a part of time,[a] Bultmann has moved skillfully in locating eschatology there.

But how can the believer be freed from the past and yet remain in the past? How can he belong to the future and yet not stand in the future? Bultmann replies that, since freedom of the will is not an abiding quality, the future can never become a secure possession, but must ever be laid hold of anew.[b] He thus transposes future as an object of the intellect into future as the self-fulfillment of man by the vital operation of his own will. Every successive moment of empirical time becomes the *now* of responsibility over against the past as well as over against the future.[c] By 'past' is meant one's own individual past (of sin), which is retained as pardoned; by 'future' is meant one's own individual future.[d]

60. Since time is the dimension of change considered as before and after, there can be no conception of time which does not contain the notes of past and future. Nor can there be any conception of the future which does not contain the characteristic of time.[a] Bultmann, in order to present a future which does not depend upon time, has taken refuge in the idea of an ever-repeatable present: because it is ever-repeatable, it lies ahead as the future does; because it is the present, it cannot be the future. The fallacy lies in this: If there has been no change, there has been no act, and the thrill of selfhood is a pure illusion; if there has been a change, then the present has thereby become past and the future has become present. But the latter alternative supposes time, and is thus excluded by definition. Therefore, the theology of Bultmann is ultimately based upon an illusion.

61. The fact that the dechronologized time of Bultmann is a pure contradiction does not destroy its functional utility. Bultmann, in fact, merely uses it to unify his approach to many questions. It enables him to interpret Christianity as an historical religion, in the sense that it takes place within the 'historicity' of the believer, and yet

---

Proclamation of the Kingdom of God (Philadelphia: Fortress, 1971).

I.56d *JCM*, 11-12.
I.56e *HE*, 38.
I.57a *BEF*, 96-97. Bultmann acknowledges (*BEF*, 102) his indebtedness to Heidegger and Gogarten for what he has learned from them.
I.57b *BEF*, 93.
I.58a *KuM* I, 129-132 (*KaM* I, 112-115).

I.59a *Supra*, 4.2:16.
I.59b R. Bultmann, "Der Mensch zwischen den Zeiten nach dem Neuen Testament," in *Man in God's Design* (Valence, 1952) [39-59], 44-45; ET: "Man between the Times according to the New Testament," in *BEF* [248-266], 254-255.
I.59c *HE*, 143. Cf. *BEF*, 207-208.
I.59d *BEF*, 254-255. Cf. *JCM*, 80-84.
I.60a *Supra*, 4.3:3.

to attribute to Christianity an exclusively eschatological character.[a] It recognizes the aura of mystery and the sense of confrontation with the extramundane which is so essential to the native Christian impulse. Bultmann admits that the Old Covenant was founded upon an event of real history (*Historie*), but "the new people of God has no real history, for it is the community of the end-time, an eschatological phenomenon," possessed of an awareness of having been taken out of the real world.[b]

62. *The Presuppositions of Demythologizing.* A complete interpretation of demythologizing would require a thorough analysis of its presuppositions in their bearing upon its conclusions. However, it may suffice for a preliminary description to list four chief presuppositions and to indicate their general influence upon its conclusions taken as a whole. These four presuppositions are contained in the following notions: rationalism, subjectivism, materialism, and the principle of literary genres.

63. *Demythologizing and Rationalism.* Considered in relation to the conclusions of the rationalist critics of the nineteenth century, demythologizing represents a reaction. Hence, in contrast to them, Bultmann is not a rationalist. However, there are certain underlying principles of rationalism which Bultmann implicitly accepts. His conclusions, mild and conservative in the view of a writer such as Fritz Buri,[a] appear extreme to one who is not viewing them in reference to the preceding German theological tradition. A common underlying source in this tradition is the liberal Lutheran doctrine of faith and justification, with its radical separation of faith and reason.[b] A liberal interpretation of this doctrine gives unlimited scope to Bultmann's private judgment, permitting him to prefer his own conclusions even to those of Jesus Christ.[c] Furthermore, although Bultmann is reacting in a limited area to the conclusions of the nineteenth-century rationalists, he nevertheless accepts the substance of their doctrine.[d] This doctrine appears implicitly in his postulates, although in inverted fashion because of the anti-concepts upon which the postulates are based. Thus, for instance, 'non-world' is derived from a full acceptance of the rationalist notion of world; 'pre-understanding' is derived from a full acceptance of the rationalist notion of science; 'non-time' is derived from a full acceptance of the rationalist notion of time (which Bultmann calls 'empirical time'); and nonhistory (*Geschichte*) is derived from a full acceptance of the rationalist notion of history (*Historie*) with its consequent disintegration of the 'historical Jesus.'[e]

64. Bultmann's existentialism is also derived from rationalism. It is occasioned by the esteem amounting almost to awe with which he regards the world of empirical science. The overpowering effect of this esteem leads him to adopt a theology which professes to abdicate entirely the rôle of the intellect in the exercise of faith. In this sense the theology of Bultmann may be called inverted rationalism. The imprint of the liberal Lutheran notion of faith is impressed upon the whole of Bultmann's theological production. Bultmann accepts this notion and he expressly states that he thinks his system to be a faithful interpretation of it.[a]

65. *Demythologizing and Subjectivism.* The profound influence of subjectivism upon demythologizing has been indicated in the preceding pages. It should be sufficient here to summarize its influence upon the five postulates. 1) 'Non-world' is neither another world nor an extension of the real world. It is not a product of 'objectifying thought,' whether mythological or scientific. It cannot exist 'out there' in any sense. Hence, it is a purely subjective being. 2) 'Pre-understanding' is an embodiment of the idea that the principle by which understanding arises does not come from the world of objectivity, but rather grows up

---

I.61a *HE*, 152-153.
I.61b *HE*, 36.
I.63a *Supra*, 1.1:4a.
I.63b *JCM*, 83-84.
I.63c *Das Urchristentum* (*supra*, 1.2:4e), 102; *Primitive Christianity*, 92.

I.63d R. Bultmann, "Autobiographical Reflections," in *BEF* [283-288] 287-288.
I.63e Cf. A. Schweitzer, *Von Reimarus zu Wrede: Eine Geschichte der Leben-Jesu-Forschung* (Tübingen, 1906); ET, *The Quest of the Historical Jesus* (London, 1954).
I.64a "Faith in God the Creator," in *BEF*, 180-181. Cf. R. Bultmann, "The Sermon on the Mount and the Justice of the State," in *BEF* [202-205] 204-205.

from within the subjective consciousness. 3) 'Non-time,' as the ever-repeated 'now' of decision is obviously a subjective being completely out of continuity with the time of the real world. The 'future' of Bultmann's eschatology has no objective existence, but is conceived of as perpetually potential being whose impossibility of ever attaining actual being (even of the purely subjective sort) is based upon the very nature of the human will, which, in Bultmann's conception, must ever decide and decide anew. 4) 'Non-history' has no continuity with the real (objective) world. The distinctive note of non-history is identified with pure existential understanding, in independence from the limitations imposed by fact in the case of real history (*Historie*). Furthermore, all 'genuine knowledge' is self-knowledge. 5) 'Genuine existence' in the theology of demythologizing is based upon a methodical reference of all human activity to the human subject in the fullest conceivable way.

66. *Demythologizing and Materialism.* Considered in relation to the conclusions of the dialectical and other extreme materialists, the theology of Bultmann again represents a reaction.[a] But materialism, too, is contained in inverted fashion in his postulates and in his conclusions, for materialism is nothing else than a deliberate and exaggerated focus of the mind upon the material side of reality. Such a focus may be plainly seen in the postulates of demythologizing. 1) The notion of 'world' which is presupposed by demythologizing is an overly material world. The existence and the activity of purely spiritual beings, such as God and the angels, are excluded from it by very definition. A miracle, for instance, is simply incompatible with Bultmann's notion of the world. 2) 'Pre-understanding' derives its plausibility from a suppression of the formal and spiritual nature of science and understanding. The objective intellectual medium which makes true science possible is pushed entirely out of focus, so that science comes to be identified with the scientific method and eventually to be immersed entirely in matter.[b] 3) 'Non-time' is derived from the conviction that the afterlife and the whole cosmic picture of the Last Things are mythological. 'Time' in the notion of demythologizing is identified with material and secularized time. 4) 'Non-history' proceeds from the dismay of a mind thoroughly convinced of the truth and the logical consequences of the historical conclusions (*Historie*) of liberal scholars in the German theological tradition. This history does not include within its focus the formal intelligibility of historical science; it is restricted to the distorting level of exclusive reference to the material side of reality. 5) 'Genuine existence' is a counter-notion to the overwhelming presence of material reality, which has become the criterion of all *real* existence (as distinguished from 'genuine' existence). The overwhelming reality of the corpses of the dead precludes any serious possibility of the immortality of the human soul. The 'legend of the Empty Tomb', e.g., simply cannot be conceived of by Bultmann as the result of a *real Resurrection* of Jesus from the dead. Hence, the 'authentic fulfillment' of the existentialistic future concedes by its very notion a substantial victory to the materialism which it is intended to oppose.

67. *Demythologizing and the Principle of Literary Genres.* The principle of literary genres, as presupposed by demythologizing, states that a type of literature is constituted according to the intention of its author.[a] It is precisely this principle which gives to the form-criticism of Bultmann its distinctive character. It should be clearly realized that the method of demythologizing is the form-criticism of Bultmann modified by a more subtle notion of myth. A major weakness of form-criticism is its questionable criteria for distinguishing the literary forms. Bultmann and Dibelius distinguish and classify forms in terms of the purpose in the mind of the author. Thus, polemic, apologetic, homiletic, and disciplinary intents become criteria for designating literary forms.[b] This method overlooks the fundamental literary distinction between the *finis operis* (intrinsic purpose of the things produced) and the *finis operantis* (extrinsic purpose of the producer). The method of form-criticism, here as around the entire circle, operates by a suppression of natural qualities based upon essence and by a confusion of

---

I.66a *HE*, 68-70.
I.66b *Supra*, 2.6:7; 2.6:15.

I.67a Cf. *GST*, 260 (*HST*, 245).
I.67b *GST*, 2-7 (*HST*, 2-6).

formal with final causes. The extrinsic purpose of an author has nothing *per se* to do with the literary genre in which he writes. Furthermore, if he is an honest man, the fact that he wishes to refute an adversary or convert a friend or receive a sum of money will not lead him to distort the facts. Such purposes pertain to the *finis operantis* and are extrinsic to the literary genus itself. If this were not so, then our first task in taking up any of the works of the form-critics would necessarily be to determine for what extrinsic ends they have composed their works, whether apologetically in order to defend their system against attack, or polemically, or for purposes of indoctrinating their disciples, and thus to classify the writings of the critics under different genera. A literary work has an intrinsic purpose too; to discover this is to know *why* it is. But literary theory which distinguishes forms by final cause alone, as form-criticism does, is woefully inadequate. One has not really defined a literary form until he can tell *what* it is. Hence none of the literary forms proposed according to the method of form-criticism has ever adequately been defined.[c]

68. It is precisely this weakness that Bultmann has sought to rectify by his method of demythologizing and, in particular, by his subtle notion of myth. The reality genres of historical-scientific thought are circumscribed by their own natures and limited to objective truth. They are not constituted at pleasure by objectifying reverie. But it is not so with the genres of literature (*Dichtung*), which are a product of human creative thought and are *per se* not limited by reality or objective truth. Mythology is such a genre. Since, then, for Bultmann, all religion is necessarily mythological, the task which he sets for theology is not the elimination of mythology from the Gospels, but the interpretation of them in keeping with a mythological vision which is acceptable to modern man. The Gospels, however, whether taken as units or dismembered into myriad sub-genres, remain forever expressions of mythology, genres of literature. Theologians, too, are producers of religious literature, manufacturers of mythological thought. They are not interested in raising the question of truth.[a]

69. Let it be noted that according to demythologizing the immediate and concrete form in which the kerygma encounters the existential believer is in the spoken and written word. Its mode of being is literary being. Moreover, if one wishes to encounter the word of the kerygma, he is told that he will find it most immediately and most acceptably in the written and oral production of Bultmann and his co-workers, for it is they who, as Bultmann declares, are the mediators and the constructors of the modern Gospel. The writings, then, of Bultmann and his colleagues are a literary genre, a species of literature similar in some respects to other forms of literature. Each genus of literature in its concrete and individual particularity has certain rules which apply only to it and a certain special configuration stemming from the different motive and different interpretation of life which has prompted its creators to write. There is in the theology of Bultmann no ultimate distinction having significance for genuine understanding between the Gospel according to Mark or the Gospel according to John and the Gospel according to Rudolf Bultmann. Consequently, the writings of Bultmann must be interpreted in accordance with the principle of literary genres, a fact which Bultmann seems to have been aware of from the start. By using this awareness consistently, he has made his approach a basically realistic one. Not only has it given him a point of contact with the Gospels themselves and with historic Christianity, it has also provided a logical motive for his approach to the reading and listening public. In the scientific order, reality is the touchstone of authenticity, while sub-

---

I.67c I am pointing here to a *double* error in form-criticism's notion of literary forms. The first error is that of not distinguishing between the *finis operis* and the *finis operantis*. The second error is in trying to define the forms in terms of their final rather than their formal cause.

I.68a Bultmann maintains that his purpose as a Christian theologian and preacher is not to eliminate the mythology of the Christian faith, but only to interpret it (1.2:1b). But the serious historian *is* interested in raising the question of truth, because he has to understand things in their relation to reality. The aim of the poet (*Dichter*) is to communicate experiences of the soul; the aim of the historian is to express historical truth.

jective existence is not a proof for anything. But in the *literary order*, reality is not the ultimate criterion of genuineness. Literary creation (*Dichtung*) is a law unto itself, and the writer (*Dichter*) may appeal to his listeners to enter its embrace for no better reason than its own attractiveness.

70. In the light of the principle of literary genres, understood in its full application, the whole theology of demythologizing becomes clear. The encounter with non-world is in no way restricted by the objectivity of the real world; it depends rather upon a deliberate departure from the real world through the mediation of faith. What, then, is faith, if not the romance of make-believe whereby one derives motivation to perpetuate the encounter? But such a resolve requires an ever-repeated act of the will, for, as soon as one ceases to make-believe, he falls back to dull reality. Mythology, moreover, is the language of the soul in its search for self-fulfillment, and Christianity is a form of mythology which has the attractiveness to woo even the modern mind away from the prosaic insufficiency of the real world.

71. In the theology of Bultmann, the notion of myth is the central idea, containing within it the ultimate explanation of mystery and paradox. The encounter with God and man is the predominant element, the fundamental law of existence and self-fulfillment: heart speaks to heart through the encounter of love. Literary intent brings a new dimension to the flat and rigid surface of objective truth. Reality, taken as a mere point of departure for the journey into subjectivity, along pathways not fenced in by the rigid rules of logic, assumes exciting new forms and meanings. Bultmann has recognized the intrinsically fictional character of demythologizing, and he has exploited his recognition to the full. He has constructed a non-world of anti-concepts. He has made the letter of the Scriptures his point of departure. He has covered the letter of his interpretations with a dazzling garment of plausibility. He has anticipated almost every objection which would normally occur to the believer tempted to give up the journey and return to the real world, and he has parried these objections skillfully. He has, in short, realized that demythologizing is a form of biblical science-fiction, and he has expressed this realization with consummate art.

72. The encounter, then, is really a paradox, and not a pure contradiction. The escape from reality is not into irrationality, but rather into romance. Since, moreover, the lone event of the encounter is exempted from the dissolving influence of subjectivism, demythologizing is, in the ultimate analysis, realistic. It is the intuitive realism of a poet who believes that all that is important to man has not been included in every sense within the narrow framework of scientific thought. The myth of God, in the Bultmannian interpretation, is a science-fiction romance whereby religious thought gradually and fascinatingly divests the traditional image of God of all of its divine characteristics until only its act remains. Meanwhile, the image of reality is gradually divested of all of its objectivity until only the awareness of self remains. Since for Bultmann religious thought is not scientific thought, the demythologizer may prescind from logic in the narration as well as in the derivation of his conclusions. This is the beauty of pre-knowledge, by which he already knows when his study begins the conclusions at which he intends to arrive. Eschatology becomes the enthralling image of a vanishing history which through the skill of the preacher gradually dissolves all of the barriers standing between the reality of the Scriptures and the romance of the kerygma. Like the Cheshire cat, which gradually disappears until only its smile remains (even the lips are gone),[a] so also, in the epic of demythologizing, the factual basis of biblical history is gradually dissolved until only its meaning remains.

I.72[a] L. Carroll, "Alice's Adventures in Wonderland," in *The Complete Works of Lewis Carroll* (London, no date), pages 66-67.

Appendix II

CONCERNING W. H. WALSH'S CONCEPTION OF HISTORY

History and Science Compared

EXPOSITION

1. In his *Introduction to Philosophy of History*,[a] W.H. Walsh brings out the intellectual character of historical thinking and its uniqueness as a form of thought, showing also that it has features making it comparable to science. Walsh observes that the equation of knowledge proper with knowledge derived from the methods of natural science was made by most leading philosophers from Descartes and Bacon to Kant (*IPH*, 12), but from about the close of the nineteenth century some philosophers of a more 'idealistic' propensity, for example, Wilhelm Dilthey in Germany and Benedetto Croce in Italy, have advanced that history offers a form of knowledge that is concrete and individual, in contrast with the 'abstract' and general knowledge afforded by the natural sciences (*IPH*, 14-15).

2. A common view of history makes it coordinate with perceptual knowledge, that is, with the knowledge of objects on the sensory level of individual facts about the past. By this view, the results of historical research simply provide data on the sensory level for the social scientist to use in constructing the science of man, in a manner similar to that in which the natural scientist uses the data of sensory perceptions of the present to derive the conclusions of natural science. But historical practice shows that most historians do not stop with conclusions on the sensory level; they strive to produce a 'significant' account in which they can show, not only what happened, but also why it happened. Historians regard history as a significant record in which events of the past are shown to be connected with one another. The explanation which historians tend to give takes the form, not of deduction in terms of general laws of history, but of understanding of the detailed course of individual events, and this fact suggests the possibility that historical thinking may be intellectual in a way that is distinct from the intellectuality of the natural sciences (*IPH*, 18).

3. The primary object of the historian's study is the human past (*IPH*, 32). The essential concern of history is with human experiences and human actions. History records a great number of natural happenings in the past, but it is interested in these events only as a background to human activities. "The historian is not concerned, at any point of his work, with nature for its own sake" (*IPH*, 31).

4. A science is "a body of knowledge acquired as the result of an attempt to study a certain subject-matter in a methodical way, following a determinate set of guiding principles." The scientist asks questions from a definite set of presuppositions, and his answers are connected just because of that (*IPH*, 35).

5. Anyone experienced in ordinary historical techniques knows that the reconstruction of events is not merely an operation of common sense (*IPH*, 38). The aim of the historian is to recount and render intelligible the precise course of individual events (*IPH*, 39). Historians study the past for its own sake, not for the light their study might hope to shed on the future course of events, but they do seek also by their study to illuminate the present, for it is the bearing on the present that makes their study relevant, and they do derive some capability of anticipating the future without going to the point of prediction (*IPH*, 40-41).

6. 'What the past was really like' is not a function of historical evidence alone, as Bury supposed, "but also of the minds of those who work at the problem of discovering it." Not only

---

II.1[a] W.H. Walsh, *An Introduction to Philosophy of History*, 3d rev. ed. (London: Hutchinson and Co., 1970).

evidence is an important factor of historical reproduction, but also the terms by which it is addressed and "the general framework of questions inside which we seek to exploit both it and the conclusions we draw from it. And these are presupposed by the historian..." (*IPH*, 195).

CRITIQUE

7. In his comparison of history with natural science, Walsh makes many observations that are useful for an understanding of what the science of history is. However, he does not identify the precise locus of historical science in intellectual consciousness. To say that history is an attempt to give a significant account of past events by relating them to what the historian feels to be important in his view of present human nature or of the laws of human behavior is not sufficient to locate history in scientific consciousness, unless this view of the present has been shown to be scientifically arrived at and scientifically applied. Otherwise, historical explanation becomes a non-scientific pursuit, taking its point of departure from what has perhaps been seriously established about facts of the past. To say that an account is 'significant' in relation to one's idea of human nature does not necessarily mean that it is *scientifically* significant; it might be merely aesthetically significant.

8. In comparing and contrasting history with science, Walsh is dealing with two insufficiently defined terms, the one being 'history' and the other 'science.' Historians have not been prone to speculate much on the nature of their study (cf. *IPH*, 11-15), while natural scientists are notoriously amateurish when it comes to defining clearly what science is. Most monographs on the subject of 'science' simply skip over the need of definition and evade the issue by declaring that 'science is identified with the scientific method,' which is obviously a contradiction in terms. This lacuna in the field of science is so glaring that some large modern encyclopedias have no entry at all under the heading of 'science' (e.g., *The Encyclopedia Americana*, Internat. Ed., New York, 1967), while others (e.g., *Encyclopaedia Britannica*, article by Charles Singer, Vol. 20, pp. 114-124) begin by expressing the doubt that science can be defined and proceed to characterize it in terms of a description of the 'scientific method.' Walsh is thus constrained to make a comparison between two unknowns.

9. The characterization of historical knowledge as being 'concrete and individual' in contrast with the general knowledge of classical science is a clue to its place in consciousness. Walsh is probing the right area. The clue can be followed up successfully after the third dimension of reality has been recognized. The science of the historian consists, not simply in recounting the precise course of individual events, which is the material object of his study, but also in the recognition of historical meaning in the course of events, which is the formal object of his study. While the object of historical interest is materially the past, the medium in which the past becomes meaningful lies somewhere in the present (4.2:2). I have shown above that for all science the locus must be the present of reality (2.4:1-3) and that the formal growth of all science must be along the third dimension of reality, that is, along the intellectual dimension of the verified (2.6:15). Historical understanding must therefore be a species of understanding within the third dimension of reality that is recognizably different from the kind of understanding that characterizes classical science, whether empirical or philosophical. Walsh has correctly suggested, in keeping with the theory of a whole line of thinkers experienced in history, that the special character of historical understanding and of historical reasoning lies in its adherence to the concrete and the particular. But if this adherence to the particular consisted primarily in the determination of material details, history would be nothing more than common sense of a most inferior sort, despite the artistry with which the historian went about his work. And experienced historians know that history involves something deeper than that.

10. To say that history is concerned only with the human past is to mistake convention for necessity. In order to understand why 'historians' have tended to devote their attention to the human past it is necessary first to distinguish the several dichotomies involved. The first is the dichotomy between the knowing self and the object of his knowledge. Historians have recognized the immense value of finding

meaning in that area of human consciousness that lies 'closer' to the knowing self than do the material things investigated by natural scientists. The second is the dichotomy between the human and the subhuman. The historical interest in the human is so great that historians have not cared to claim for their study those areas of the past occupied by subhuman developments. But this absence of claim does not deprive the subhuman areas of the past of their potentially historical character. The third is the dichotomy between the present and the past. The intellectual movement of the historian is formally towards the present, in the sense that the deepening of his understanding of events in the past requires the recognition that the past lies *formally* in the present. This is not a contradiction in terms. The past does not formally exist in itself. It exists within the presence of the mind of the knowing subject.

11. History and classical science are alike in the fact that both require a general framework of 'principles' that must be operative in the mind of the thinker, if the characteristic type of conclusion is to be reached. Both history and classical science are specialized types of science. The difference lies in this, that the characteristic presuppositions of the classical scientist are principles of the general type, while the characteristic presuppositions of the historian are principles of the concrete type.

Subjectivity in History

EXPOSITION

12. With regard to the question of objectivity in history, Walsh observes that while positivist thinkers make the natural sciences the sole repositories of human knowledge, reducing history to "something other than a cognitive activity," certain idealist philosophers place history alongside or above the natural sciences, maintaining that "it is an autonomous branch of learning, with a subject-matter and methods of its own, resulting in a type of knowledge which is not reducible to any other." He argues that history is more in the order of science than of simple sense perception, and it goes beyond common sense. Thus, he says, it is a kind of science in its own right (*IPH*, 93-94).

13. There seems to function in historical thinking a subjective element not found in scientific thinking, and many historians of today would argue that "every history is written from a certain point of view and makes sense only from that point of view. Take away all points of view, and you will have nothing intelligible left, any more than you will have anything visible if you are asked to look at a physical object, but not from any particular point of view" (*IPH*, 97).

14. The historian brings to his studies a set of interests, beliefs, and values which influence what he takes to be important (*IPH*, 98). While the distinction between fact and interpretation does not seem to be tenable in the ultimate analysis, disagreement among historians as to what conclusions to draw from a given body of evidence seems largely a technical matter, but disagreement about the proper interpretation of the conclusions drawn is what calls into question the scientific status of history. The main factors leading to this disagreement in interpretation are grouped under four headings: a) personal bias (personal likes and dislikes, whether for individuals or classes of persons); b) group prejudice (assumptions associated with the historian's membership in a certain group); c) conflicting theories of historical interpretation; d) underlying philosophical conflicts (basically different moral beliefs, conceptions of the nature of man, or *Weltanschauungen*) (*IPH*, 99-103).

15. "Differences between historians are in the last resort differences of philosophies." By philosophies Walsh means "moral and metaphysical beliefs." Moral beliefs are "the ultimate judgments of value historians bring to their understanding of the past." Metaphysical beliefs are "the theoretical conception of the nature of man and his place in the universe with which these judgments are associated" (*IPH*, 103). F.H. Bradley, in his *Presuppositions of Critical History*,[a] avers that we can believe about the past only that which bears some analogy to what we know in our own experience. If we

---

II.15[a] London, 1874. Reprinted in F. H. Bradley, *Collected Essays*, 2 vols. (Oxford, 1935), Vol. I.

accept Bradley's formula for history, we must understand by 'experiences' not merely experience of physical nature, but experience of human nature as well, and this experience is not all given, but includes an *a priori* element, for the 'hard core' of our experience of human nature is connected with our moral and metaphysical beliefs (*IPH*, 105).

CRITIQUE

16. Walsh sees elements of truth and of error in both the positivist and the idealist approaches to the question of objectivity in history. In arguing that history is a kind of science in its own right, he stands closer to the idealist thesis, but he is also keenly aware of the subjective factor in history. What he does not clearly bring out is that, if history is or can be a science, all four heads of disagreement among historians in interpreting conclusions may stem ultimately from their imperfect grasp of the essential presuppositions of their science. In other words, history may be a hitherto largely unexploited science whose contradictory expressions are attributable to the prescientific thinking of historians.

17. Not only historical accounts, but all expositions of natural science are written from some point of view. Take away the point of view, and you have nothing intelligible left. The reason why the viewpoint of history seems subjective and that of natural science does not is that natural science has advanced much further in recognizing its medium and making an object of it. It is the unfulfilled task of historians to isolate and study their medium of thought and draw from it objective conclusions upon which they can all agree. This formal development of the science of history will eliminate personal and group bias. By producing a clearly analyzed set of inescapable presuppositions, this work will also eliminate those conflicts of historical interpretation that arise from an unscientific use of the historical medium. 'Moral and metaphysical beliefs,' in the sense intended by Walsh, do not pertain to the historical medium as such; they are conclusions belonging to a different area of human consciousness and serve as general presuppositions of historical thought.

18. Walsh cites the importance of intellectual as well as sensory experience in our association of the past with our own experience. The *a priori* element is not primarily our subjective feelings about moral or metaphysical objects; it is the potentiality of the intellect to grow in understanding as well as the understanding itself inborn or achieved. The emergence of historical science requires that the objective principles of human understanding, especially as regards the unique kind of understanding characteristic of history, be recognized and formulated objectively.

Objectivity in History

EXPOSITION

19. Walsh points out that a 'point of view' is a name given to something that is by no means homogeneous. It seems possible for the historian to abstract from some things in his point of view (e.g. his personal likes and dislikes) but not from others. Historical skepticism is based on the principle that all historical points of view ultimately express subjective attitudes about which argument is futile. The perspective theory, on the other hand, accepts the existence of irreducibly different viewpoints among historians, but allows for a 'weakened' or 'secondary' objectivity in the sense that the historian can depict the facts accurately from his own point of view, so that different histories can be considered to complement rather than necessarily contradict each other. A third theory admits the possibility at least in principle that an objective viewpoint having universal acceptance among serious historians could be developed (*IPH*, 106-107).

20. Historical skepticism is supported by the view, "now so common that it has almost become an article of philosophical orthodoxy," that metaphysical and moral statements are not descriptions of real features of fact. While such extreme skepticism seems unwarranted, it can be conceded that the historian has, also a practical purpose in pursuing his primary aim of discovering truth about the past for its own sake. "The conclusion we must draw is that history throws light not on 'objective' events, but on the persons who write it; it illuminates, not the past, but the present. And that is no doubt why each generation finds

it necessary to write its histories afresh" (*IPH*, 108-109).

21. According to the perspective theory of interpretation, any developed history is the product of two essential factors: the subjective element contributed by the historian (his point of view) and the evidence from which he begins and which is accepted as given. Just as a portrait painter sees his subject from his own viewpoint, so the historian regards the past in keeping with his own presuppositions, but that does not preclude insight into the 'real' nature of his subject (*IPH*, 111). The perspective theory recognizes some points of continuity between history and the sciences (for instance that both are primarily cognitive activities), while not admitting that history can be absolutely objective in the way that natural science claims to be (*IPH*, 112-113).

22. We should hope for the achievement eventually of a single historical point of view, a set of presuppositions acceptable to all historians (*IPH*, 115). Metaphysical disputes may be soluble, in principle at least, and the possibility of general agreement on moral principles is not to be ruled out. Until this is accomplished there is no alternative but to fall back on the perspective theory (*IPH*, 116).

CRITIQUE

23. The word 'viewpoint' is often used, even by natural scientists, as a vague representation of an awareness that has not been sufficiently analyzed. Walsh has the insight to realize the need of bringing the historical 'point of view' into view for the sake of examining it, and he makes a beginning of its analysis. The observation that there is an area of the historical viewpoint from which the historian can abstract seems especially important, because it opens up an area of objectivity in historical understanding. Distinguishing thus between objective 'beliefs' and subjective 'attitudes,' Walsh goes on to allow for the possibility, at least in principle, of unanimity of moral and metaphysical belief. The hope expressed here by W.H. Walsh for the ultimate attainment of a single set of historical presuppositions can be realized if explicit attention is given to their graduated objectivity. Along the line of formal abstraction the recognition of results in the absolute past in terms of general conceptions of the concrete type is most specific to historical understanding and most liable to common agreement among historical scientists. For example, the way in which improved technology was actually applied to farming in a particular region provides a medium in which the development of farming in that region can be historically understood. This type of viewpoint occurs in all cases of historical reasoning, but agreement becomes more difficult where human acts and situations are the object of study. However, a systematic separation of these specifically historical 'presuppositions' (historical syntheses) from what Walsh calls 'moral and metaphysical beliefs' would seem to make agreement possible even in the case of human events.

24. Moral and metaphysical beliefs pertain to the general present of the historian. It is my contention that they, too, can be products of scientific thinking, but they are embodied in sciences distinct from both history and natural science. Agreement concerning the presuppositions of these sciences is also possible.

25. Walsh does not seem to distinguish adequately between the 'primary aim' of the historian and his 'practical purpose' in pursuing his study. The instance of history does not differ essentially in this regard from that of natural science. The practical purpose of any scientist is important under certain aspects, but it does not enter into the intrinsic finality of the science as such, where science is itself being regarded as a cognitive activity. Where natural science stands out is in its marked success in isolating the objectivity of its medium, so that it provides an area of recognized concentration distinct from personal 'attitudes' of the scientists. History needs to develop its field in similar fashion. The historical medium, both special and general, does throw light on 'objective' events; it is an illumination standing formally in the present, but the objects it illuminates are in the past. By the 'past' I mean here the sensory images of the past (standing presently in the imagination) in their formal intelligibility as 'past.' Historical understanding illuminates the mind of the historian by giving him present insight into the past, but this kind of illumination of the present should not be called a practical purpose. In point

of fact, as an illumination of the understanding it does not differ essentially from the illumination of natural science, even though it is a different kind of illumination than that of natural science. We are dealing here with two species of genuine scientific understanding.

26. Each generation rewrites its history, but so does each generation rewrite its natural science. The common reason is that as the science develops (or decays) the scientist will tend to view it differently. What is specific to history in this regard is that the special medium of history, the recognized results of the historical process, is itself moving forward in time, and it is the latest stage of the process that constitutes historical perspective in the specific and absolute sense. Therefore, the perspective of every generation will necessarily differ in important respects. But this fact does not make the science of history fundamentally relative to constantly changing viewpoints, because the movement of the perspective is not simply a change; it is a development with its own continuity and intrinsic unity. It is therefore a principle of historical science that the entire picture of the past cannot be seen at any stage of the historical process for the simple reason that part of it has not yet emerged, but the portion that has emerged has meaning in its own right. As more of the picture comes into view, fuller understanding of the past becomes possible. Thus the essential difference between the viewpoints of successive generations is the difference between the lesser and the greater.

27. In comparing historical science with a game to be played according to the conventional rules and to the art of painting (*IPH*, 110), Walsh overlooks its essentially scientific content. The basic rules of historical interpretation are to be found in the structure of the facts being studied and in the laws of human understanding. These are not a matter of convention, and their employment affords an understanding that can be purely cognitive, not simply aesthetic as in the case of fine art or poetry. The example of the portrait painter can be used in a different fashion. To the extent that the portrait painter depicts in his subject insights that are merely suggestive and ambiguous, vague and indefinable traits of character, he remains a simple artist. But to the degree that his intent is to depict through the medium of his art in a clear and unambiguous way his real understanding of the person whose portrait he paints, he becomes a documenter of fact. The portrait artist could find aesthetic meaning in creating a caricature of his subject; the historian who does the same is not a producer of scientific history. The error of the 'perspective theory' lies in its failure to recognize that viewpoints are subject to reasonable argument to the extent that the principles upon which they are based can be identified and made objects of intelligent examination.

Historical Certainty

EXPOSITION

28. According to the theory of *correspondence*, truth means the correspondence of a statement with fact. Facts designated as 'hard' or 'given' are distinguished from theories in that the latter are more tenuous, being at best 'well-grounded' (*IPH*, 74). A difficulty inherent in the correspondence theory is that experiences in themselves cannot be used to test theories. Direct experiences have to be expressed, given conceptual form, and raised to the level of judgment before they can become a criterion. In this process of being interpreted, the original experiences are transformed; they are modified by being brought into relation with previous experiences and classified under general concepts (*IPH*, 75). In the case of history the difficulty of correspondence is increased by the fact that direct experience of the past is not possible, and dependence upon the *ipse dixit* of authorities is regarded as a failure to exercise the universally critical attitude that modern historians maintain with regard to statements from the past (*IPH*, 81).

29. According to the theory of *coherence*, truth means the coherence of any statement with all other statements we are prepared to accept. No statement is made in entire isolation; every statement depends upon certain presuppositions or conditions (*IPH*, 76). A 'fact' is thus conceived, not as an independent existent, but as a conclusion of a process of thinking. Facts are not simply apprehended; they are established. The division of fact from theory vanishes,

and a fact becomes a theory that has been established (*IPH*, 76-77).

30. The coherence theory can be modified to the position that, while every historical judgment is probable in the sense of being subject to revision as knowledge accumulates, yet some historical statements are considered to have a greater degree of probability than others (*IPH*, 86). History deals only with practical, not mathematical, certainty, since no matter-of-fact statement can ever be raised to the status of a logically necessary truth. The belief of many working historians that the past is a world of its own, stretching out behind the present as a fixed, finished, and independent reality, awaiting only discovery, is a philosophical absurdity (*IPH*, 87).

31. The historical past cannot be identified with the remembered past in the sense of being limited to the range of living memory. But historical thinking depends upon memory in a special way; if there were no such thing as memory, it is doubtful if the notion of the past would make sense to us at all. Memory must be a form of knowledge in the strict sense (*IPH*, 83-84). We should be no more expected to prove that there have been past events than that we have experience of an external world. The contents of acts of thought are not mere constituents of the temporal flow of consciousness. Two persons can think the same thoughts and can know that they do, where 'thought' is equated with what the persons think. Memory gives no direct vision of the past, but it does afford a point of contact, enabling us to divine to some degree the true shape of past events. Apart from this, the sole available criterion of truth, in history as in other branches of factual knowledge, is the internal coherence of the beliefs erected on that foundation (*IPH*, 90-92).

CRITIQUE

32. If the past is defined as the 'remembered absent,' the act of remembering becomes its specific characteristic. But this definition does not completely delimit the past of history, for history is also characterized by temporal relationship, while the past is somewhat broader. By the past is meant what has 'passed.' Formally, it has passed through the mind of someone in the form of recognized experiences, either direct or reconstructed. The theorems of Euclid are 'past' for a mathematician; they have passed through his mind and he preserves them in some way in his memory. He also uses them in advancing to higher problems, but their relationship to the elements of the higher problems is not temporal. There are many other cases in which the remembered absent stands in a temporal relationship but is not strictly historical. For instance, the perception of a musical rendition requires the experience of separate sounds presented consecutively in time and their retention in the memory so as to form a conception of larger units. Another instance is the apprehension of the speech of another, in which separate syllables achieve meaning as they are retained in memory and built into words, phrases, and sentences. These primary units of music and speech belong to the remembered absent in its most general sense. The past of history includes the notion of development in the sense of genetic relationship.

33. Memory may be defined as "the power or process of reproducing or recalling what has been learned and retained, especially through associative mechanisms" (*Webster*). It is important to note in the context of the present discussion that memory can be intellectual as well as sensitive. In the case of history, and especially of historical science, the memory involved is based upon the sensitive memory but is formally intellectual. The historian remembers (directly or by the suggestion of a source or by his own reconstruction) a series of sensitive experiences, just as the musician or the mathematician does, but he synthesizes them into units which he calls 'events' and which are events because they stand in genetic relationship to one another. When history is called the 'remembered past,' the adjective 'remembered' is taken broadly to include, not only what is preserved in the memory of direct living experience, but everything that has passed in the form of events and yet remains as past within human awareness.

34. Walsh points out that the need to view the past "through the eyes of the present" means that we cannot escape seeing it in terms of our present conceptual scheme. He does not affirm with sufficient clarity that this 'conceptual scheme' is both general and

specific for historical science and is itself factual when it has been isolated as an intellectual object and shown to conform to the intellectual dimension of reality. We may say that 'facts' are essentially intellectual objects, for they are the conclusions resulting from the apprehension of sensory objects within the field established by the notion of reality, and this in turn means the recognition of intellectual objects ('meanings') not formally identical with the sensory objects but spread out along the intellectual dimension of the things that the sensory objects represent.

35. Walsh affirms that historians cannot know any absolutely certain facts about the past. I agree only in the sense that every human being is subject to illusion, deception, and error to one degree or another. In this sense, natural scientists do not know any absolutely certain facts about the present either, and no one can be absolutely certain that he has just put sugar into his coffee, even though he is presently tasting the sweetness. The admission of the fallibility of the human subject of knowledge is a factor of the scientific possession of knowledge, and it does not infringe upon intellectual insight into the existence of absolute truth as long as human weakness is not made into a principle used to deny the objective character of intellectual objects.

36. The correspondence theory becomes tenable when it is taken to mean correspondence, not merely with the presentations of sense, but also with the reality of human intelligence. The conceptual form given to sensory experiences pertains to the intellectual dimension of reality when it is arrived at rationally. The problem of history is to distinguish the completely rational conceptions from those that are defective.

37. The question of how to treat authorities of the past deserves close and honest study in its own right, since the rationality of doubting all authorities may not then look as self-evident as it does on the surface of things. The attitude of universal doubt assumed by many historians of the modern era to the writings of the past has not been matched by the same quality of criticism of their own medium of thought. Often the rejection of statements for lack of 'evidence' has been based on reference to material evidence alone, with little or no regard given to the intellectual objects implied. Walsh points out that two persons can share the same thought as a content. It is the content of the thought as intellectually discernible to which I refer in speaking of an 'intellectual object.'

38. Walsh's distinction between practical and mathematical probability does not represent a complete analysis. Empirical science recognizes mathematical probability as practical certainty, since the odds of error are so small. The more apt distinction is between logical certainty, which is had only by deduction from necessary truths, and 'moral' certainty, which takes possession where the degree of concrete certainty is sufficient to preclude rational doubt in the practical realization of one's thinking about a problem. This latter kind of certainty is 'practical' in the sense of allowing a premise to be used in the process of thinking, but the premise itself may represent what is scientifically known to pertain to external reality and objective truth.

39. The past with the events of which it is composed is not an independent substance. However, it is a reality independent of the immediate present in which it is anchored. What is real about the past is not the physical base of the remembered happenings. Obviously, this does not exist any longer. The reality stands formally in the objective meaning apprehended by the intellect of the one reviewing the past through the medium of his present. It is not reasonable to suppose that such meaning is imaginary or unreal, simply because it inheres in objects that no longer physically exist. It is reasonable to note that the things underlying the events no longer exist substantially in the same form, but as known through memory they carry real meaning with them. Granted that the meaning coalesces entirely in the present of the historian. The past is simply a formal distinction within that present. The object of historical thought remains really distinct from the formality in which it is viewed. I have just mentioned the parallel example of a musical rendition. We must grant that at any stage of the rendition only one sound is actually being heard. But the succession of sounds carries with it a meaning that goes beyond the sound of the moment and is appre-

hended by synthetic understanding. The same is true of history, except that in the case of music the meaning is essentially aesthetic, while in the case of historical science it is essentially realistic. There is a real difference between intellectual objects of the past and other intellectual objects. It is not a difference of two worlds, the past and the present. It is rather a difference of fields within intellectual consciousness.

Historical Causation

EXPOSITION

40. In his analysis of historical causation, W. H. Walsh observes that Robin G. Collingwood limits historical causality to the affording to a conscious and responsible agent of a motive for performing a free and deliberate act (*IPH*, 191), and Oakeshott claims that searching for historical causes means either seeking to explain historical events in terms of things entirely outside of them (e.g., geographical or economic expressions) or falsely turning past events into independent existents by artificially detaching them from the background in which they inhere (*IPH*, 192). Walsh points out that Oakeshott's view is defective in not allowing for the fact that the historian does present the persons he describes as conscious and responsible agents who are caused to do things in the way identified by Collingwood (*IPH*, 197).

41. The historian examines two other types of cause as well. He answers the question of 'Who caused what?' by fixing responsibility and assessing the amount of an agent's contribution to a given end. He also studies the circumstances which influenced the outcome from the point of view of the agent concerned. Problems about historical causes could be easily settled if these three types of cause were all recognized as relevant to historical inquiry and no other type of causal question were allowed to enter the field (*IPH*, 198-200).

CRITIQUE

42. The conception of history in itself should be systematically distinguished from the bearing of the past upon the practical life of the historian. The separation of the two should follow the lines of the analytical distinction between the *finis operis* (purpose of the work) and the *finis operantis* (purpose of the worker). The historian can have any number of remote or proximate aims in producing his historical narrative, some of which may touch upon his most cherished values, but his work is scientific only to the extent that he is able to prescind from them and write his history in terms of its own intrinsic purpose. That intrinsic purpose is not distinct from the known outcome of the events described and from the intrinsic meaning contained in that outcome. The true historian respects the past for the reality which it contains, in no way distorting or violating that reality; he loves the past for the meaning contained in past reality. The reality of the past requires an exact and a sufficiently detailed narrative as its material base. However, the reality of the past is not sufficiently described in a direction of thought which simply aims at greater and greater detail, ignoring or suppressing the intellectual dimension embodied in these details. History has not explained itself until the intellectual dimension has been recognized and at least implicitly set forth. The definition of meaning as the immersion in material details is the error of historical materialism.

43. Various kinds of causes appear in historical descriptions, but the distinctive type of causality in terms of which historical explanations are expounded is genetic causality. The unclarity in historical circles as to whether or not there are historical causes is largely due to the posing of the question in terms of efficient causality or of formal causality, neither of which is proper to history as such. The historian uses formal causality in determining what the things are that underlie the events he is studying. He must know whether a particular object was a wagon or a sled, a knife or a letter opener. But this kind of knowledge is not proper to the domain of history; indeed, the historian is dependent upon savants of other fields for much of it. The historian uses also efficient causality in describing the genesis of events. He must ascertain, for instance, whether a tree fell because it was broken by the wind or because it was cut down with an axe; he must learn whether a ship sank because it struck a reef or because it was blown up. But, here

again, the arriving at this kind of knowledge is not proper to the domain of history; it is shared by other fields. It is important to the historian to know whether a person struck by a car had been run over accidentally or assailed deliberately, but the determination of such motives pertains more properly to courts of law than to the cerebrations of historians. For the assigning of many efficient causes, the historian makes use of conclusions arrived at in other fields, not accepting them uncritically, but admitting them as evidence unless he has other evidence to oppose them.

44. The kind of historical cause admitted by Collingwood, that is, the motivating of a free and responsible agent, adds an element of purpose to the notion of efficient causation, but the finality is backwards as far as historical understanding is concerned. The determination of subjective purpose is important for many historical descriptions, but is not proper to the historian as such. To the extent that the historian succeeds in reproducing a situation exactly from the viewpoints of the original protagonists he is producing a copy of their views, and these views are directed towards the present actuality of their situation; hence they are blind to any historical knowledge of their situation. The view of history is always towards the past, not towards the future. What gives history its formal intelligibility is the knowledge which the historian has of the actual outcome of the situation; it is distinct from the intentions of the agents; it is historical reality as distinguished from the hopes and desires of the agents. The very notion of history demands that the standpoint of the historian not be the standpoint of the agents involved in the events.

45. The three kinds of cause enumerated by W. H. Walsh are additional to history; they do not pertain to its essence. Once they have been determined, the task of the historian is not finished; it still remains for him to explain the events in terms of the causality proper to his own field.

Historical Understanding

EXPOSITION

46. Walsh observes that historians seek inner connections of events with one another on the assumption that they constitute a single process (*IPH*, 24). This process, proper to historical thinking, may be called 'colligation,' to use a term of the nineteenth-century logician Whewell. The historian does not aim at the formulation of a system of general laws, although he does presuppose general propositions about human nature (*IPH*, 25).

47. In Dilthey's view history is concerned with human thoughts in the wider sense of all human experiences; these thoughts are reached and understood by others through the mediation of external expressions, such as joy or grief, not by a process of inference, but by direct passage from awareness of the expression to awareness of what is expressed and by having in ourselves the same experience (*IPH*, 49-50). Collingwood sought to avoid reducing historical understanding to shrewd guessing about the experiences of other people, present and past, by arguing that we can share only their intellectual operations, since there is no reason why the same act of thought in this sense cannot fall within two different mental series (*IPH*, 51). He proposed that the central concept of history is the concept of action, meaning thought expressing itself in external behavior. For Collingwood the rethinking of the thoughts of another in a given situation is immediate and non-discursive; it is not an application of general principles to the particular case (*IPH*, 52).

48. Walsh maintains that Collingwood's main thesis will not stand up, because we need more than a single act of intuitive insight to discover what people were thinking; we need to interpret the evidence before us by at least implicit reference to general truths (*IPH*, 58). Actions are, broadly speaking, the realization of purposes; since a single purpose can find expression in a whole series of actions, some historical events are intrinsically related in an intelligible sense. The series of actions forms a whole, of which the later members are determined by the earlier and the earlier members are affected by the fact that the later ones were envisaged. This is a situation which is not found in nature, for even if, in the case of organic bodies, as mechanist biologists would deny, the concept of purpose is needed to explain their mode of behavior, it is not purposive action *in the same sense* as human behavior is (*IPH*, 59-60).

49. Walsh notes that "a straightforward teleological explanation" is justified for some historical events, especially where the pursuance of coherent policies is concerned, and a semi-teleological explanation is sought for the other events. The historian colligates events under appropriate conceptions so as to show in a significant account how detailed facts become intelligible in the light of certain dominant concepts or leading ideas whose connections can be traced (*IPH*, 61). This does not mean that history is a rational process in any disputable sense; the 'dominant concepts' concerned are not Hegel's concrete universals in disguise, for it says nothing about the *origin* of the ideas apprehended by the historian; it is sufficient that they were influential at the time of which he writes (*IPH*, 62).

50. In addition to colligation, the historian has recourse to a quasi-scientific type of explanation involving the application of general principles to particular cases (*IPH*, 63). Positivists maintain that there is no such thing as history in the generic sense; it is real only in its species; but this diffusionist notion of history overlooks the single overriding aim of all historical work "to build up an intelligible picture of the human past as a concrete whole, so that it comes alive for us in the same way as the lives of ourselves and our contemporaries." Furthermore, in addition to the specific generalizations of individual historians, each has a fundamental set of judgments to guide his thinking. "These judgments concern human nature: they are judgments about the characteristic responses human beings make to the various challenges set them in the course of their lives..." What the historian takes to be credible depends upon what he conceives to be humanly possible; it is in the light of his conception of human nature that the historian must ultimately decide both what he will accept as fact and how to understand what he does accept (*IPH*, 64-65).

51. The fact that we think that we can understand past ages is an indication of something constant in human nature, even though it does not prove it. Thus a science of human nature may be considered possible in principle at least, in spite of obvious variations in behavior and in beliefs from one age to another (*IPH*, 68). Since our knowledge of human nature rests on experience, it is subject to constant revision as our experience is widened (*IPH*, 69). Collingwood's aim was to bring to light the peculiar character of historical knowledge; his doctrine could be reconstructed without any reference being made to intuition. His dictum that all history is the history of thought could be understood as an attempt to bring out the conceptual structure of historical knowledge rather than as an account of what historians do (*IPH*, 70-71).

CRITIQUE

52. Walsh expresses the recognition that 'colligation' under general terms in history is a different form of thought from the application of universal laws to individual cases, as is done in classical science. He brings out the insight that explanation of events by tracing their intrinsic relations to each other within a context of development is a procedure distinctive of historians. He sees cases in which a frankly teleological explanation is justified, observing that for the others a quasi-teleological explanation cannot be entirely avoided. He sees the insufficiency of the 'diffusionist' notion of history, which denies that there is any general concept of history at all. A very important observation seems to be the judgment that the historian's ultimate decisions as to what he accepts as fact and how he will understand a fact he has accepted depend upon his conception of human nature. His better judgment refuses to let him admit that there could be nothing constant in human nature throughout the ages, although he does not feel in a position to show that there is anything constant. Perhaps most important of all is his drawing attention to the "conceptual structure of historical knowledge," a notion placing the focus of attention upon the historical medium and opening the door to one's entry into historical science of the advanced and genuine type.

53. Walsh's hesitancy about whether historical explanation is scientific or not seems to stem from an insufficiently clear and broad idea of science. History is not a science in the sense that it copies empirical science in a quasi-scientific way. When understood properly, history is a science in its own right, a separate species of knowledge of the real. The unique character of

historical explanation lies in its presentation of meaning in a manner that is different from deduction from general principles. It utilizes the intellectual apprehension of concrete, individual things to build syntheses that are themselves concrete and individual. 'Europe,' 'the Second World War,' 'chivalry' are concrete concepts having a generality quite diverse from logical universals and mathematical terms of the classical type. I would disagree with Walsh's assertion that "to say that we explain historical events by referring to the ideas they embody is not to hold that history is a rational process in any disputable sense," for, if this were true, history could not be a science except in a very limited degree. A lexicographer might be called a scientific worker to the extent that he sticks to the level of words and does not try to construct any sentences of his own. A historian whose history is not a rational process in any disputable sense might be called a scientific (or quasi-scientific) worker to the extent that he simply ascertains details after the fashion of a detective of the past and supplies them to scientists true and proper. For the rest he would be a writer of a special kind of fiction. If he could be said to possess a branch of knowledge, it would be knowledge whose base is factual but whose form is fanciful. History would not escape the essential ambiguity of simple art, and every serious work of history would be looked upon as mere art whose point of departure was real on the material level. Walsh seems almost reconciled to the locating of history in the realm of the artistic and the literary, basing its claim to wisdom upon common experience and the genius of poetic insight. But history can be more genuinely scientific than that.

54. The 'inner' connection of historical events needs to be understood on the intellectual level by a visualization that goes beyond sense imagery. It is not nature itself that is all on the surface, but the sensory phenomena of nature. The natural scientist must penetrate beyond the 'outside' of material phenomena in order to understand the objects of his study. He understands by arriving at meanings, and meanings are the intellectual dimension of nature. The study of 'thoughts' is not peculiar to history; it is a pursuit shared with other sciences. We can clarify Collingwood's difference from Dilthey on the question of which thoughts are the object of history by the following distinction. All kinds of human experiences are the object of history in the prescientific sense. But the formal object of historical science is intellectual thought alone. The 'inside' of events should not be overly restricted by the image of 'getting inside' of someone else's mind. The fact that historical studies have centered around human subjects has led idealist historians to the false conclusion that the 'inside' of historical events is to be found inside the minds of the human subjects. It has led also to the misunderstood dictum that 'history is intelligible because it is a manifestation of mind.' Certainly, in a true sense, everything intelligible is so because it is a manifestation of mind — fundamentally, the 'mind' of the Creator. But all that is essential in the intelligibility of historical events as such is that they have meaning along the intellectual dimension.

55. That which grounds the essential intelligibility of historical events is the finality in which they are realized. Again, idealists have been misled by the fact that human beings act with conscious purpose, thus adding a secondary finality to the essential finality of all historical events as such. The confusion of the primary and secondary finalities of human events has falsified much historical interpretation and deprived it of its base in reality. What the historian as a scientist aims to do is not to judge the acceptability of past events according to what in his theory of human nature the human subjects were capable of doing. That would be making up history rather than ascertaining it. The serious historian simply finds out what the evidence indicates has happened in the past, using the most reliable methods available, and bases his historical explanation upon that. The historian must be a *realist* in his approach to the past; his mind must be open to any eventuality. Much bad history has been written by historians who rejected significant events simply because their minds were not sufficiently open to the facts as they presented themselves.

56. In addition to his own distinctive type of reasoning, the historian makes use of classical reasoning as a subordinate form. Within the direct confrontation of past with present, he uses classical reason to determine what

is past and what is present. In this way historical reason becomes a transcendent form by which spheres of classical reason are brought into intelligible relationship with one another. The remote meaning of classical conclusions is elevated to concrete and immediate meaning by historical insight, not simply on the level of common sense, but even on the advanced level of formal abstraction. Positivists who are blind to this kind of intellectual development have simply suppressed an important area of meaning from their lives.

57. Does it pertain to the historian to deduce particular conclusions from universal principles? The answer is complex. In order to be able to penetrate deeply into his subject-matter, in order to be able to recognize and explain more advanced and abstract historical meaning in the events he is examining, the historian must have a developed mind, and he must be able to use the understanding in which the development of his mind is embodied. Therefore, in the assessing of any situation, he is using general principles of the classical type and applying them to concrete cases. To the extent that his explanation involves meanings transcending the commonsense situation, he must employ specialized reasoning of the classical type. For instance, he cannot give an historical explanation of a Supreme Court decision if he knows nothing about law. Technical reasoning can appear in an historical explanation either incidentally or *ex professo*. Those who write the history of law, those who write the history of biology, and so forth, are just as much historians as those who write the history of baking or baseball. A general history attempts to bring events of both the familiar world and the technical world under general but concrete concepts. The fact that this confrontation is made implies the need of a form of reasoning that transcends classical categories, and, when the comparison stands within relationships of time, the comprehensive form of reasoning becomes distinctively historical.

58. General principles about human nature are important to the historian for the main reason that most historians tend to center their explanations around human events. But it is a mistake to identify the general mental formation of the historian with his knowledge of human nature. He must have good general knowledge of many fields. The significance of knowing human nature lies in the fact that one cannot ascertain what a human being did or experienced if he does not know what a human being is. But the knowledge of human nature is not used by the historian principally to be able to anticipate characteristic human responses; such anticipation merely supplies clues. The chief use of knowledge about human nature is for the historian the fitting of human happenings into developmental meaning in terms of the whole conscious life that persons of the past have had, whether of goals or aspirations, of perfections or attainments, of success or failure, of scandal or inspiration. Mental life presents meaning to various sciences of man. The historian must make use of the findings of all of these sciences in deriving full historical meaning from the human events of the past.

59. The evidence that human nature contains unchanging and constant elements is studied in the sciences of psychology, epistemology, and others. The findings of these sciences form part of the general present of the historian, and they help him to interpret correctly the changing elements.

60. The basic historical meaning of past human events stands outside of the conscious intentions of the agents; a secondary historical meaning stands within their conscious intents. A certain amount of empathy is necessary to establish a link with the past; furthermore, the good and noble deeds of others will excite esteem and appreciation in ourselves. We tend to love what is good and reject what is evil, if our own moral attitude is sound. But the historian should realize that he can never arrive fully at the concrete outlook of other persons, past or present, and that should not be his ultimate aim. What the historian ultimately aims to do is to reach fuller theoretical and practical insight into the meaning of reality by bringing to light those areas of reality that are visible only through the medium of history. The final aim is not to understand himself, or to understand others, but to understand reality.

61. The central concept of history is development in the sense of genetic relationship. Human actions are a species of the general concept, and their central rôle in the constructs of history

stems from genetic relationship as known from what actually took place, not as efficient causes looking forward to what has not yet taken place. The dominant concepts underlying historical explanations are retrospective generalizations, to which ideas about human nature, physical nature, and all their parts are essentially subordinate.

62. The judgments of historical science are universally valid, in the sense that they must be accepted by all competent and honest historians; they are timeless, even though their intelligibility is based upon genetic time. It is true that these judgments are set in new patterns as time progresses and historical science grows, but in this respect they are not essentially different from the case of judgments of empirical science and philosophy. There is common confusion in the use of the word 'universal.' Because of the long ascendency of classical thinking in the realm of science, the adjective 'universal' is spontaneously associated with open and unlimited classes of things. But this association is not fully rational; it tends to discredit other kinds of universal concepts without an adequate hearing. The fact is that the universe is not itself an open class of things; it is a concrete object. The total content of what anyone knows is not an open class; it is a concrete content. The recognition of any situation is concrete and particular; there is a whole world of judgments made about situations that are concrete and particular. The 'classifying' of all such judgments as belonging to 'common sense' or to prescientific intuition is not based on a scientific and rational investigation into the nature of these judgments; it is largely based on oversight and bias. It is my contention that any serious investigation on the level of logic and theory of science will show that many of these judgments are intellectual, that they have distinctive intellectual value, that they can be as abstract as judgments of the classical type, that they have a solid theoretical fundament, and that ignorance of them creates a gap in the intellectual life of man.

APPENDIX III

CONCERNING R.G. COLLINGWOOD'S CONCEPTION OF HISTORY

The Philosophical Science of History

EXPOSITION

1. In his *Idea of History*,[a] Robin Collingwood draws attention to the fact that the philosophizing mind, in thinking about an object, necessarily "thinks also about its own thought about that object," since philosophy is "thought of the second degree" (*IH*, 3). The establishment of a new philosophical science necessitates a revision of all previous philosophical science. Just as by the seventeenth century the philosophy of (empirical) science had ceased to be a particular branch of philosophical science and had permeated the whole area of philosophy to become itself a complete philosophy conceived in a 'scientific' spirit, so in the present instance the redimensioning of all philosophical questions in the light of the results achieved by the 'philosophy of history' in the narrower sense will produce a complete new philosophy of history in the wider sense, conceived from the historical point of view (*IH*, 6-7). "For the philosopher, the fact demanding attention is neither the past by itself, as it is for the historian, nor the historian's thought about it by itself, as it is for the psychologist, but the two things in their mutual relation" (*IH*, 2) For the philosopher the thought of the historian is not a complex of mental phenomena but a system of knowledge; the past is not a series of events but a system of things known (*IH*, 3).

2. History is 'for' human self-knowledge. "Historians nowadays think that history should be (a) a science, or an answering of questions; (b) concerned with human actions in the past; (c) pursued by interpretation of evidence; and (d) for the sake of human self-knowledge" (*IH*, 10-11).

III.1a R.G. Collingwood, *The Idea of History* (Oxford: Clarendon Press, 1946) (hereinafter referred to as *IH*).

3. Collingwood observes that the historian does not investigate mere events, having only an outside and no inside, but actions, each of which is the unity of the outside and the inside of an event. "By the outside of the event I mean everything belonging to it which can be described in terms of bodies and their movements ... By the inside of the event I mean that in it which can only be described in terms of thought ..." The historian "is interested in the crossing of the Rubicon only in its relation to Republican law, and in the spilling of Caesar's blood only in its relation to a constitutional conflict" (*IH*, 213). Collingwood claims that in the case of nature the distinction between the outside and the inside of an event does not arise. "To the scientist, nature is always and merely a 'phenomenon,' not in the sense of being defective in reality, but in the sense of being a spectacle presented to his intelligent observation; whereas the events of history are never mere phenomena, never mere spectacles for contemplation, but things which the historian looks, not at, but through, to discern the thought within them."

4. "All history is the history of thought," Collingwood avers. The historian discerns thoughts by re-thinking them in his own mind. "The historian of philosophy, reading Plato, is trying to know what Plato thought when he expressed himself in certain words." But the historian, as he re-enacts past thought in his own mind, also criticizes it, forms his own judgment of its value, and corrects the errors he espies. This criticism is an indispensable condition of historical knowledge. "All thinking is critical thinking; the thought which re-enacts past thoughts, therefore, criticizes them in re-enacting them" (*IH*, 215-216).

5. In Collingwood's view, history is not "a story of successive events or an account of change." The historian is not concerned with events as such at

all. His fundamental concern is with thoughts alone; their outward expression simply reveals to him the thoughts which are his objectives (*IH*, 217). Historical inquiry reveals to the historian the powers of his own mind. All memory and personal experience belong to historical thinking (*IH*, 218-219). Science cannot generalize from historical facts as it does from sensory data. "(T)he facts, in order to serve as data, must first be historically known; and historical knowledge is not perception, it is the discerning of the thought which is the inner side of the event. The historian, when he is ready to hand over such a fact to the mental scientist as a datum for generalization, has already understood it in this way from within. ... But if he has done so, nothing of value is left for generalization to do." Statements that similar things have happened elsewhere have value only where the particular fact cannot be understood by itself (*IH*, 222-223).

6. The novelist has the single task of constructing a coherent picture. The historian must construct a coherent picture of things as they really were and of events as they really happened (*IH*, 246).

7. "The scheme upon which the exact sciences have been traditionally arranged depends on relations of logical priority and posteriority: one proposition is placed before a second, if understanding of the first is needed in order that the second should be understood; the traditional scheme of arrangement in history is a chronological scheme, in which one event is placed before a second if it happened at an earlier time." History is a special kind of science whose business it is to study events inferentially, arguing to them from 'evidence' accessible to our observation (*IH*, 251-252).

8. Collingwood maintains that the word 'authority' no longer belongs to the vocabulary of historical method. 'Critical history,' he says, as it was worked out from the seventeenth century onwards, and "officially acclaimed in the nineteenth as the apotheosis of the historical consciousness," is now of interest to the student of historical method "only as the final form taken by scissors-and-paste history on the eve of its dissolution" (*IH*, 257-260).

9. Collingwood epitomizes his theory in the statement that the proper object of historical knowledge is thought: not things thought about, but the fact of thinking itself. History is distinct from 'natural science' in that the study of a given or objective world is distinct from the act of thinking it; history is distinct from psychology in that the study of immediate experience, sensation, and feeling, although being the activity of the mind, is not the activity of thinking (*IH*, 304).

CRITIQUE

10. What the establishment of a new science necessarily requires is a revision of all formal knowledge as possessed by human understanding from the special point of view of that science. Collingwood has rightly observed (III.1) that the emergence of historical science requires a revision of knowledge as gathered and understood from the historical point of view. Historical science is the elevation of the prescientific view of history to the scientific viewpoint. Any science, in regarding its object, "thinks also about its own thought about that object," not necessarily on the level of cognitional theory, but always and necessarily in the conscious interposition of its proper medium of thought. The medium becomes the center of focus as the direct object of study when the science scrutinizes it and formulates it in terms of postulates; the medium is then used in the science as the set of proximate presuppositions of the reasoning of the science. Collingwood sees (III.1) that past and present are the object and medium of historical science, but his explanation of their relationship to one another is not clear.

11. Collingwood's assertion that history is 'for' human self-knowledge is true but ambiguous (III.2). What he is talking about is not all history, but that area of history called historical science. Since the intrinsic purpose of any science is the informing of the intellect, the aim of historical science is to produce human self-knowledge in that sense. Collingwood's assertion lacks stress upon the objective character of the science of history. Historical science produces insight vested in human selves through the instrumentality of an objective medium in their minds. History is 'for' human self-knowledge in this objective and intelligible sense, not in the sense that the 'self' is the object of historical knowledge.

12. Collingwood notes (III.8) that 'critical history' is not the definitive form of emerging historical science; it was simply a precursor. The earlier 'critical approach' carried its 'universal methodic doubt' to all of the sources of history, but it was not fully critical in the examination of its own medium of thought. The new scientific history is critical of its own medium and presuppositions.

13. The 'outside' of an event is the event on the level of sense perception. The 'inside' of an event is the event on the level of intellectual apprehension. Collingwood's claim that the distinction does not arise in the case of the physical sciences (III.3) is not factual. Physical scientists do derive meaning on the level of intellectual observation from phenomena on the level of sensory perception. The difference between empirical science and historical science lies essentially in the difference in the kind of meaning derived. Collingwood's confusion concerning the wider incidence of intellectual 'insight' seems to have arisen from an insufficiently critical notion of what insight into meaning is.

14. The dictum, "All history is the history of thought" (III.4), is ambiguous and misleading. The dictum is most true in the reflexive sense that the 'thought' in which history formally consists has its own history, or development, in the mind of each individual thinker about history. The dictum is misleading in the sense that it is not usually taken in the reflexive sense. In the originary sense all history is the history of past events; it is events that constitute the material object of history, and events as such are not to be identified with human thought.

15. Some history is the history of thought in the direct and originary sense. The historian of empirical science, the historian of poetic literature, the historian of philosophical science constructs an account of thought as the subject of his inquiry. This thought is objective; it consists in objective meanings expressed in conventional signs and thus communicated to others. In this case, the material object of the historical inquiry is the set of conventional signs and verbal sentences together with the sensory phantasms which these signs are apt to arouse in the mind of the historian. The remote formal object is the intellectual meaning embodied in the material symbols, and the proximate formal object is the conscious medium personally possessed by the historian himself.

16. Collingwood claims that history is not a story of successive events or an account of change (III.5). His concern is with history as 'thought.' The correct distinction to be made is that material historical science is the account of change, while formal historical science is the understanding of reality contained potentially within the change (3.2:10).

17. Collingwood notes (*IH*, 215) that the only way in which the historian of philosophy can know what Plato thought is by thinking for himself what Plato has expressed in words, that this is what is meant by 'understanding' the words. But thinking for oneself can be understood as more than a mere subjective experience; it can be regarded as the vision of an intellectual object. And it is the recognition of thoughts as intellectual objects that puts them on the scientific plane. The apprehension of the intellectual object consciously intended by the words of Plato may consist formally in the recognition of an ontological fact or truth. In other words, the fact of Plato's having recorded his recognition of an ontological truth or his conception of an ontological problem is concrete as an event of recording and it is concrete as a previous event of recognizing, but it is abstract in the universalizing sense in the content of the recognition and of what has been recorded. What elevates this concrete fact to the level of historical understanding is its recognized intelligible genetic relationship with another prior or subsequent concrete fact Therefore, in assessing the value of past thoughts for present sciences, we must distinguish between the thoughts as materially past and formally present on the one hand and as both materially and formally past on the other.

18. Collingwood (III.7) is referring to a fundamental aspect of the problem of historical science when he distinguishes between the pattern of logical priorities of the 'exact' sciences and the pattern of chronological priorities of history. His reduction of the problem to the manner of juxtaposing the elements of the reasoning is right to the point, although the use of terms is inexact. Classical science juxtaposes two

propositions in a relationship of 'logical' priority or posteriority, deriving conclusions from the greater or lesser distribution of the medium common to both. This is the process of deduction from the universal to the particular. Historical reasoning, on the other hand, is derived from the juxtaposition of two concrete elements (propositions) in a relationship of chronological priority or posteriority, deriving conclusions from the greater or lesser fulfillment of the identical thing that serves as the medium common to both elements (propositions). The thinking of historical science is rational, but not 'logical' in the sense of classical logic. It is inferential in the sense that it does derive conclusions of understanding by a comparison of two terms on the basis of an identical common element. We call this kind of inference 'immediate' to distinguish it from the 'mediate' inference of reasoning based upon a quasi-mathematical distribution of the common middle term. Collingwood affirms (III.7) that history is inferential in the sense that it argues to the structure of the past from evidence accessible to our observation in the present. This is the inference of verification in reality, and I do not see how it differs essentially from the verification requisite to any science. The empirical scientist argues also to the validity of his constructs from his observations in present experience. Actually, empirical scientists are quite accustomed to basing arguments upon facts of the past. For instance, when a chemist writes down what he has observed in the laboratory, he is describing the past as he remembers it from his experience. The fact that historians usually verify a different series of constructs of the past does not change the essential conformity of their kind of verification with all scientific verification. Hence, what constitutes the essence of historical thinking as such is what is done with the facts after they have been verified: they are juxtaposed in genetic order.

19. Collingwood's rejection (III.8) of all authority in the construction of historical accounts is a prescientific attitude. Every science will doubt authorities where it can find solid reason to do so, but every science is based upon the principle of authority both in the secondary sense that scientists have to trust one another and in the primary sense that reality itself is the ultimate authority of science. A critical attitude is necessary in the true scientist, but what has unfortunately accompanied the universal rejection of authority in the realm of history up to the present time is a hypercritical attitude towards the testimony of the past and an uncritical attitude towards the presuppositions of the modern critical process.

Science Is History

EXPOSITION

20. In a short essay published in 1922,[a] Robin Collingwood maintains that the distinction between science as knowledge of the universal (what Aristotle calls 'philosophy') and history as knowledge of the particular is illusory.[b] He notes the consensus of opinion of many twentieth-century philosophers in the schools of Mach, Bergson, James, and Croce that "science is not knowledge at all but action, not true but useful, an object of discussion not to epistemology but to ethics." For these same philosophers "any cognition must be of the particular and must therefore be history: what is called a cognition of the universal cannot be a cognition at all but must be action."[c] It is Collingwood's contention that a scientist is effectively a scientist only when he is in the act of interpreting concrete facts in the light of his general concepts. The framing of these general concepts is not the end of science but is only the means. One who possesses the concepts apart from applying them is not actually but only potentially a scientist.[d] These general concepts are not like a previously constructed tool to be used upon separately given material (the facts). "Neither the concept nor the fact is 'possessed' (*thought* and *observed* respectively) except in the presence of the other. To possess or think a concept is to interpret a fact in terms of it; to possess or observe a fact is to interpret it in terms of a concept.

III.20[a] R.G. Collingwood, "Are History and Science Different Kinds of Knowledge?" in his *Essays in the Philosophy of History* (New York: McGraw-Hill, 1966),

pages 23-33 (reprinted from *Mind*, 1922).
III.20[b] *Ibid.*, 23-26.
III.20[c] *Ibid.*, 25-26.
III.20[d] *Ibid.*, 27.

Science is this interpretation." Collingwood claims that knowledge is not of the particular or of the universal but only of the individual: the sense-datum (pure particular) and the concept (pure universal) are false abstractions when taken separately, even though, as elements in the one concrete object of knowledge (the individual interpreted fact) they are capable of being analytically distinguished. The object which the scientist knows is not a universal, but is always particular fact, "a fact which but for the existence of his generalizing activity would be blank meaningless sense-data." Collingwood does not except mathematics from this rule. "Even mathematics does not consist of abstract equations and formulae but in the application of these to the interpretation of our own mathematical operations."[e] The aim of the scientist is to know the individual, that is, "to interpret intuitions by concepts or to realize concepts in intuitions."[f] But this is also the work of the historian, since the reconstruction of historical narrative is an interpretation of individual facts.[g] "The analysis of science in epistemological terms is thus identical with the analysis of history, and the distinction between them as separate kinds of knowledge is an illusion."[h]

CRITIQUE

21. Collingwood defines science as the interpretation of a fact in terms of a concept. Since this definition implies a vague recognition of the scientific medium, it is correct but imprecise. Collingwood does not explain what it means on the scientific level to interpret a fact *in terms of* a concept; he distinguishes between fact and concept, but he does not explain their functional relationship to one another. Facts are recognized identities that fit under the concept of reality. If these identities do not pass the reality-test, they are not facts. This means that facts are the standard product of all scientific thinking. Collingwood's definition of science is therefore correct for the reason that one's thinking recedes from fact to the same extent that it recedes from science, and one's thinking recedes from science to the same extent that it recedes from fact. But the definition is imprecise, because it does not make clear that the one concept with which all science is necessarily concerned is the concept of reality. The phrase 'in terms of' can mean nothing else but that the concept is a medium through which (or in which) the object of thought is viewed. Obviously, one cannot view the mediated object of thought without also viewing the medium. But the object and the medium are functionally, as well as analytically, distinct.

22. It is vaguely correct to say that general concepts are the means and not the end of science. They are the means in the precise sense that they are the objective medium in which the science exists. Science itself consists formally and essentially in this medium. It is incorrect to say that the science cannot be possessed except in the presence of practical material objects upon which the mind is concentrating, because science arises and grows through concentration upon its own medium as distinct from the remote objects that form it subject matter. In one sense a science is not distinct from the matter that forms its object. Astronomy is not distinct from the celestial objects that it studies. But in another sense a science is distinct from its subject matter. The astronomer can concentrate upon the telescope that constitutes the material medium of his observation. He can also concentrate upon the mathematical concepts and formulae that constitute the formal medium of his observation. Collingwood errs in saying that mathematics does not consist of abstract equations and formulae but only of their application. The truth is that in one sense mathematics consists purely of the concepts and in another sense it consists of the concepts and their application. It is the task of scientific reflection as such to differentiate the two senses and understand the relation between them. In not differentiating these two senses of science and their function, Collingwood stands on the prescientific level of thought.

23. Because Collingwood does not clearly distinguish between the dynamic and the static aspects of human understanding, he reverts to a theory of cognition that is inferior to the

---

III.20[e] *Ibid.*, 28-29.
III.20[f] *Ibid.*, 30.

III.20[g] *Ibid.*, 31.
III.20[h] *Ibid.*, 32.

insights already possessed by the ancient Greek philosophers. Those who say that science is not knowledge at all but is only action have lost sight of that recognition of the stable element of knowledge which gave rise to science in the first place, and the conclusions that they draw from this erroneous presupposition can only be unscientific. To say that a scientist is not a scientist apart from the process of scientific thinking that happens to be taking place at the moment in his consciousness is a false conclusion drawn from the failure to recognize the stable aspects of science. While the act of insight is transitory, the act of understanding abides. A scientist retains his understanding even when he is asleep. When he awakens he is able to bring his mind as a tool differentiated by the insights of his past reflection to bear upon new problems. Collingwood's theory overlooks the scientific character of human understanding as the abiding element in science. He fails to see the formal factor that makes science be what it is.

24. It is equally false to say that science is not true but useful, since the *finis operis* of science as such is the perception of truth, while the *finis operantis* of science may be the useful. Even when the satisfaction and personal fulfillment that come from the understanding of truth are included under the category of the 'useful,' the *finis operis* of science remains the perception of truth. It is no less false to say that science is not an object of epistemological but only of ethical discussion. It is the object of both, but for different reasons. In its intrinsic character as the perception of truth, science is necessarily an object of epistemology alone. As is evident from the final quotation of the exposition above (III.20), Collingwood himself develops his discussion of science from the viewpoint of epistemology. Again, while it is true that every cognition is in some sense history, it is precisely the task of the historical scientist to discover and explain in what sense every cognition is history without attempting to suppress the distinctive intelligibility of every other science, as would be implied in the statement that knowledge of universals is not knowledge at all but action. Since knowledge is the perception of objects, the knowledge of universals is knowledge, for what is here meant by 'universals' is nothing other than those objects of intellectual perception constituted by open classes of things. The true reason why every cognition is in some sense history is because in the process of cognition there is necessarily included a genetic relationship of mental events whose recognition constitutes an area of historical understanding. But the understanding of the act of cognition need not be limited to the genetic viewpoint; it has in fact been understood from other viewpoints as well. Concepts considered separately from the sense-data are by no means 'false abstractions'; the distinction of concepts from the data of sense is what abstraction means. While all scientific thinking involves an intervening intellectual object and a more remote object ultimately traceable to the data of sense, the more remote object need not itself be a datum of sense; it can be a meaning (2.6:10); and even a material fact is more than the data of sense, as has been show above (III.21). According to the insights long established by classical science the object which the scientist knows is in a crucial sense the universal. It therefore devolves upon the historical scientist to show that his science is a science nevertheless. Such a demonstration will center upon the ambiguity of the term 'universal.' While distinctively historical conceptual objects are not 'universals' in the sense of open classes of things, they are universals in the sense of intellectually recognized closed and complete objects of understanding.

Appendix IV

CONCERNING SOME REPRESENTATIVE CURRENT CONCEPTIONS OF HISTORY

A. *The Historical Theory of W.B. Gallie*

EXPOSITION

1. W. B. Gallie agrees with the contention of philosophers like Windelband, Rickert, Croce, and Collingwood that "the way we understand history is basically the same as that in which we understand all purposive thought and action, and radically unlike the way in which we understand natural phenomena as instances of some scientific law or theory." Gallie contributes to this thesis the belief that the concept of a story, which is the proper starting point and basis of all historical thought and knowledge, is a *sui generis* form of human understanding.[a] A story is a teleologically guided form of attention in which "we are pulled along by our sympathies toward a promised yet always open conclusion, across any number of contingent, surprising events, but always on the understanding that these will not divert us hopelessly from the vaguely promised end." An element of story is essential to all history. While a great deal of what is called historical literature (lists, accounts, diary jottings, etc.) but which is actually ancillary to or parasitic upon history of a more central and substantial kind, may not be story or narrative, nevertheless, the kind of history that interests all of us *most* (including historians) is that which treats of some major achievement or failure of men living and working together.[b] These central works of history can be considered as primarily narratives, that is, as stories. "Granted that every genuine work of history is also a work of reason, of judgment, of hypothesis, of explanation: nevertheless every genuine work of history displays two features which strongly support the claim that history is a species of the genus Story." First, to use a book or a chapter of history means to follow it "in the light of its promised or adumbrated outcome through a succession of contingencies, and not simply to be interested in what resulted or could be inferred as due to result from certain initial conditions. ... History, like all stories and all imaginative literature, is as much a journey as an arrival, as much an approach as a result." Secondly, every genuine work of history is read in this way because its subject-matter is felt to be worth following.[c]

2. Gallie points out that the historian, in taking up a problem of research, knows broadly how his narrative will go, but, once he is immersed in the relevant materials, there is almost no limit to the surprises that they may hold in store for him. In historical research there is always a story.[a] There is also a 'sensation of pastness,' the experience of touching right back to a past lived reality, which it is perhaps natural to expect in reading any work of history, although the successful expression of this sensation is no less within the capability of the poet and the novelist as it is of the historian.[b] It is, moreover, the primary task of the historian "to provide a followable (and on the evidence an acceptable) account of such social changes as are characteristically human interest because they depend upon the ideas, choices, plans, efforts, successes and failures of individual men and women."[c] The crucial moments in any major historical work consist in indications of how this or that individual or group reacted in certain institutional rôles. "In such

---

IV.1[a] W.B. Gallie, *Philosophy and the Historical Understanding*, 2d ed. (New York: Schocken Books, 1968), 1.
IV.1[b] Ibid., 64-65.

IV.1[c] Ibid., 66-67.
IV.2[a] Ibid., 72.
IV.2[b] Ibid., 74.
IV.2[c] Ibid., 83-84.

moments we see and feel the true growing-points — or dying-points — of history." [d]

3. Gallie claims that historical understanding is "the exercise of the capacity to follow a story, where the story is known to be based on evidence and is put forward as a sincere effort to get at *the* story so far as the evidence and the writer's general knowledge and intelligence allow." Explanations in history are no more than aids to this basic capacity or attitude of following.[a] The characteristic function of historical explanations is pragmatic and ancillary.[b] Historical narrative, as it goes along, does need to be 'righted' and logically endorsed by helpful explanations, but more characteristically every historical narrative is *self-explanatory*. Hence the deductivist model of historical explanation should be replaced by the new model of the philologist's gloss, in a sense broad enough to include large-scale reinterpretations.[c]

CRITIQUE

4. Gallie (IV.2) implicitly defines history as "a followable (and on the evidence an acceptable) account of such social changes as are of characteristically human interest because they depend upon the ideas, choices, plans, efforts, successes and failures of individual men and women." This definition is inadequate on both the common-sense and the scientific levels. It is inadequate on the level of common sense because it excludes from the category of history (rightly and properly so called) every account that does not describe those social changes that are of 'characteristically' human interest in that they depend upon the purposive thought and action of individual men and women. According to this definition any account of changes that are not social in character does not belong to history, and any account of changes that are not wrought by the purposive activity of individual men and women does not belong to history. If such accounts are not history, even though they are narratives of change, then what is the right word to designate them? Gallie's definition appears confused, not only because it does not state clearly what history is, but especially because it casts a cloud over the nature of those forms of history which are not "central and substantial," and which are therefore classified as ancillary or parasitic. On the scientific level Gallie's definition shows lack of reflection upon the terms involved. What makes an historical account 'acceptable,' and what is acceptability? No reason is given or implied for the limiting of history to accounts of social changes, except that this is what historians are characteristically interested in. But why are they characteristically interested in this area of change? Furthermore, to limit history to the results of the purposive intent of men and women is to falsify history and to mistake the true nature of its essential causality (5.1:2). A certain dam may have burst because it was sabotaged or because it was weakened by an earthquake. In either case the description of the event pertains equally to history. It is subjective and arbitrary to say that the former case pertains more to history because it has more 'human interest.' For some humans there may be more interest in the latter case. It is prescientific to define a field in terms of the interest that the researchers have without explaining why they should have that interest and how that interest is related to the interest of workers in other fields. Gallie's definition of history is restricted to one species of history and falsified by the inclusion of human intent, a causality that, however great may be its importance, is not essential to it.

5. It is unscientific to restrict the definition of history to "the kind of history that interests all of us *most*" (IV.1). It is an open contradiction to define a genus in terms of one of its species, as Gallie here does. In the context of scientific thought, *any* species of history can happen to be the one that interests us most. But it is important to realize that I am speaking of scientific interest and not of aesthetic or pragmatic interest. It is of great interest to historical science to recognize and examine those species of history whose historical character has been overlooked or neglected by many theoreticians. Attention to these forms will render a clearer and deeper understanding of what history is, and will therefore make our interest in history more characteristically human in the

---

IV.2[d] *Ibid.*, 85.
IV.3[a] *Ibid.*, 105.

IV.3[b] *Ibid.*, 107.
IV.3[c] *Ibid.*, 113.

sense that this interest will represent our own success as individual men and women in grasping the idea and import of history. Now, granting that Gallie's definition (IV.2) is overly restricted by the adjective 'social' and falsified by the introduction of the wrong kind of causality, can we admit that history may at least be defined as "a followable (and on the evidence an acceptable) account of changes"? Reflection will show that the definition is still too restricted. An account of change that is acceptable on the evidence pertains only to history as science; it does not account for history in the form of fiction (6.3:2). Gallie consciously intends to speak only of historical science (as I have defined it), but he does not provide an explanation for the common elements in fictional and in real accounts of change. Nor is the adjective 'followable' specific enough to designate what is distinctive about history. Historical narrative is an account of *genetic* change (5.1:3). The reason why an historical account is followable is precisely because of the genetic relationship that gives it intelligibility (5.1:4). One could give a followable account of the changes in a turning kaleidoscope, but the account would not have historical character. May we, then, in the final analysis, define history as 'an account of genetic change'? I think that even this last definition is too restricted, for it applies only to history as the art of historiography, and not at all to history as the *knowledge* or understanding from which historiography springs (6.2:1). Gallie's definition of history misses completely the cognitive aspect of history, which is actually the most central and substantial area of interest for any scientific discussion about history. On the level of scientific discussion it is not exact to say that the primary task of the historian is to provide an account (IV.2). The primary task is to achieve the insight and understanding that will make the account possible. The account is but the artistic expression of the understanding achieved (6.2:4). The *finis operantis* of the historian is characteristically the composition of an account of genetic changes, but the *finis operis* of history is the historical understanding implied in the account (6.2:5).

6. There is more to history than a 'sensation of pastness' (IV.2). To the extent that the historian or the reader of history understands what history is about there is an intellectual grasp of the concept of the past (4.3). To the extent that the historian understands the science of history, "the experience of touching right back to a past lived reality" will be elevated to the intellectual experience of grasping the meaning of that past reality in the present reality of his own understanding (4.2:6). Such an intellectual experience cannot be characteristically afforded by the poet or the novelist, because the medium of their accounts is not reality (2.4:1). When Gallie says (IV.2) that the successful expression of the experience of touching right back to a past lived reality pertains as much to the poet and the novelist, he destroys the scientific character of the history he is describing and confuses the two areas of reality and imagination (3.2:1). Gallie states that there is an element of story in all history, calling history a species of the genus 'story' (IV.1), but he does not succeed in distinguishing story as a real account from story as a fictional account. By attributing to both the historian and the writer of fiction the aim of expressing successfully the experience of past reality, he ascribes to fiction the one essential element by which real history differs from it, since real events are the proper object of serious historical study. And Gallie fails also to express the formal object of historical science. He says that historical understanding is "the exercise of the capacity to follow a story," in the sense of a sincere effort to get at *the* story in so far as the evidence and the knowledge and intelligence of the writer will allow (IV.3). If by *the* story Gallie means the exact material facts as inexactly reported in the earlier account, he is confusing historical understanding with verification in sense-experience (3.2:6). The reason why every historical narrative is *self-explanatory* is because genetic insight occurs in the very act of following the account. This means that what Gallie calls historical explanation is not characteristically historical explanation at all (5.2:13). He has missed the true identity of historical explanation, of historical understanding, and of history itself.

7. The conception of a story is the proper starting point and basis, not of all historical thought and knowledge, as Gallie asserts (IV.1), but only of the art of historical expression (except to

the extent that historical thought and knowledge remain on the merely descriptive level of the story). Story is a *sui generis* form of human thought, but not of human understanding. Gallie's definition of story in terms of a vaguely promised end (IV.1) is incorrect. The end is simply what it is, and this cannot be promised. Story in the form of fiction holds the attention of the reader because of the aesthetic beauty and attractiveness which it offers, above all in the artistry of its conclusion. Story in the form of serious history holds the attention of the reader because of the genetic insight it offers to those who view the succeeding events in the light of their conclusions. Historical knowledge (serious history) is thus a *sui generis* form of human understanding, while literary knowledge (fictional history) is not characteristically (*per se*) a form of human understanding. The stories of imaginative literature are "as much a journey as an arrival" for the reason that their purpose lies in the aesthetic satisfaction of following them. The following of serious history may also afford aesthetic satisfaction, and in this it does not differ from its fanciful counterpart. But this is a secondary (*per accidens*) effect. Serious history is "as much a journey as an arrival" in its primary function for the reason that its meaning lies formally in the conclusion and materially in the antecedents. As the reader reviews the antecedents, he sees more and more meaning in them in the light of a growing awareness of the conclusion. And, as the reader reviews the antecedents, he sees more and more meaning in the conclusion because of his growing knowledge of the antecedents. The crucial moments in any work of serious history (IV.2) are the major transitions from antecedents to results. In order to express these transitions in a way that communicates true historical insight, the historian must be aware also of the metaphysical problems involved.

B. *Morris Cohen's Theory of History*

EXPOSITION

8. Morris Cohen points out that originally the word 'history' denoted any account of the nature of a thing. "When Aristotle said that poetry was more true and serious than history, he was using the latter word in its original sense, his idea being that mere description leaves out the universal essence." Cohen observes that in current usage the word is restricted to an account of past events, although by a natural extension its meaning refers also to what happened, as distinct from the knowledge or account of it.[a] The impulse to poetry and fancy is basic in human nature, and history oscillates between poetry as a product of the imagination and the desire to know what actually happened.[b] The almost irresistible temptation to tell a fascinating story makes us introduce imaginary elements to fill out into some unity our account of what happened, but the fact that history involves imagination does not mean that it is arbitrary. The difference between history and fiction is that the historian is not only docile to facts but is actively interested in searching for them through scientific investigation and verification.[c] "If 'science' means knowledge based on the most careful examination of all the available evidence, the scientific historian certainly aims at such knowledge, and his work can be judged by the extent to which he attains his aim." In the popular view historical narrative is considered a branch of literature, but the primary work of the historian is finding out what actually did happen at a given time and place. "This differs not only generically in aim and method from works of pure fiction, but also to a degree from the common knowledge that has not developed any special techniques either to guard us against our natural credulity and our disinclination to pursue an inquiry or to teach us to withhold final judgment until all the available evidence is critically examined."[d]

9. "History is concerned with establishing specific events that occurred at a definite time and place, whereas the facts or laws which general physical science seeks to establish deal with repeatable elements and assert that whenever and wherever A then B." Time and space enter into general science, not in the form of dates and locations, but as repeatable intervals,

---

IV.8[a] M.R. Cohen, *The Meaning of Human History* (La Salle, Illinois: Open Court, 1961), 8.

IV.8[b] *Ibid.*, 31.
IV.8[c] *Ibid.*, 32.
IV.8[d] *Ibid.*, 36.

notwithstanding the fact that historical elements are present in all sciences and notwithstanding the correlative fact that laws cannot be eliminated from history.[a] History is an applied science, although the historian seldom explicitly states the laws of human events that he assumes. "Yet implicitly he does make such assumptions, as when he asserts that a given ruler acted in a given way because of ambition or from a desire to please his people. In these cases it is, of course, assumed that if anyone is ambitious he is likely to act in that way."[b] As regards the social sciences, what is distinct about human history is the focus or perspective which makes description or understanding of individual happenings in time and place central. Similarly, the difference between 'natural history' and natural science is that "the former focuses on description, while the latter stresses theory, or systematic deduction from assumed principles or hypotheses."[c] Natural history is "a knowledge that certain things exist." Science, on the other hand, aims "not only to recognize things but to comprehend and understand them." What Cohen means is that "science aims to understand the laws or grounds which determine these existences and their connections."[d]

10. Cohen notes that positivists like Mach, Duhem, and Vaihinger have asserted that only particular sensations are real, and consequently that all universals or mathematical entities are fictions. But nature does not consist only of elements corresponding to sensation, for, just as a sentence does not consist of words alone, so intellectual consciousness does not consist of sensations alone.[a] No historian who is conscious of his task can avoid the problem of evaluation. In the maze of human events it is the viewpoint from which the historian selects his material that determines what is important. "The safeguard against bias in the writing of history, as in the natural sciences, is not to indulge in useless resolutions to be free of bias but rather to explore one's preconceptions, to make them explicit, to consider their alternatives, and thus to multiply the number of hypotheses available for the apprehension of historical significance."[b]

11. Cohen is of the opinion that Aristotle's doctrine of the four causes is "not much different from the modern principle of sufficient reason except that in the interest of clarity we restrict the causal relation to changes of phenomena in time and space."[a] A scientific explanation is a kind of description, a description in which the phenomenon is related to other phenomena in accordance with certain laws.[b] "(I)n its most rigorous form causality denotes the sum of the necessary and sufficient conditions for the occurrence of any event." Only the circumstance that is sufficient to bring about the given event can be the cause of the event. But in the popular view there can be a plurality of causes, and the historian under pressure to tell a coherent story may not interrupt his narrative here and there to indicate the inadequacy or inconclusiveness of his evidence. For this reason most historians "select from the vast conglomerate of determinants which form the necessary and sufficient conditions of a given event some element or elements to which they attach special importance and this they call 'the cause,' classifying all other elements as 'conditions.' "[c]

CRITIQUE

12. Morris Cohen's theory of history is based upon extensive acquaintance with historical literature and activity. However, it is confused with regard to the nature of historical science and historical causality.[a] His dictum that history oscillates between poetry as a product of the human imagination and the desire to know what actually happened (IV.8) may be true for nonscientific forms of history, but it does not apply characteristically to scientific history. The active interest of the serious historian to find out the facts is not basically what distinguishes him from the writer of historical fiction. A good historical novelist may be just as interested in finding out the facts regarding the events he is going

---

IV.9[a] *Ibid.*, 36-37.
IV.9[b] *Ibid.*, 38.
IV.9[c] *Ibid.*, 40-41.
IV.9[d] *Ibid.*, 68.
IV.10[a] *Ibid.*, 79.

IV.10[b] *Ibid.*, 80.
IV.11[a] *Ibid.*, 95-96.
IV.11[b] *Ibid.*, 99.
IV.11[c] *Ibid.*, 112-113.
IV.12[a] *Supra*, 3.2; 5.1.

to depict imaginatively. He may know the facts better than any serious historian. The real difference between fiction and historical science is that the scientist must adhere to reality in his mode of describing the events, while the novelist and the poet are free to roam into creative imagination without this restriction. The 'fascinating story' in the sense of aesthetic appeal is the genre of fiction, not of science. For the historical scientist there is only the fascination of the truth, which is a different kind of fascination, but is no less fascinating in its own right. The adept historical scientist does resist the temptation to introduce imaginary elements in order to fill out the unity of his account. Cohen's distinction between historical narrative as a branch of literature and the aim of the historian as finding out what really did happen is therefore not adequate. Two distinctions are actually involved. 'Literature' is an ambiguous word that can mean fanciful writing (*Dichtung*) or serious writing. There is poetic literature and there is scientific literature. There are poetic forms of historical literature, and there are scientific forms of historical literature. The uncritical viewer is confused about this distinction, even though he accords to history more of the factual than of the fanciful. But the scientific viewer knows this distinction and carefully separates scientific historical literature from all of the fanciful and prescientific forms (3.2:1). Again, while the primary aim of the historian as a scientist is to perceive and understand the historical truth, his practical aim in writing history is to express the historical truth. In this sense history is an applied science.

13. Cohen's teaching about history as description is confused. Natural history, he says, is a knowledge that certain things exist, while natural science aims both to know that things exist and to understand them. History, he says, focuses on description, while science stresses theory, or systematic deduction (IV.9). But Cohen's distinction between description and explanation is not clear-cut. A scientific explanation, he says, is a description according to causal relationship (IV.11). Again, he says that history focuses upon "the description or understanding of individual happenings in time and space" (IV.9). The difference between history and general science is in this wording not clear, for science is thus the description *and* understanding (by deductive reasoning) of individual phenomena, while history is the description *or* understanding of individual phenomena (happenings in time and space). What Cohen leaves too vague in his exposition is the true character and function of understanding in history. He adds the *or understanding* too casually to his notion of history as description (one is not sure whether it is disjunctive or conjunctive), considering that he also admits that science is description plus understanding. I would clarify Cohen's distinction as follows. Both historical science and general science are based upon correct descriptions of concrete, individual phenomena. General science understands its phenomena by viewing them in the light of universal concepts and general principles. Historical science understands its phenomena by viewing them in the light of their genetic relationship to results that are concepts in the sense that they are intellectually abstracted and grasped, but are not universals in the sense of open classes of things (5.2:14). I find Cohen to be correct in affirming that intellectual consciousness does not consist of sensations alone, and I extend this affirmation to include a place for distinctively historical understanding within intellectual consciousness.

14. I find Cohen's idea of historical causality to be incorrect. Aristotle advanced the notion of the four causes to correct the popular misconception that there is only one cause of any particular effect, namely, the efficient cause. To restrict Aristotle's doctrine of the four causes to changes of phenomena in time and space, as Cohen does (IV.11), is to destroy it entirely. The formal cause disappears, and the final cause never comes into sight. Aristotle's point was that the sum of the necessary and sufficient conditions for the occurrence of any event (as Cohen puts it) requires four causes and not just one. But the central cause of historical concern is the final cause, and the mistake of many theoreticians of history as well as of many practicing historians is that, when they select what they call *the* cause, they are concentrating upon an efficient causality (known, postulated, or imagined) which is secondary to history, and they are overlooking the final causality from which historical understanding springs.

C. Gustaaf Renier's Notion of History

EXPOSITION

15. Gustaaf Renier maintains that history should be defined as "the story of the experiences of men living in civilized societies."[a] He quotes Acton to the effect that "history is a generalized account of the personal actions of men, united in bodies for any public purposes whatever." While acknowledging that biography contains elements that "cannot be described as experiences of men living in societies," he considers it wise to admit that even these private aspects of human existence do have an influence over the public rôle of human beings, and therefore that "in all its aspects biography belongs to history." In theory, a human being living entirely by himself is of no interest to the historian. "Did Robinson Crusoe on his island before the arrival of his Man Friday stay outside history? Only if no traces of his sojourn had ever been found, if his life on the island had not been conditioned by his previous experiences as a social being."[b] Renier claims that his definition of history grasps it in action, not in its essence; as an operation, not as a thing.[c] He considers it unfortunate that the word 'history' is often used for the content of the story, as synonymous with the past experiences of human beings living in civilized societies, or even simply as 'the past.' The *Oxford English Dictionary*, he notes, includes this sense of the word when it calls 'history' "the whole train of events connected with a particular country, society, person, thing, etc., and forming the subject of his or its history," and again "the aggregate of past events in general, the course of events or human affairs." Renier observes that this "new acceptation" of the word is sometimes called 'objective history,' while the "older sense" represented by his definition is then called 'subjective history.' He recalls that Hegel takes delight in the union of an objective and a subjective aspect in the one word 'history.' Yet nothing that Hegel says shows any real advantage in the fact that the word has two different meanings. The subject matter of history could just as well be called 'the human past' and not 'history' at all. The double use of the word merely gives rise to equivocations. "It is by chance, or, more precisely, as a result of causes which have nothing to do with the nature of history that its name acquired a second acceptation." Some have tried to escape from the equivocation, says Renier, by using the word 'historiography' for the discipline that he identifies with 'history.' But 'historiography' is an ill-starred word, because "it commands the affection of the extremists on the right" who "would reserve the name 'history' for research, on the ground that the Greek word *historia* meant investigation." Renier considers etymology to be a poor guide for the correct use of language.[d]

CRITIQUE

16. Renier's definition of history is similar to that of Gallie (IV.2) and is roughly subject to the same criticism (IV.4). Acton's definition is no better. When Acton speaks of history as a "generalized account," it may be added, he is raising the problem without answering it of what 'generalization' in history is, as distinct from the generalization of classical science. Renier recognizes that the twofold meaning of 'history' has given rise to much equivocation, but he does not see why this is so. What he calls "subjective history" and "objective history" are but the two (incorrectly defined) species of a single genus, which is history. It is the lack of correct definition of both genus and species that gives rise to the equivocations. The problem of equivocation cannot be solved by eliminating one of the two species or by calling it a different name. The solution lies only in recognizing in what sense each species of history is history and using it only in its proper sense. If the subject matter of history could just as well be called 'the human past,' as Renier remarks, so also could the literary form of narrative be called 'the story of the past' and not 'history' at all. That would solve nothing. It would only suppress the valid suspicion that both of the species have

---

IV.15a G.J. Renier, *History: Its Purpose and Method* (London: George Allen and Unwin, 1950), 38.

IV.15b *Ibid.*, 36-37.
IV.15c *Ibid.*, 79.
IV.15d *Ibid.*, 82-84.

something in common, which suspicion is already a confused look in the right direction. Renier's refusal to look for the essence of history is the source of his trouble. He would define history as action, as an operation, and not as a thing. But not to define history as a thing means not to know what he is talking about (2.6:1; 6.3:4). And he admits this implicitly when he says that "in theory" a human being living entirely by himself is of no interest to the historian, but nevertheless "in all its aspects biography belongs to history." It is clear that here he is comparing biography to the general idea of history — to a vague essence which he has never taken trouble to define and which he says he refuses to define. In terms of Renier's definition of history (an 'essence' in its own right) it is also clear that biography *in one sense* is history and in another sense is not. That is the force of Renier's own argument. Therefore it is illogical for Renier to say in the terms of his own argument that *in all its aspects* biography belongs to (comes under the essence of) history. This inattentiveness to logic, this lack of concern for the essence of the problem he is treating explains the absence of logical coherence in his conclusion.

17. Since biography is a species of history, it is not true to say that a human being living entirely by himself is of no interest to the historian. The past of a man living entirely by himself is of great interest to himself as his own biographer, that is, as the historian of his own life. It is obvious that happenings which by definition are strictly personal and proper to one individual man and cannot ever be or at least never are communicated to other human beings do not belong to those species of history which by definition regard interpersonal happenings. But it is not logical to exclude biography from history on that ground, since the concept of history itself does not include the qualification 'interpersonal.' Renier's example of Robinson Crusoe is indeed curious. Traces of his sojourn have certainly never been found, since Robinson Crusoe is a completely fictitious character. Daniel Defoe had a "firm hold on fact," and he remains one of the greatest of English journalists, but *Robinson Crusoe* is pure fiction in the form of "realistic make-believe."[a] It is likely that Defoe got the idea for the book from accounts of the strange adventures of the Scottish sailor Alexander Selkirk, but Defoe's book does not narrate the adventures of Selkirk.[b] It has been observed that "Defoe's books all sound like fact, and several of them have passed as such."[c] As an historian, Renier must have known that Robinson Crusoe was a fictitious character on an imaginary island. But what is curious is that Renier would use him as an example of how a man who lived entirely alone could yet be of interest to the historian. He might better have chosen one of the anchorites of the desert. The life of Robinson Crusoe is of no direct interest to the historical scientist as such. His life is of literary interest alone. But indirectly (*per accidens*), as a character whose life has been read by a vast reading public, as an embodiment of Defoe's masterful writing technique, as an influence upon the serious history of poetic literature, etc., there is interest for the serious historian as well. With the concept of 'history' correctly defined, it becomes clear that Robinson Crusoe's life on the deserted island is not in itself subject matter for the science of history (is not factual in that sense), but nevertheless in another sense it is historical: it belongs to the historical species of the imaginary story and to the subspecies of imaginary biography. It should be recognized for what it is.

D. *Bernard Norling's Notion of History*

EXPOSITION

18. Bernard Norling insists that "history is not a science." Natural scientists deal only with objects whose characteristics and reactions are known and invariable. Biologists seek information about one individual in order to generalize about a species. But history studies only the unique, the one situation for itself. "The substance of history is the deeds of men in time." Time never ceases to move, and men

---

IV.17[a] *The Encyclopedia Americana* (international ed., 1967), Vol. 8, p. 593a.

IV.17[b] *Ibid.*, Vol. 23, p. 585.

IV.17[c] J.W. Cunliffe, K. Young, M. van Doren eds., *Century Readings in English Literature* (New York: Appleton-Century-Crofts, 1940), 397.

never cease to act as free agents. Thus history is not a science; it is a kaleidoscope. If history were a science, we could predict the future in detail. From the knowledge of history we cannot predict what particular individuals or nations will do at particular future times. "If one is determined to foretell the course of human affairs in detail he should consult astrologers, not historians." [a]

CRITIQUE

19. The fact that history as such does not tell us what will happen in the future does not exclude it from the realm of science. Norling's definition of science is too narrow; it extends only to what natural scientists and biologists do. Science is knowledge of the real; it is knowledge of the existence of the real and understanding of the meaning of the real (2.6:9). There can be knowledge of the real that does not include prediction of the future; there can be understanding of the real that does not carry with it (*per se*) prediction of the future. History can therefore be a science, even though it does not enable us to predict the future (6.1:3). History is a science to the extent that it gives us knowledge of the existence of the real and understanding of the meaning of the real. Genetic insight enables us to derive understanding from unique situations; it perceives a causal relationship and intelligible order that is far from being a kaleidoscope. History may or may not be a science, but historical science is definitely a science (5.1:4).

IV.18[a] B. Norling, *Towards a Better Understanding of History* (Notre Dame, Indiana: University of Notre Dame Press, 1960), 43.

BIBLIOGRAPHICAL INDEX

(Authors of works mentioned in this volume and first footnote references)

Abbagnano, N., 2.4:12[a]
Abell, G., 2.6:2[b]
Amann, E., 2.1:1[a]
Aquinas, T., 2.1:1[a], 2.1:1[c], 2.1:2[a], 2.2:7[a], 2.3:10[b], 5.1:1[a], 7.2:2[a], I.45[b]
Aristotle, 2.1:1[c], 2.1:2[a], 2.3:10[b], 5.1:1[a], 5.2:19[a]

Barnes, H. E., 3.1:1[a], 3.1:1[c]
Barthel, P., 1.1:1[b]
Bartsch, H. W., 1.1:1[b], 8.0:1[a]
Berdyaev, N., 6.3:4[a]
Bernheim, E., 3.1:1[a]
Bradley, F. H., 2.2:2[b], 2.6:17[a], II.15[a]
Brandenburger, E., 1.1:1[b]
Brinkmann, B., 1.1:1[b]
Bultmann, R., 1.1:1[a], 1.2:1[a], 1.2:2[a], 1.2:3[a], 1.2:4[a], 1.2:4[b], 1.2:4[e], 1.2:6[e], 1.3:1[a], 1.3:10[a], 8.0:1[a], 8.2:5[d], 8.2:15[c], 8.2:21[a], 8.3:5[a], 8.3:11[c], 8.3:12[f], I.1[b], I.17[a], I.32[a], I.32[d], I.38[b], I.41[a], I.41[b], I.42[c], I.44[b], I.50[b], I.54[b], I.59[b], I.63[d], I.64[a]
Buri, F., 1.1:4[a]

Carroll, L., I.72[a]
Castelli, E., 1.1:1[b]
Cattell, R. B., 2.1:1[a]
Chollet, A., 2.1:1[a]
Cohen, M. R., IV.8[a]
Collingwood, R.G., 3.1:1[a], III.1[a], III.20[a]
Conant, J. B., 2.5:2[d]
Cullmann, O., 1.1:3[a]
Cunliffe-Young-van Doren, IV.17[c]

Dahl, N. A., 1.1:4[d]
Danto, A., 2.5:2[a]
Delafresnaye, J. F., 2.1:2[b]
Denzinger, H., 8.1:2[a]

Forcellini, A., 2.1:1[a]
Frank, E., 1.2:7[h], I.54[a]
Funk and Wagnall, 3.1:5[a]

Gallie, W. B., IV.1[a]
Garrigou-Lagrange, R., 2.5:2[c]
Gloege, G., 1.1:4[e]
Greig, J. C., 1.2:4[c]
Grundmann, W., 1.1:4[b]

Harré, R., 2.6:5[a]
Hartlich-Sachs, 1.3:8[a]
Heidegger, M., 1.2:6[b]
Hermann, W., 1.3:8[c]
Holloway, E., 2.1:1[c]

James, W., 2.2:2[b], 2.3:14[b]
Jones, G. V., 1.1:3[a]

Kantor, J. R., 2.6:2[c]
Kegley, C. W., 1.1:1[a]
Kevane, E., 6.3:4[a]
Kittel, G., I.32[d]
Kolping, A., 1.1:1[b]
Kubie, L. S., 2.1:2[b]

Latourelle, R., 7.7:5[b]
Lattanzi, U., I.43[a]
Locke, J., 2.1:1[a]
Lonergan, B., 2.2:7[b], 2.5:4[a]

McCarthy, J. F., 2.6:13[b]
Macquarrie, J., 8.2:5[d]
Margenau, H., 2.3:19[b]
Marlé, R., 1.1:1[b]
Menzel, D., 2.3:7[a]
Moore, T. V., 2.1:1[b], 2.2:5[a]

Newman, J. H., I.42[k]
Norling, B., IV.18[a]

Oepke, A., 1.1:4[b]
Ogden, S., 1.2:4[c]

Palazzi, F., 2.1:1[a]
Petruzzellis, N., 6.2:1[a]

194

BIBLIOGRAPHICAL INDEX

Piaget, J., 2.1:4[b]

Renier, G. J., IV.15[a]
Rienecker, F., 1.1:4[d]
Riesenfeld, H., 1.1:4[d]
Rigaux, B., 1.1:1[b]

Salcedo - Fernández, I.46[a]
Sasse, H., 1.1:4[b]
Scheid, J. E., 1.1:1[a]
Schlink, E., 1.1:4[d]
Schmidt, E., 1.1:4[b]
Schmithals, W., 1.1:1[a]
Schweitzer, A., I.63[e]
Selvaggi, F., 2.4:3[a], 2.6:5[a]
Sheen, F. J., 2.6:25[a]
Shotwell, J. T., 3.1:1[a]

Shotwell - Jacob, 3.1:1[b]
Simmons, E. D., 2.4:9[b]
Smart, J. D., 1.1:4[c]
Stauffer, E., 1.1:4[d]
Swift, J., 2.1:9[a]

Teggart, F. J., 3.1:9[c]
Theunis, F., 1.1:1[a]
Tübingen School, 1.1:4[e]

Van Laer, H., 2.4:9[b]
Vögtle, A., 1.1:1[b]

Walsh, W. H., 3.1:1[a], II.1[a]
Webster, N., 2.5:2[b]
Weiss, J., I.56[c]
Wernz - Vidal, 2.6:13[b]
Wilder, A., 1.3:6[e]

If you have enjoyed this book, consider making your next selection from among the following . . .

Life Everlasting. Garrigou-Lagrange, O.P.	12.50
The Three Ways of the Spiritual Life. Garrigou-Lagrange, O.P.	4.00
Eucharistic Miracles. Joan Carroll Cruz	13.00
The Four Last Things—Death, Judgment, Hell, Heaven	5.00
The Facts About Luther. Msgr. Patrick O'Hare	13.50
Little Catechism of the Curé of Ars. St. John Vianney	5.50
The Curé of Ars—Patron St. of Parish Priests. Fr. B. O'Brien	4.50
St. Teresa of Ávila. William Thomas Walsh	18.00
The Rosary and the Crisis of Faith. Msgr. Cirrincione & Nelson	1.25
The Secret of the Rosary. St. Louis De Montfort	3.00
Modern Saints—Their Lives & Faces. Book 1. Ann Ball	15.00
Modern Saints—Their Lives & Faces. Book 2. Ann Ball	20.00
The 12 Steps to Holiness and Salvation. St. Alphonsus	6.00
The Incorruptibles. Joan Carroll Cruz	12.00
Raised from the Dead—400 Resurrection Miracles. Fr. Hebert	13.50
Saint Michael and the Angels. Approved Sources	5.50
Dolorous Passion of Our Lord. Anne C. Emmerich	13.50
Our Lady of Fatima's Peace Plan from Heaven. Booklet	.75
Divine Favors Granted to St. Joseph. Pere Binet	4.00
St. Joseph Cafasso—Priest of the Gallows. St. J. Bosco	3.00
Catechism of the Council of Trent. McHugh/Callan	20.00
The Sinner's Guide. Ven. Louis of Granada	11.00
Padre Pio—The Stigmatist. Fr. Charles Carty	12.50
Why Squander Illness? Frs. Rumble & Carty	2.00
The Sacred Heart and the Priesthood. de la Touche	7.00
Fatima—The Great Sign. Francis Johnston	7.00
Heliotropium—Conformity of Human Will to Divine. Drexelius	11.00
Purgatory Explained. (pocket, unabr.). Fr. Schouppe	5.00
Who Is Padre Pio? Radio Replies Press	1.00
Child's Bible History. Knecht	4.00
The Stigmata and Modern Science. Fr. Charles Carty	1.00
The Life of Christ. 4 Vols. H.B. Anne C. Emmerich	55.00
The Glories of Mary. (pocket, unabr.). St. Alphonsus	8.00
Is It a Saint's Name? Fr. William Dunne	1.50
The Precious Blood. Fr. Faber	11.00
The Holy Shroud & Four Visions. Fr. O'Connell	1.50
Clean Love in Courtship. Fr. Lawrence Lovasik	2.00
The Prophecies of St. Malachy. Peter Bander	4.00
The History of Antichrist. Rev. P. Huchede	2.50
The Douay-Rheims Bible. Leatherbound	35.00
St. Catherine of Siena. Alice Curtayne	11.00
Where We Got the Bible. Fr. Henry Graham	5.00
Imitation of the Sacred Heart of Jesus. Fr. Arnoudt	13.50
An Explanation of the Baltimore Catechism. Kinkead	13.00
The Way of Divine Love. Sr. Menendez	17.50
The Curé D'Ars. Abbé Francis Trochu	18.50
Love, Peace and Joy. St. Gertrude/Prévot	5.00
St. Bernadette Soubirous. Trochu	15.00
Rhine Flows into the Tiber. Wiltgen	12.00
St. Anthony, the Wonder-Worker. Stoddard	3.50
Hell Plus How to Avoid Hell. Schouppe/Nelson	10.00
Golden Arrow—Autob. Sr. Mary of St. Peter	10.00

At your bookdealer or direct from the publisher.

Prices guaranteed through December 31, 1992.